Centenarians—A Model to Study the Molecular Basis of Lifespan and Healthspan

Centenarians—A Model to Study the Molecular Basis of Lifespan and Healthspan

Editors

Annibale Puca
Calogero Caruso

MDPI • Basel • Beijing • Wuhan • Barcelona • Belgrade • Manchester • Tokyo • Cluj • Tianjin

Editors
Annibale Puca
Universita di Salerno
Italy

Calogero Caruso
University of Palermo
Italy

Editorial Office
MDPI
St. Alban-Anlage 66
4052 Basel, Switzerland

This is a reprint of articles from the Special Issue published online in the open access journal *International Journal of Molecular Sciences* (ISSN 1422-0067) (available at: https://www.mdpi.com/journal/ijms/special_issues/centenarians).

For citation purposes, cite each article independently as indicated on the article page online and as indicated below:

LastName, A.A.; LastName, B.B.; LastName, C.C. Article Title. *Journal Name* **Year**, *Volume Number*, Page Range.

ISBN 978-3-0365-0980-8 (Hbk)
ISBN 978-3-0365-0981-5 (PDF)

Contents

About the Editors

Annibale Puca is an M.D. and board-certified in neurology. He started as a postdoc at TIGEM (Naples) in 1995 with Prof. Brunella Franco. In 1998, he was a postdoc at Harvard Medical School in Prof. Luo Kunkel's lab working on the genetics of exceptional longevity. In Cambridge, USA, he scientifically directed and co-founded Centagenetix Inc. in 2001. In 2006, he started a group at Multimedica (Milan) where he focused its efforts on the discovery of genetic risk factors for exceptional longevity. At the academic level, he became Associate Professor in 2011 and Full Professor in Clinical Pathology in 2020 at the Department of Medicine of University of Salerno. He is the winner of grants as PI (Ministero della salute, Cariplo Foundation) and unit leader (Ministero della Ricerca, Ministero della Salute). He has a strong track record of publications (Scopus H index of 37 with more than 130 publications) about the therapeutic use of a longevity-associated variant (LAV) of BPIFB4 and he started a spin-off (LGV1) in 2015 around its potential therapeutic applications.

Calogero Caruso , MD, formerly Full Professor of General Pathology, is Professor Emeritus of the University of Palermo. In his academic career, he was Coordinator of the PhD course in Molecular Medicine and Biotechnology, Director of the Postgraduate School in Laboratory Medicine, and Dean of the Medical School Graduate Course. He directed the Immunopathology Laboratory and the Transfusion Medicine Unit of "Paolo Giaccone" University Hospital. He is the author of 382 scientific papers with a total of 13,093 citations (H-index 59), mainly on topics of immunogenetics (in 1981 he was a fellow at the Department of Immunohematology of Leiden University) and immunopathology, immunosenescence, aging and longevity. Founder and Editor of Immunity & Aging from 2004 to 2018, he is currently Academic Editor of PlosOne, Mediators of Inflammation, and the International Journal of Molecular Sciences. He recently edited the book "Centenarians. An example of Positive Biology" for Springer Nature, and with Prof. Giuseppina Candore, he edited for Elsevier the book "Human Aging: From Cellular Mechanisms to Therapeutic Strategies" to be published in May. Coordinator of an Italian national project on centenaries and longevity (2017–2020), with Prof. Giuseppina Candore, he is responsible for the local unit of the European project "Improved Vaccination Strategies for Older Adults".

International Journal of
Molecular Sciences

Editorial

Special Issue "Centenarians—A Model to Study the Molecular Basis of Lifespan and Healthspan"

Calogero Caruso [1,*] and Annibale Alessandro Puca [2,3]

1   Laboratory of Immunopathology and Immunosenescence, Department of Biomedicine, Neuroscience and Advanced Diagnostics, University of Palermo, 90134 Palermo, Italy
2   Department of Medicine, Surgery and Dentistry "Scuola Medica Salernitana", University of Salerno, 84081 Baronissi, Italy; apuca@unisa.it
3   Cardiovascular Research Unit, IRCCS MultiMedica, 20138 Milan, Italy
*   Correspondence: calogero.caruso@unipa.it

Citation: Caruso, C.; Puca, A.A. Special Issue "Centenarians—A Model to Study the Molecular Basis of Lifespan and Healthspan". *Int. J. Mol. Sci.* **2021**, *22*, 2044. https://doi.org/10.3390/ijms22042044

Received: 6 February 2021
Accepted: 10 February 2021
Published: 19 February 2021

Publisher's Note: MDPI stays neutral with regard to jurisdictional claims in published maps and institutional affiliations.

1. State of Art

People are living longer, not, as was previously the case, due to reduced child mortality, but because we are postponing the ill-health of old age. The global population aged 60 years or over numbered 962 million in 2017, more than twice as large as in 1980 when there were 382 million older persons worldwide. The number of older persons is expected to double again by 2050, when it is projected to reach nearly 2.1 billion; the number of persons aged 80 years or over is projected to increase more than threefold between 2017 and 2050, rising from 137 million to 425 million. [1–3].

Ageing is a component of life, which derives from the breakdown of the self-organization system and the reduced ability to adapt to the environment. As we age, harmful changes accumulate in the molecules, cells and tissues responsible for the decline of normal physiological functions. This inexorably leads to a reduced ability of the individual to maintain adequate homeostasis, resulting in a greater susceptibility to different types of stressors. Hence, ageing processes are defined as those that amplify the vulnerability of individuals, as they age, to the factors that ultimately lead to death [4].

This extraordinary increase in older people emphasizes the importance of studies on ageing and longevity and the need for a prompt dissemination of knowledge on ageing and longevity with the aim to satisfactorily diminish the medical, economic and social problems associated with advancing years, problems caused by the continuous increase in the number of older people at risk of frailty and age-related diseases. In fact, the increase in the duration of life (lifespan) does not coincide with the increase in the duration of health (healthspan), that is the period of life free from serious chronic diseases and disabilities [1,5]. Therefore, improving the quality of life of older people must be a priority. This makes studies of the processes involved in ageing and longevity of great importance.

In particular, understanding why some individuals have escaped from neonatal mortality, infectious diseases in the pre-antibiotic era and the fatal outcomes of age-related diseases, thus living 100+ years, might allow the factors involved in the attainment of healthy ageing. For scientists, centenarians are, in fact, the paradigmatic example of healthy ageing [6].

Long-lived individuals should refer to people belonging to the 5th percentile of the survival curve, that is, in the Western world, to those over ninety. However, it is often the canonical age of 100 that is regarded as the threshold of exceptional longevity [4,7]. Worldwide, the number of centenarians fluctuates between half a million and a million with five women for one man. Over the course of human history their number has grown [2,8].

Human beings are the product of their genes, but they would not be the same without the experiences they have/had since those in the womb. It has been shown that the month of birth, a proxy for early environmental influences (i.e., epigenetics), affects the possibility of reaching 100 years [9]. Living organisms are subject to nature laws and genetic programs where both Brownian random motions, i.e., the erratic random movement of particles in a fluid, as a result of continuous bombardment from molecules of the surrounding water molecules in the fluid, and crossing over contribute to leave space for the chance. There is evidence of the inherent stochastic nature of both gene expression and macromolecular biosynthesis. Several genes are in fact transcribed in minimal amounts of mRNA, which can cause large fluctuations in macromolecular biosynthesis. Chance is just that, the random occurrence, i.e., an event happening not according to a plan [2,10].

Asking whether ageing and longevity depend on the environment or genetics, even if legitimate, is too simplified. To respond fully means to consider all that is important, the case and circumstances of life, diet, physical activity, environmental exposure and lifestyle (which also affect epigenetics), the burden of natural life pathogens, stress management and social networking, education and gender and, obviously, DNA (genetics and epigenetics) in the forefront [10–13]. However, taking into account the risk factors we can modify lifestyle, particularly diet, that is, beyond doubt, the most important.

The deepening of this knowledge could allow modulating the ageing rate by providing valuable information on how to achieve healthy ageing.

2. Diet and Healthy Ageing

A long life in a healthy, vigorous and young body has always been one of humanity's greatest dreams. Anti-ageing strategies aimed not at rejuvenating but at slowing ageing and delaying or avoiding the onset of age-related diseases are welcome. It has to be emphasized that the goal of ageing research is not to increase human longevity regardless of the consequences, but to increase active life free from disability and functional dependence.

As previously stated, the role of diet in the attainment of longevity by slowing ageing and fighting age-related diseases is well known. A particular aspect of diet concerns some compounds principally contained in fruit and vegetables, called nutraceuticals, i.e., "naturally derived bioactive compounds that are found in foods, dietary supplements and herbal products, and have health promoting, disease preventing, and/or medicinal properties". Several nutraceuticals exhibit anti-ageing features by acting on the inflammatory status and on the prevention of oxidative reaction. Similar effects are thought to be obtained with the use of probiotics [14,15]. In this special issue six papers addressed these aspects in vitro or in vivo.

In an in vitro study [16], it was examined in pheochromocytoma 12 (PC12) cells, whether *Hericium erinaceus* (HE), a medicinal plant, could exert a protective effect against oxidative stress and apoptosis induced by di(2-ethylhexyl)phthalate (DEHP), a plasticizer known to cause neurotoxicity. The authors demonstrated that pre-treatment with HE significantly attenuated DEHP induced cell death. This protective effect was attributed to its ability to reduce intracellular reactive oxygen species levels.

Two studies [17,18] focused on the role played by some olive oil polyphenols [19] on ageing pathophysiology with special attention to Parkinson's disease like symptoms in the well-known invertebrate model organism *Caenorhabditis (C) elegans*. The studies demonstrate that polyphenolic extract treatment has the potential to partly prevent or even treat ageing-related neurodegenerative diseases and ageing itself. Future investigations including mammalian models and human clinical trials are needed to uncover the full potential of these olive compounds.

The review of Rosselli et al., [20], analyzed, instead, the impact of probiotics on *C. elegans*. The picture emerging from their analysis highlights that several probiotic strains are able to exert anti-ageing effects in nematodes by acting on common molecular pathways, such as insulin/insulin-like growth factor-1 and p38 mitogen-activated protein kinase. So, *C. elegans* appears to be advantageous for shedding light on key mechanisms involved in host pro-longevity in response to probiotics supplementation.

Then, in their review [21], the authors discuss the possible effect of Royal Jelly and its components on healthy ageing and longevity in animal models as well as its positive effects on health maintenance and age-related disorders in humans. The findings pave the way to inventing specific Royal Jelly as anti-ageing drugs.

Finally, a small clinical trial was instead performed on another model of neurodegeneration, Meniere's disease (MD) [22]. The authors evaluated systemic oxidative stress and cellular stress response in MD patients in the absence and in the presence of treatment with a biomass preparation from *Coriolus versicolor*, endowed with various biological actions, including antioxidant ones. It was concluded that systemic oxidative stress was reduced in MD patients treated with *Coriolus versicolor*, which was paralleled by a significant induction of vitagenes, known to encode survival and anti-oxidant proteins. Vitagene up-regulation after *Coriolus versicolor* supplementation indicates a maintained response to counteract intracellular pro-oxidant status. Thus, searching innovative and more potent inducers of the vitagene system can allow the development of pharmacological strategies capable of enhancing the intrinsic reserve of vulnerable neurons, such as ganglion cells to maximize antidegenerative stress responses and thus providing neuroprotection.

3. Hallmarks of Ageing

Some aspects fighting three hallmarks of ageing, i.e., characteristic features considered to contribute to the ageing process, hence determining the ageing phenotype [23], are instead treated in the last four papers.

Concerning the hallmark altered intercellular communication [23], e.g., in this series, Adverse Food Reactions (FAs) [24] have been treated as example of immunosenescence [25]. FAs show peculiar characteristics in older people that concern both the pathogenesis and the clinic. FAs in older people are driven by immunosenescence, as well as the cell ageing and tissue modifications that characterize advanced age. The aged gastrointestinal mucosa is central in the development of FAs in older people through its compromised digestive properties and structural changes, as well as the alteration of its immune functions linked to immunosenescence and age-related microbiota remodeling. Among the risk factors for the sensitization to food allergens in older people, in addition to chronic damage and inflammation of gut epithelia due to the ageing process, there are chronic alcohol consumption, chronic infections, multimorbidity, polymedication, and drug side effects.

Data of Jimenez-Gutierrez et al., [26] regard the prevention of the hallmark cellular senescence [23]. Their data are consistent with the paradigm that interfering with function of Nuclear β-dystroglycan (β-DG) involved in the maintenance of nuclear architecture and function results somehow in aberrant multipolar mitoses. That, in turn, evokes a p53-dependent DNA-damage response, arresting the cell cycle progression and thereby inducing senescence, to avoid the propagation of damaged genomes. That supports a role for DG in protecting against senescence, through the maintenance of proper lamin B1 expression/localization and proper mitotic spindle organization.

Epigenetics, another hallmark [23] is treated in the last two papers. In the first paper [27], the authors used next-generation sequencing to identify differentially expressed microRNAs (miRNAs) in serum exosomes isolated from young (three-month-old) and old (22-month-old) rats and then used bioinformatics to explore candidate genes and ageing-related pathways. Taken together, their findings suggest that changes in the makeup of circulating exosomal miRNAs with age not only can be considered as a potential predictor of age but also may contribute to ageing via several key signaling pathways that regulate ageing and lifespan.

In the paper of Gutman et al. [28], exceptionally long-lived individuals (ELLIs) demonstrated juvenile performance in DNA methylation age clocks and overall methylation measurement, with preserved cognition and relative telomere length. The findings suggest a favorable DNA methylation profile in ELLIs enabling a slower rate of ageing in those individuals in comparison to controls. It is possible that DNA methylation is a key modulator of the rate of ageing and thus the ELLIs DNAm profile promotes healthy longevity.

4. Conclusions

This series of papers is noteworthy because the identification of the factors that predispose to a long and healthy life is of enormous interest for translational medicine.

Author Contributions: Both authors (C.C. and A.A.P.) wrote. Both authors have read and agreed to the published version of the manuscript.

Funding: The original work of the Authors in the topic was supported by 20157ATSLF project (Discovery of molecular, and genetic/epigenetic signatures underlying resistance to age-related diseases and comorbidities), granted by the Italian Ministry of Education, University and Research.

Conflicts of Interest: The authors declare no conflict of interest.

References

1. Christensen, K.; Doblhammer, G.; Rau, R.; Vaupel, J.W. Ageing populations: The challenges ahead. *Lancet* **2009**, *374*, 1196–1208. [CrossRef]
2. Caruso, C. (Ed.) *Centenarians*; Springer: Cham, Switzerland, 2019; pp. 1–179. [CrossRef]
3. World Population Ageing. Available online: https://www.un.org/en/development/desa/population/publications/pdf/ageing/WPA2017_Highlights.pdf (accessed on 2 February 2021).
4. Avery, P.; Barzilai, N.; Benetos, A.; Bilianou, H.; Capri, M.; Caruso, C.; Franceschi, C.; Katsiki, N.; Mikhailidis, D.P.; Panotopoulos, G.; et al. Ageing, longevity, exceptional longevity and related genetic and non genetics markers: Panel statement. *Curr. Vasc. Pharmacol.* **2014**, *12*, 659–661. [CrossRef]
5. Crimmins, E.M. Lifespan and Healthspan: Past, Present, and Promise. *Gerontologist* **2015**, *55*, 901–911. [CrossRef] [PubMed]
6. Engberg, H.; Oksuzyan, A.; Jeune, B.; Vaupel, J.W.; Christensen, K. Centenarians—A useful model for healthy aging? A 29-year follow-up of hospitalizations among 40,000 Danes born in 1905. *Aging Cell* **2009**, *8*, 270–276. [CrossRef]
7. Villa, F.; Ferrario, A.; Puca, A.A. Genetic Signatures of Centenarians. In *Centenarians*; Caruso, C., Ed.; Springer: Cham, Switzerland, 2019; pp. 87–97. [CrossRef]
8. Herm, A.; Cheung, S.; Poulain, M. Emergence of oldest old and centenarians: Demographic analysis. *Asian J. Gerontol. Geriatr.* **2012**, *7*, 19–25. Available online: http://ajgg.org/image/module/ajgg_issue/11/v7n1_SA3_A%20Herm.pdf (accessed on 2 February 2021).
9. Gavrilov, L.A.; Gavrilova, N.S. Season of birth and exceptional longevity: Comparative study of american centenarians, their siblings, and spouses. *J. Aging Res.* **2011**, *2011*, 104616. [CrossRef] [PubMed]
10. Accardi, G.; Caruso, C. Causality and Chance in Ageing, Age-Related Diseases and Longevity. In *Updates in Pathobiology: Causality and Chance in Ageing, Age-Related Diseases and Longevity*; Accardi, G., Caruso, C., Eds.; Palermo University Press: Palermo, Italy, 2017; pp. 13–23. Available online: https://pure.unipa.it/en/publications/updates-in-pathobiology-causality-and-chance-in-ageingage-related (accessed on 2 February 2021).
11. Caruso, C.; Passarino, G.; Puca, A.; Scapagnini, G. "Positive biology": The centenarian lesson. *Immun. Ageing* **2012**, *9*, 5. [CrossRef]
12. Passarino, G.; De Rango, F.; Montesanto, A. Human longevity: Genetics or Lifestyle? It takes two to tango. *Immun. Ageing* **2016**, *13*, 12. [CrossRef]
13. Lutz, W.; Kebede, E. Education and Health: Redrawing the Preston Curve. *Popul. Dev. Rev.* **2018**, *44*, 343–361. [CrossRef]
14. Aiello, A.; Accardi, G.; Candore, G.; Carruba, G.; Davinelli, S.; Passarino, G.; Scapagnini, G.; Vasto, S.; Caruso, C. Nutrigerontology: A key for achieving successful ageing and longevity. *Immun. Ageing* **2016**, *13*, 17. [CrossRef]
15. Aiello, A.; Accardi, G.; Candore, G.; Gambino, C.M.; Mirisola, M.; Taormina, G.; Virruso, C.; Caruso, C. Nutrient sensing pathways as therapeutic targets for healthy ageing. *Expert Opin. Ther. Targets* **2017**, *21*, 371–380. [CrossRef]
16. Amara, I.; Scuto, M.; Zappalà, A.; Ontario, M.L.; Petralia, A.; Abid-Essefi, S.; Maiolino, L.; Signorile, A.; Trovato Salinaro, A.; Calabrese, V. Hericium Erinaceus Prevents DEHP-Induced Mitochondrial Dysfunction and Apoptosis in PC12 Cells. *Int. J. Mol. Sci.* **2020**, *21*, 2138. [CrossRef]
17. Di Rosa, G.; Brunetti, G.; Scuto, M.; Trovato Salinaro, A.; Calabrese, E.J.; Crea, R.; Schmitz-Linneweber, C.; Calabrese, V.; Saul, N. Healthspan Enhancement by Olive Polyphenols in *C. elegans* Wild Type and Parkinson's Models. *Int. J. Mol. Sci.* **2020**, *21*, 3893. [CrossRef]
18. Brunetti, G.; Di Rosa, G.; Scuto, M.; Leri, M.; Stefani, M.; Schmitz-Linneweber, C.; Calabrese, V.; Saul, N. Healthspan Maintenance and Prevention of Parkinson's-like Phenotypes with Hydroxytyrosol and Oleuropein Aglycone in *C. elegans. Int. J. Mol. Sci.* **2020**, *21*, 2588. [CrossRef]
19. Gambino, C.M.; Accardi, G.; Aiello, A.; Candore, G.; Dara-Guccione, G.; Mirisola, M.; Procopio, A.; Taormina, G.; Caruso, C. Effect of Extra Virgin Olive Oil and Table Olives on the Immune Inflammatory Responses: Potential Clinical Applications. *Endocr. Metab. Immune Disord. Drug Targets* **2018**, *18*, 14–22. [CrossRef] [PubMed]
20. Roselli, M.; Schifano, E.; Guantario, B.; Zinno, P.; Uccelletti, D.; Devirgiliis, C. *Caenorhabditis elegans* and Probiotics Interactions from a Prolongevity Perspective. *Int. J. Mol. Sci.* **2019**, *20*, 5020. [CrossRef]
21. Kunugi, H.; Ali, A.M. Royal Jelly and Its Components Promote Healthy Aging and Longevity: From Animal Models to Humans. *Int. J. Mol. Sci.* **2019**, *20*, 4662. [CrossRef] [PubMed]

22. Scuto, M.; Di Mauro, P.; Ontario, M.L.; Amato, C.; Modafferi, S.; Ciavardelli, D.; Trovato Salinaro, A.; Maiolino, L.; Calabrese, V. Nutritional Mushroom Treatment in Meniere's Disease with *Coriolus versicolor*: A Rationale for Therapeutic Intervention in Neuroinflammation and Antineurodegeneration. *Int. J. Mol. Sci.* **2019**, *21*, 284. [CrossRef] [PubMed]
23. López-Otín, C.; Blasco, M.A.; Partridge, L.; Serrano, M.; Kroemer, G. The hallmarks of aging. *Cell* **2013**, *153*, 1194–1217. [CrossRef]
24. De Martinis, M.; Sirufo, M.M.; Viscido, A.; Ginaldi, L. Food Allergies and Ageing. *Int. J. Mol. Sci.* **2019**, *20*, 5580. [CrossRef]
25. Aiello, A.; Farzaneh, F.; Candore, G.; Caruso, C.; Davinelli, S.; Gambino, C.M.; Ligotti, M.E.; Zareian, N.; Accardi, G. Immunosenescence and Its Hallmarks: How to Oppose Aging Strategically? A Review of Potential Options for Therapeutic Intervention. *Front. Immunol.* **2019**, *10*, 2247. [CrossRef]
26. Jimenez-Gutierrez, G.E.; Mondragon-Gonzalez, R.; Soto-Ponce, L.A.; Gómez-Monsiváis, W.L.; García-Aguirre, I.; Pacheco-Rivera, R.A.; Suárez-Sánchez, R.; Brancaccio, A.; Magaña, J.J.; CR Perlingeiro, R.; et al. Loss of Dystroglycan Drives Cellular Senescence via Defective Mitosis-Mediated Genomic Instability. *Int. J. Mol. Sci.* **2020**, *21*, 4961. [CrossRef]
27. Zhang, H.; Jin, K. Peripheral Circulating Exosomal miRNAs Potentially Contribute to the Regulation of Molecular Signaling Networks in Aging. *Int. J. Mol. Sci.* **2020**, *21*, 1908. [CrossRef]
28. Gutman, D.; Rivkin, E.; Fadida, A.; Sharvit, L.; Hermush, V.; Rubin, E.; Kirshner, D.; Sabin, I.; Dwolatzky, T.; Atzmon, G. Exceptionally Long-Lived Individuals (ELLI) Demonstrate Slower Aging Rate Calculated by DNA Methylation Clocks as Possible Modulators for Healthy Longevity. *Int. J. Mol. Sci.* **2020**, *21*, 615. [CrossRef]

International Journal of
Molecular Sciences

Article

Hericium Erinaceus Prevents DEHP-Induced Mitochondrial Dysfunction and Apoptosis in PC12 Cells

Ines Amara [1,2,†], Maria Scuto [2,†], Agata Zappalà [2], Maria Laura Ontario [2], Antonio Petralia [3], Salwa Abid-Essefi [1], Luigi Maiolino [3], Anna Signorile [4,*], Angela Trovato Salinaro [2,*] and Vittorio Calabrese [2]

1   Laboratory for Research on Biologically Compatible Compounds, Faculty of Dental Medicine, University of Monastir, Rue Avicenne, Monastir 5019, Tunisia; ines.amara15@yahoo.fr (I.A.); salwaabid@yahoo.fr (S.A.-E.)
2   Department of Biomedical and Biotechnological Sciences, University of Catania, Torre Biologica, Via Santa Sofia n. 97, 95125 Catania, Italy; mary-amir@hotmail.it (M.S.); azappala@unict.it (A.Z.); marialaura.ontario@ontariosrl.it (M.L.O.); calabres@unict.it (V.C.)
3   Department of Medical and Surgery Sciences, University of Catania, 95125, Via Santa Sofia, 78, 95123 Catania, Italy; petralia@unict.it (A.P.); maiolino@policlinico.unict.it (L.M.)
4   Department of Basic Medical Sciences, Neurosciences and Sense Organs, University of Bari, Piazza G. Cesare, 11, 70124 Bari, Italy
*   Correspondence: Anna.signorile@uniba.it (A.S.); Trovato@unict.it (A.T.S.)
†   These authors contributed equally to this work.

Received: 10 February 2020; Accepted: 17 March 2020; Published: 20 March 2020

Abstract: *Hericium Erinaceus* (HE) is a medicinal plant known to possess anticarcinogenic, antibiotic, and antioxidant activities. It has been shown to have a protective effect against ischemia-injury-induced neuronal cell death in rats. As an extending study, here we examined in pheochromocytoma 12 (PC12) cells, whether HE could exert a protective effect against oxidative stress and apoptosis induced by di(2-ethylhexyl)phthalate (DEHP), a plasticizer known to cause neurotoxicity. We demonstrated that pretreatment with HE significantly attenuated DEHP induced cell death. This protective effect may be attributed to its ability to reduce intracellular reactive oxygen species levels, preserving the activity of respiratory complexes and stabilizing the mitochondrial membrane potential. Additionally, HE pretreatment significantly modulated Nrf2 and Nrf2-dependent vitagenes expression, preventing the increase of pro-apoptotic and the decrease of anti-apoptotic markers. Collectively, our data provide evidence of new preventive nutritional strategy using HE against DEHP-induced apoptosis in PC12 cells.

Keywords: Di (2-Ethylhexyl) pthalate; *Hericium erinaceus*; vitagenes; oxidative stress; apoptosis; mitochondrial respiratory complexes

1. Introduction

Phthalates are common plasticizers, used in a large variety of household and medical products to confer flexibility to many polyvinyl chlorides (PVC)–based plastics [1]. Di(2- ethylhexyl)phthalate (DEHP) is one of the most extensively used phthalates, which have a variety of applications including food packages, cosmetics, clothing, children's toys, and medical devices such as blood storage bags [2]. DEHP is highly hydrophobic compound and it is well absorbed after oral exposure [3], being estimated that human absorption of DEHP could be as high as 25%. Due to its liposolubility, DEHP leaches from plastics following lipophilic fluids [4], making it possible that DEHP crosses the blood-brain barrier into the central nervous system (CNS) tissue, which could lead to neural toxic effects.

Recently, an increasing number of studies have provided evidence of a significant association between DEHP exposure and neuronal disruption. The toxicity of DEHP in mammalian cells has been investigated, and it has been demonstrated that this phthalate is cytotoxic and induces apoptosis in Neuro-2-a cells, a neuroblastome cell line [5]. Other investigators have studied the in vivo administration of DEHP and demonstrated that this plasticizer causes neurodegeneration in the brain of rats [6]. Other lines of evidence indicate that prenatal and postnatal exposure of DEHP affect CNS [7]. Indeed, in utero exposure, DEHP caused disruption of rat brain development [8], while a reduction in the number of mid brain dopaminergic neurons and a motor hyperactivity have been demonstrated after postnatal exposure to this phthalate [9].

An often-mentioned mechanism of chemical-induced neuronal damage is oxidative stress. Excess of reactive oxygen species (ROS), mainly associated with mitochondrial dysfunctions, disrupts antioxidant defense system, generating a vicious cycle resulting in further damage to cellular functions, energy insufficiency, and eventually leading to neuronal cell death [10].

Apoptosis is the most common and well-defined form of programmed cell death which is a genetically directed process of cell self-destruction that is essential for embryonic development, immune-system function, and the maintenance of tissue homeostasis in multi-cellular organisms [11]. It is marked by different changes in the mitochondria such as the release of caspase activators, changes in electron transport, loss of mitochondrial membrane potential (MMP), and involvement of pro and anti-apoptotic Bcl-2 family proteins [12]. In recent years, considerable research has been carried out on identifying naturally occurring substances endowed with neuroprotective properties, impacting on apoptotic processes, and thus capable to prevent or delay neurodegenerative processes. Mushrooms, which have been used in traditional medicine for thousands of years [13,14], have been reported to possess various biological actions, including antitumor, immunomodulatory, antioxidant, antiviral, antibacterial, and hepatoprotective effects [15,16]. Some of the most potent immunostimulatory molecules derived from mushrooms are β glucans, which activate many types of immune cells and stimulate cytokine responses [17–20]. Administration of complex mixtures of molecules to unknown concentrations is difficult to reconcile with current pharmaceutical practices involving highly purified compounds, and hence, as the active ingredients may be unknown, to patent mushroom extracts is a very difficult task.

Moreover, mushroom-derived polysaccharides are complex molecules that cannot be synthesized, as the mass production of these compounds would require timely and costly extraction processes. As a result, many research efforts have focused on low molecular weight compounds, such as cordycepin, which is a cytotoxic nucleoside analog inhibitor of cell proliferation. Of the mushroom-derived therapeutics, polysaccharopeptides obtained from *Hericium erinaceus* (HE) are commercially the best established. HE is known to have a neuroprotective property evidenced by the regulation of inflammation in relation to the pathology of Alzheimer's disease. In another modality of protection against Alzheimer's disease, HE has been shown to stimulate the synthesis of nerve growth factor (NGF) in cultured astrocytes [21]. This growth factor, acting on cholinergic neurons by modulating the activity of two enzymes: cholineacetyltransferase and acetylcholinesterase in Alzheimer's disease, the activity of these two enzymes is inhibited since the dysfunction of cholinergic neurons is an initial event in Alzheimer's disease [22].

The ability to sense oxidative and proteotoxic insults and to coordinate defensive stress response are basic elements for cellular adaptation and survival [23,24]. With regard to this, medicinal mushrooms, including HE may have beneficial protective effects in low doses [25–27]. Consistently, hormesis dose response is characterized by low dose stimulation and a high dose inhibition. The biphasic dose-response phenomenon is characterized by a U-shaped or inverse dose response curve, depending on the different measured endpoints [28–30]. The hormetic dose response results from either a direct stimulation or through an overcompensation stimulatory response following disruption in homeostasis [31]. Such hormetic dose responses provide a quantitative description of the bounds of biological plasticity, and a measure of the extent to which adaptive processes may be upregulated, which is especially relevant to the comprehension of protective effects induced by plant and fungal species [32–36]. It has been known that mushrooms activate the heat shock protein (Hsp) pathway in different brain regions of rats, which plays a crucial role in the cellular stress response [22,37,38], and hence, *Hericium erinaceus biomass* preparations can have neuroprotective effects through modulation of inflammatory processes associated with the neuropathology, as well as through regulation of brain stress response plasticity mechanisms [39]. In addition, antioxidant activity of HE has also been demonstrated in other tissues, including liver [40]. In light of the above-mentioned evidence in the present study, we have investigated in vitro the neuroprotective role of HE biomass preparation against DEHP-induced neurotoxicity.

2. Results

We first evaluated whether DEHP alone or HE alone treatment was toxic to pheochromocytoma 12 (PC12) cell line. Cells were treated for 24 h with increasing concentrations of DEHP (20, 40, 60, 80, and 100 µM) and HE (0.5, 1 1.5, and 2 mg/mL) (Figure 1a) and cell survival was determined by 3-(4,5-dimethylthiazol-2 yl)-2,5-diphenyltetrazolium bromide (MTT) assay (Figure 1a,b). DEHP treatment induced a dose-dependent reduction in cell viability with approximately IC50 observed at 85 µM (Figure 1a). Consequently, cytotoxic induction with 85 µM DEHP for 24 h was used in the subsequent experiments. Besides, while HE alone exhibited no toxicity towards PC12 cells (Figure 1b), pretreatment with this mushroom at 0.5 mg/mL, significantly decreased DEHP-mediated cytotoxicity (Figure 1c).

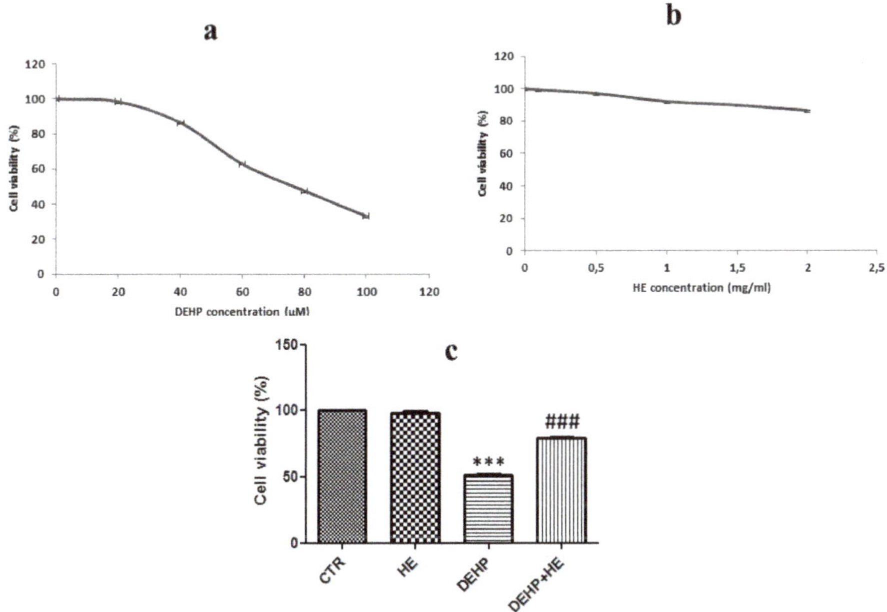

Figure 1. (**a,b**) Cytotoxic effect of di(2-ethylhexyl)phthalate (DEHP) and *Hericium Erinaceus* (HE) on pheochromocytoma 12 (PC12) cells. Cells were treated with DEHP or HE at the indicated concentrations for 24 h. (**c**) HE reduces DEHP-induced cytotoxicity in PC12. Cells were pretreated for 2 h with HE (1 µM) before DEHP treatment for 24 h (85 µM). Cell viability was determined using 3-(4,5-Dimethylthiazol-2 yl)-2,5-diphenyltetrazolium bromide (MTT) assay. Data are expressed as the mean ±SD of three separate experiments. *** $p \leq 0.001$ vs. control and ### $p \leq 0.001$ vs. DEHP alone.

Indeed, the DEHP treatment results in 51.16 % cell viability, but pretreatment for 2 h with HE at 0.5 mg/mL increase cell viability at 78.73 % (Figure 1c). Thus, we asked whether DEHP exposure affects mitochondrial ROS formation. Therefore, ROS production in PC12 cells was measured using fluorescent dye DCFDA. The levels of intracellular ROS markedly increased after treatment with DEHP. However, pretreatment with HE (0.5 mg/mL) significantly decreased the intracellular ROS generated by DEHP in PC12 cells (Figure 2).

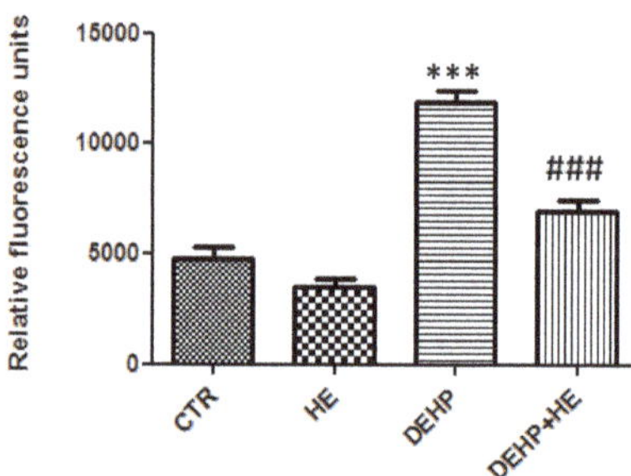

Figure 2. Effects of HE on DEHP-induced reactive oxygen species (ROS) generation. PC12 cells were pretreated with HE (1 µM) for 2 h before DEHP treatment for 24 h (85 µM). The relative intracellular ROS production was evaluated by recording the fluorescence of DCF, the product of 2, 7 - Dichlorofluorescein diacetate (DCFH) oxidation. Data are expressed as the mean ±SD of three separate experiments. *** $p \leq 0.001$ vs. control and ### $p \leq 0.001$ vs. DEHP alone.

It has been reported that electron transport chain (ETC) complexes are important sources of mitochondrial reactive oxygen species, and their inhibition has been associated with elevated levels of ROS [41]. Given the observed induction of ROS after DEHP treatment, we hypothesized that this excess of ROS may be mediated by mitochondrial dysfunction. To assess this, we analyzed the activity of respiratory chain complexes in PC12 cells treated with DEHP and the effect of HE pretreatment. The data presented in Table 1 shows that the enzymatic activities of complexes I, II-III, IV, as well as ATP synthase drastically decreased after DEHP treatment. However, pretreatment with HE reduces DEHP-induced alterations in mitochondrial respiratory complex activities and significantly restored activity of complex I and II+III ($p < 0.05$).

Table 1. Effect of HE on di(2-ethylhexyl)phthalate (DEHP) treatment on the activity of Complex I, II–III, and IV and ATP Synthase in PC12 cells.

	NADH-UQ Oxidoreductase Complex I µmol Oxidized NADH/min/mg Protein	Succinate-Cytochrome c Oxidoreductase Complex II-III µmol Reduced Cyt c/min/mg Protein	Cytochrome c Oxidase Complex IV µmol Reduced Cyt c/min/mg Protein	ATP Synthase Complex V µmol Oxidized NADH/min/mg Protein
CTR	0.053 ± 0.0036	0.050 ± 0.0013	0.1882 ± 0.0152	0.0066 ± 0.0008
HE	0.0452 ± 0.0036	0.043 ± 0.0015	0.1742 ± 0.0078	0.0075 ± 0.00049
DEHP	0.0182 ± 0.0007***	0.026 ± 0.0031***	0.0961 ± 0.0173***	0.0034 ± 0.00040**
DEHP+HE	0.037 ± 0.0031#	0.035 ± 0.0037#	0.1218 ± 0.0067	0.0047 ± 0.00032

Effect of HE on DEHP treatment on the activity of Complex I, II–III, and IV and ATP Synthase in PC12 cells. # $p \leq 0.05$, ** $p \leq 0.01$, *** $p \leq 0.001$.

ROS overproduction is accompanied by increased expression of genes participating to recovery of mitochondrial function, detoxification, and cell survival, stress responsive, genes called vitagenes. Vitagenes encode for heat shock proteins (Hsps), thioredoxin, and sirtuin protein systems [42]. The effect of DEHP given alone or combined with HE on vitagenes expression is shown in Figures 3 and 4.

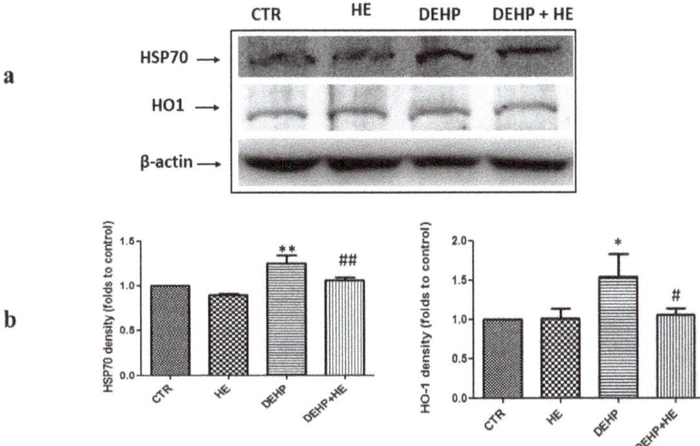

Figure 3. (**a**) The effect of DEHP and HE treatment on HSP70 and HO-1 protein levels. PC12 cells were pretreated with HE (1 µM) for 2 h before DEHP treatment for 24 h (85 µM). The cell lysate proteins were separated by SDS–PAGE, transferred to nitrocellulose membranes, and immunoblotted with the antibodies against HSP70 and HO-1. Protein loading was assessed by re-probing the blots with the β-actin antibody. (**b**) The bars represent the percentage changes of density (folds to control) SD of three independent experiments. * $p \leq 0.05$ vs. control, ** $p \leq 0.01$ vs. control, # $p \leq 0.05$ vs. DEHP alone, ## $p \leq 0.01$ vs. DEHP alone.

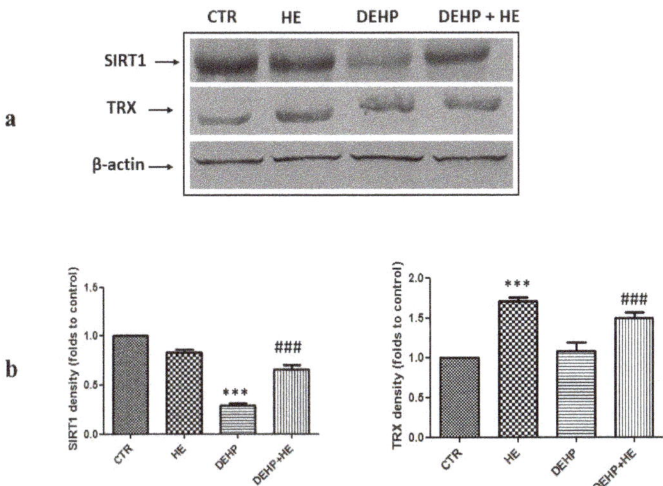

Figure 4. (**a**) The effect of DEHP and HE treatment on SIRT1 and TRX protein levels. PC12 cells were pretreated with HE (1 µM) for 2 h before DEHP treatment for 24 h (85 µM). The cell lysate proteins were separated by SDS–PAGE, transferred to nitrocellulose membranes, and immunoblotted with the antibodies against SIRT1 and TRX. Protein loading was assessed by re-probing the blots with the β-actin antibody. (**b**) The bars represent the percentage changes of density (folds to control) SD of three independent experiments. *** $p \leq 0.001$ vs. control, ### $p \leq 0.001$ vs. DEHP alone.

Figures 3a,b and 4b show an increase of HSP70, HO-1, and Trx protein levels induced by DEHP in PC12 cells, whereas this increase was not statistically significant when compared to control for Trx expression. However, DEHP decreased the expression of SIRT1 (Figure 4a). Pretreatment of cells by HE for 2 h, significantly modulates expression of stress responsive vitagenes (Figures 3a,b and 4a,b).

The transcription factor Nrf2 is activated under stressful conditions. Nrf2 binds to the ARE of DNA, leading to the transcription of cytoprotective genes, including HO-1 and Hsp60 [33]. After treatment with DEHP, an increase in the level of Nrf2 was observed, but pretreatment of cells with HE, modulated redox induced Nrf2 expression (Figure 5).

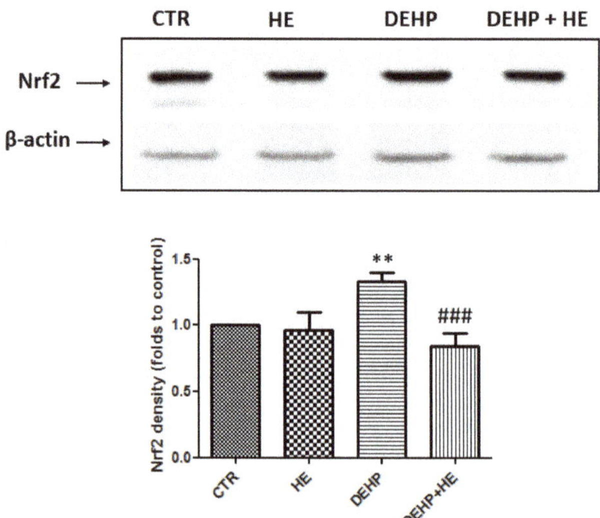

Figure 5. (**a**) The effect of DEHP and HE treatment on Nrf2 protein levels. PC12 cells were pretreated with HE (1 µM) for 2 h before DEHP treatment for 24 h (85 µM). The cell lysate proteins were separated by SDS–PAGE, transferred to nitrocellulose membranes, and immunoblotted with the antibodies against Nrf2. Protein loading was assessed by re-probing the blots with the β-actin antibody. (**b**) The bars represent the percentage changes of density (folds to control) SD of three independent experiments. ** $p \leq 0.01$ vs. control and ### $p \leq 0.001$ vs. DEHP alone.

Since mitochondrial anion fluxes are impacted by oxidative modifications, we subsequently examined whether DEHP induced dissipation of the mitochondrial membrane potential (MMP). Rhodamine 123 (Rh123) fluorescence was used to measure the MMP associated with DEHP treatment. As shown in Figure 6a, DEHP treatment reduced the Rh123 uptake to indicating a loss of mitochondrial potential, that was prevented when the cells were pretreated with HE for 2 h.

In addition, double staining cells with FITC-labeled-AnnexinV and PI (Figure 6b) allowed us to confirm by flow cytometry that DEHP induced apoptosis. Compared to the control values, DEHP at 85 µM increased the percentage of early apoptotic cells (AnnV+/PI−) to about 3.1% and late apoptotic/necrotic cells (AnnV+/PI+) to about 16.9% (Figure 6b). Nevertheless, cell pretreatment with HE for 2 h significantly reduced the rate of apoptotic cells (Figure 6b).

Neurodegenerative disorders are associated with mitochondria dysfunctions, alteration of the respiratory chain complexes, MMP decrease, and ROS increase [43]. Excessive ROS generation, MMP disturbance, and modulation of pro and anti-apoptotic proteins play a key role in neuronal apoptosis, thus we tested by western blotting analysis the effect of DEHP on apoptotic protein expression in PC12 cells.

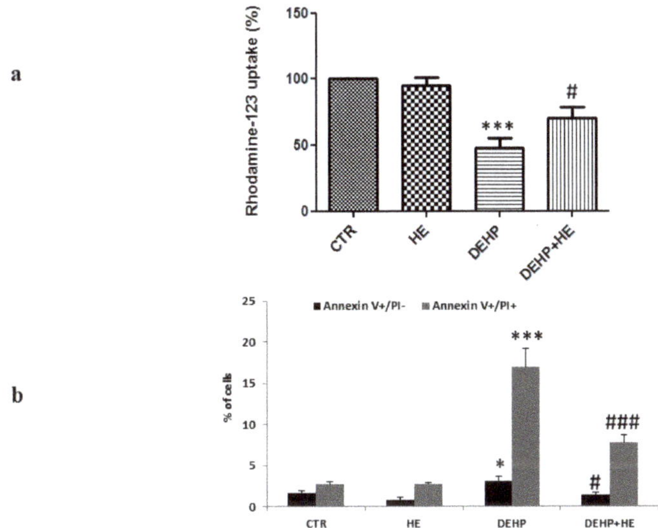

Figure 6. (**a**) Effects of HE on DEHP-induced loss of mitochondrial transmembrane potential. PC12 cells were pretreated with HE (1 µM) for 2 h before DEHP treatment for 24 h (85 µM). The mitochondrial potential was assessed by measuring the uptake of Rhodamine-123. (**b**) Effects of HE on DEHP-induced cell apoptosis. Different subsets of cells were measured by AnnexinV/PI staining after treatment with DEHP (85 µM) and/or HE (1 µM). Early apoptotic cells are positive for AnnexinV and negative for PI (AnnV+/PI-) and late apoptotic/necrotic cells are both positive for AnnexinV and PI (AnnV+/PI+). Data are expressed as the mean ± SD of three separate experiments. * $p \leq 0.05$ vs. control, *** $p \leq 0.001$ vs. control, # $p \leq 0.05$ vs. DEHP alone, ### $p \leq 0.001$ vs. DEHP alone.

The results in Figure 7 show that the expression of apoptosis biomarker, p53, increased after DEHP treatment (Figure 7a,b). This induction was associated with the overexpression of proapoptotic protein, Bax (Figure 7a,b), and with a decrease in antiapoptotic protein, Bcl2 (Figure 7a,b). All these effects were prevented by pretreatment with HE (Figure 7a,b). We next examined involvement of caspase 3 in apoptosis pathways. Western blotting analysis showed that DEHP treatment resulted in increased level of cleaved active caspase 3 (Figure 7a,b). In addition, caspase-3 activity was measured by fluorimetric assay. As shown in Figure 7c, DEHP treatment significantly increased caspase-3 activity. HE pretreatment was found to be effective to prevent DEHP-induced activation of caspase-3 (Figure 7a–c). Our results indicate the involvement of the caspase-mediated apoptosis pathway.

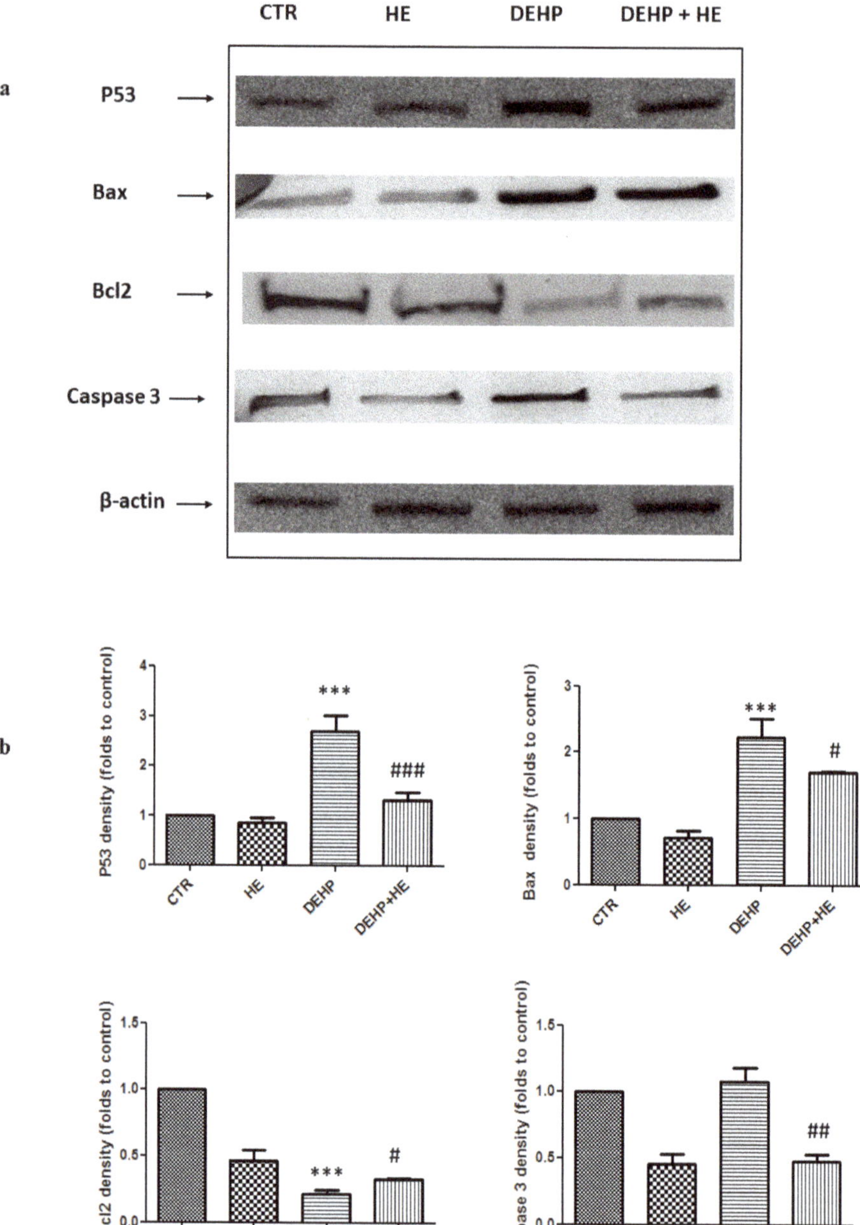

Figure 7. *Cont.*

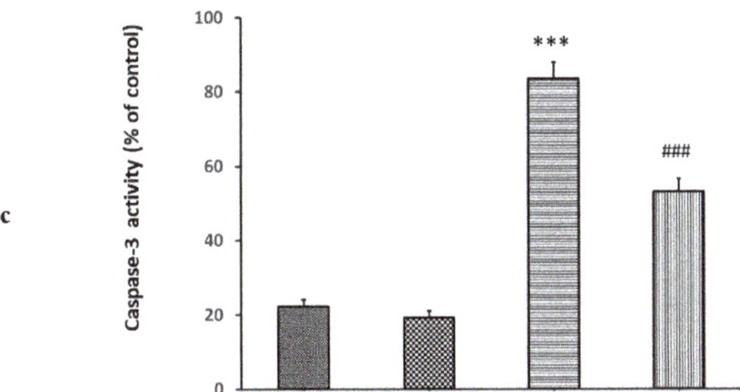

Figure 7. (**a**) The effect of DEHP and HE treatment on P53, Bax, Bcl2, and caspase-3 protein levels. PC12 cells were pretreated with HE (1 μM) for 2 h before DEHP treatment for 24 h (85 μM). The cell lysate proteins were separated by SDS–PAGE, transferred to nitrocellulose membranes and immunoblotted with the antibodies against P53, Bax, Bcl2, and activated caspase-3. Protein loading was assessed by reprobing the blots with the β-actin antibody. (**b**) The bars represent the percentage changes of density (folds to control) SD of three independent experiments. (**c**) The effect of DEHP and HE treatment on caspase-3 activity. PC12 cells were pretreated with HE (1 μM) for 2 h before DEHP treatment for 24 h (85 μM). Caspase-3 activity was measured using a commercialized kit. Data are expressed as the mean ± SD of three separate experiments. *** $p \leq 0.001$ vs. control, # $p \leq 0.05$ vs. DEHP alone, ## $p \leq 0.01$ vs. DEHP alone, ### $p \leq 0.001$ vs. DEHP alone.

3. Discussion

Although DEHP is extensively used as a plasticizer, few studies have focused on its brain tissue toxicity. In the present study, we investigated the protective effect of HE to counteract oxidative stress, mitochondrial energy deficit and cell death in PC12 cells. Our results showed that HE significantly increased PC12 cells survival against DEHP insult, an effect associated with a decrease in the level of ROS generation, and with modulation of mitochondrial respiratory complex activities, as well as reduction of apoptosis. The present data shows that DEHP induces oxidative stress as demonstrated by robust increase in ROS generation. ROS excessively produced during oxidative stress cause cell damage leading to cell death [33].

Pre-treatment of neuronal cells with HE at the dose of 0.5 mg/mL inhibits intracellular ROS formation and protects cells against DEHP-induced oxidative stress. Protective effects of HE against oxidative stress-induced injuries have been reported in various in vitro and in vivo studies [39,44]. So, we proposed that this anti-oxidative effect of HE may be partly though the possibility of direct elimination of ROS or by rescuing the efficiency of complex I given that the inhibition of this complex is related to ROS overproduction. In fact, it is known that overproduction of ROS might occur through energy dependent mechanisms, mainly a consequence of an inhibition of complex I which generally is associated with cell death [45,46].

To assess the possible alterations in energy metabolism following exposure to DEHP, we measured mitochondrial respiratory complex activities, i.e., NADH-UQ oxidoreductase (complex I), Succinate-cytochrome c oxidoreductase (complex II-III), Cytochrome c oxidase (complex IV) and ATP synthase (complex V). Our results clearly show that DEHP inhibits the mitochondrial respiratory complex I, II-III, IV and V. The mechanism by which DEHP inhibits these complexes is still unclear, however, we hypothesize that inhibition might result from an imbalance in redox status associated with alteration in mitochondrial membrane potential, which can eventually lead to structural disorganization and dysfunction of critical enzymes. Further studies are necessary, however, to determine whether DEHP induces structural and/or post-translational modifications of these enzymes. On the other hand, our results show for the first time that HE treatment could restore the activity of respiratory complexes I, II-III, IV, and V, suggesting that HE may play a direct role in preserving complex activities and eventually prevent ROS production, thus renewing the capacity of neurons to produce energy. Such alterations in respiratory complex activities could be associated with a disturbance of the oxidant/antioxidant balance equilibrium which induces cell degeneration.

Cells are endowed with cellular pathways involved in the maintenance of cellular homeostasis to recovery mitochondrial function and confer protection against oxidative stress [39]. Among these pathways, there is a complex network of the so-called longevity assurance processes, composed of several genes termed vitagenes [27], which include members of the HSP family, such as HO-1, Hsp72, but also sirtuins and the thioredoxin/thioredoxin reductase system [27,42,47]. Molecular chaperones have been known to protect cells against a wide variety of toxic conditions as well as oxidative stress, extreme temperatures or heavy metals exposure. Chaperones play also an important role in the preservation and repair of the correct conformation of the cellular macromolecules, such as proteins, RNAs, and DNA [48]. Indeed, HO-1 catalyzes the degradation of heme and produces carbon monoxide and bilirubin, which can directly scavenge free radicals and repair DNA damage caused by oxidative or nitrosative stresses [39]. Sirtuins are histone deacetylases which, in the presence of NAD^+ as a cofactor, catalyze the deacetylation reaction of histone substrates and transcriptional regulators. Sirtuins regulate different biological processes, such as apoptosis, cell differentiation, energy transduction, and glucose homeostasis [42]. Furthermore, Trx, is a major redox control system, consisting of a 12 kDa redox active protein Trx, and a homodimeric seleno-protein called thioredoxin reductase (TrxR1). TrxR1 is a flavoprotein that catalyzes the NADPH-dependent reduction of oxidized thioredoxin protein. It is usually located in the cytosol, but it translocates into the nucleus in response to various stimuli associated with oxidative stress, thereby playing a central role in protecting against oxidative stress [27].

In this work we provide experimental evidence that PC12 cells treatment with DEHP for 24 h results in upregulation of vitagenes, in particular Hsp70, HO-1, and Trx and in down regulation of SIRT1. However, our study shows for the first time that HE pretreatment modulates vitagenes expression in PC12 cells. Our results are consistent with evidence obtained in mice, showing neuroprotection by HE on Aβ25–35 peptide-induced cognitive dysfunction [49].

We also provide experimental evidence that upregulation of HO-1 might involve the transcription factor Nrf2, which was highly expressed in the nuclear fraction of cells exposed to DEHP. Nrf2 is a transcription factor that regulates the expression of genes involved in protection against oxidative stress. Nrf2-dependent transcription is under control of the amount of ROS present in cells. Indeed, under basal conditions, Nrf2 is localized in the cytoplasm in its inactive form where it is bound to its inhibitor Keap1 which promotes its degradation by the proteasome via an E3 ubiquitin ligase complex. However, under oxidative stress, Nrf2 dissociates from Keap1, moves into the nucleus, and activates AREs present in promoter regions of a set of genes [50]. Upregulation of Nrf2 observed in our study might be due to oxidative stress generated by this phthalate, which on the other hand, was modulated by HE pretreatment. In addition to this, we found inhibition of the mitochondrial enzyme complexes activities, which can be considered an event preceding the reduction of the mitochondrial membrane potential. Consistent to this finding, excess ROS production by decreasing mitochondrial complex activities and promoting the decline of mitochondrial membrane potential, can induce, as a potent mediator, cell pathway death [39].

We report here that DEHP induces apoptosis through oxidative stress. To understand whether pretreatment with HE can alleviate apoptosis following DEHP exposure, some key factors involved in apoptosis pathway were further evaluated in this study. The apoptotic process is known to be triggered in cells through either the extrinsic or the intrinsic pathway [12]. The intrinsic apoptotic pathway is regulated by the Bcl-2 family proteins which are important modulators of MMP [12]. In order to determine whether Bax, a pro-apoptotic protein, and Bcl2, an anti-apoptotic protein of Bcl-2 family, contributed in the regulation of DEHP-induced decrease of MMP, their expression levels were measured. Our results show that DEHP treatment increased Bax and decreased Bcl2 expressions. Consistent to our findings, it is known that cellular stresses lead to stabilization and activation of the p53 tumor suppressor protein. Moreover, depending on the cellular context, this results in one of two different outcomes: cell cycle arrest or apoptotic cell death. Cell death induced through the p53 pathway is executed by caspase proteinases, which, by cleaving their substrates, lead to the characteristic apoptotic phenotype such as chromatin condensation, plasma membrane asymmetry, and the formation of apoptotic bodies [51]. Our results clearly show that DEHP induces a p53 and caspase-3 activations. These findings confirm that PC12 cells, treated by DEHP, underwent a p53 and caspase-dependent apoptosis. In the other hand, when combined to DEHP (85 µM) and HE (0.5 mg/mL) significantly reduced the apoptosis induced by this phthalate by decreasing the loss of membrane mitochondrial potential, diminishing Bcl2 expression level, increasing Bax, P53, and Caspase-3 expression levels. Although, at the present we cannot precisely indicate whether the apoptotic pathway involved is intrinsic or extrinsic, however our findings support that HE induces an anti-apoptotic activity. It is also known that tryptophan is an essential amino acid and it is a precursor of 5-hydroxytryptamine (serotonin), which is involved in the physiological regulation of several behavioral and neuroendocrine functions [52]. Tryptophan pathway regulated by the rate-limiting enzyme, indoleamine-2,3-dioxygenase (IDO-1), has evolved as a therapeutic target in immunosuppression-induced cancer autoinflammatory diseases [53–56]. Consistent with this notion, the gut microbiota, which is part of a complex physiological networks, is considered an important source of tryptophan and tryptophan-derived metabolites also involved in neurotransmitter synthesis [57–59]. In addition, genetically susceptible individuals with impaired mucosal integrity, undergo escaping of microbial antigens through the epithelial barrier, presenting higher risk for inappropriate immune response and/or underlying chronic inflammation [60]. Our results are relevant to the biology of chronic inflammatory pathology, both in central and peripheral tissues, as many researches have demonstrated that polysaccharides, naturally occurring substances derived from plants or mushrooms [23,37], exhibit favorable therapeutic and health-promoting benefits [61], particularly in relation to diseases associated with inflammation [22,37–39]. HE is a medicinal fungus, with the effect of prevention and treatment of gastrointestinal disorders, particularly, owing to its potential therapeutic effect on cancer, promoting immune stimulation and improving lipid metabolism and thus gastrointestinal pathology, where it has been shown that HE supplementation can improve the immune system via regulation of metabolism and composition of gut microbiota [60], thereby reducing ulceration and providing protection against gastric mucosal damage [62]. In conclusion, the present work gives new indication on molecular mechanisms occurring in DEHP exposed PC12 cells, resulting in compensative events against energy defect, and identifies a novel activity of HE to counteract mitochondrial energy deficit, ROS generation, and apoptosis in PC12 cells.

4. Materials and Methods

4.1. Chemicals

DEHP was purchased from Sigma-Aldrich (St. Louis, MO, USA), *Hericium erinaceus* was from Mycology Research Laboratories Ltd. (Luton, United Kingdom), 3-4,5-Dimethylthiazol-2-yl, 2,5-diphenyltetrazolium bromide (MTT), Cell culture medium Dulbecco's modified Eagle medium (DMEM), horse serum, fetal bovine serum (FBS), phosphate buffer saline (PBS), trypsin-EDTA, penicillin and streptomycin mixture and l-glutamine (200 mM) were from GIBCO-BCL (UK). 2,7-Dichlorofluoresce diacetate (DCFH-DA) was supplied by Molecular Probes (Cergy Pontoise, France). All other chemicals used were of analytical grade.

4.2. Hericium Erinaceus (HE) Biomass Preparation

Hericium erinaceus is found almost worldwide; however, its bioactivity varies depending on the habitat in which it grows. To eliminate these variations, established HE-OX strain was used which demonstrates rapid and aggressive colonization. HE, containing both mycelium and primordia (young fruit body) biomass, obtained cultivating the biomass that is grown on a sterilized (autoclaved) substrate. The production process involves the inoculation of sterile organic edible grain with spawn from the mother culture. The fungus is allowed to completely colonize the growth medium aseptically. At the correct stage of development, corresponding to the maximum bioavailability, the living biomass is aseptically air-dried, granulated, tested microbiologically, and reduced in powder for tablet preparation. In comparison to *Hericium* extracts, biomass has the advantage of preserving all nutraceutical potential which is usually reduced with extracts or concentrates, including lyophilisation, and thus the activity of the product corresponds with the source mushroom, while being further intensified by utilizing the entire mycelium. Powder of *Hericium erinaceus* biomass containing mycelium and primordia of the respective mushroom, as the product commercially available, were used for experiments. Optimal concentration was chosen according to previous studies [39,63].

4.3. Cell Culture

PC12 (ATCC® CRL-1721™) rat cell line were used for experiments. Cells were cultured in DMEM, supplemented with 10% horse serum, 5% FBS, 1% L glutamine (200 mM), 1% of mixture penicillin (100 IU/mL), and streptomycin (100 µg/mL), at 37 °C in a CO_2 incubator. Before the experiments, cells were differentiated by culturing in serum-free medium containing 50 ng/mL nerve growth factor (NGF) for 5 days. Protein concentration was estimated using the Bradford assay by spectrophotometrically reading at 595 nm [64].

4.4. Cell Viability Assay (MTT)

The cell viability was determined by the MTT assay as described in [65]. This test is based on the ability of living cells to metabolize the yellow tetrazolium salt to a blue formazan via the mitochondrial succinate dehydrogenase which is a member of mitochondrial electron transfer system complex. To determine the neuroprotective effect of HE, PC12 cells (10^5 cells/well in a 24-well plate) were incubated at 37 °C after pretreatment with HE (1 µM) for 2 h and then incubated with DEHP (83 µM) for 24 h. A negative control containing only cells was also evaluated. After treatment, cells were incubated with 5 mg/mL MTT for 3 h at 37 °C, the medium was removed carefully after the incubation and the formazan crystals were dissolved in 150 µL of DMSO and absorbance of formazan reduction product was measured by spectrophotometer at 570 nm using a microplate reader (Biotek, Elx 800, USA). The results were expressed as the percentage of MTT reduction relative to the absorbance measured from negative control cells. All assays were performed in triplicate. Based on the results obtained from cell viability assay, the effective dose of HE against DEHP toxicity was utilized to study the effect of HE by assessing reactive oxygen species (ROS), mitochondrial membrane potential (MMP), and apoptotic protein marker expression.

4.5. Determination of Reactive Oxygen Species (ROS) and Oxidative Stress Status

The intracellular ROS amounts were determined using a fluorometric assay with 2,7-dichlorofluorescein diacetate (DCFHDA). After diffusion inside the cell, the probe is hydrolyzed by intracellular esterase to non-fluorescent dichlorofluorescein (DCFH) and then oxidized to fluorescent DCF in the presence of ROS [66]. PC12 cells were seeded on 24-well culture plates at 10^5 cells/well for 24 h. After incubation, cells were treated with 20 µM DCFHDA. Intracellular production of ROS was measured after 30 min incubation at 37 °C by fluorometric detection of DCF oxidation on a fluorimeter with an excitation wavelength of 485 nm and emission wavelength of 522 nm. Results are expressed as the ratio of intensity of fluorescence in treated cells to that of the HE responding fluorescence in the control.

4.6. Measurement of Mitochondrial Membrane Potential (MMP)

Changes in MMP were determined by the mitochondrial specific, incorporation of a cationic fluorescent dye Rhodamine-123 (Rh-123) [67]. In a typical experiment, the seeded cell in 96- well culture plates were treated with DEHP alone or combined to HE for 24 h. Then cells were rinsed with PBS and 100 μL of Rh-123 (1 μM) in PBS was added on the plates. Cells were incubated (37 °C, 5% CO_2) for 15 min. Next, the PBS solution containing non-uptaken Rh-123 was washed and replaced by fresh PBS and estimated by fluorimetric detection. The results were expressed as the percentage of uptaken Rh-123 fluorescence relative to the fluorescence measured from negative control cells.

4.7. Cell Death Induced by DEHP

To distinguish apoptotic versus necrotic cells, Annexin V/propidium iodide (AnnV/PI) double staining was performed. PI in combination with FITC-AnnV permit to discriminate between viable (AnnV-/PI-), early apoptotic (AnnV+/PI-), and late apopto-tic/necrotic (AnnV+/PI+) cells. The Annexin V assay was performed following the manufacturer's instructions (Annexin V-FITC kit, Bender MedSystems, Vienna, Austria). Fluorescence of at least 5000 cells was analyzed by flow cytometry.

4.8. Preparation of Mitoplasts

PC12 cells were seeded on 6-well culture plates (Polylabo, France) at 1×10^5 cells/well for 24 h of incubation, after, the cells were incubated with DEHP (85 μM) alone or combined to HE (1 μM), for 24 h at 37 °C PC12 cells were collected by trypsinization, pelleted by centrifugation at 500× *g* and resuspended in PBS (pH 7.4). The cell suspension was exposed for 10 min on ice to 2 mg of digitonin/mg cellular proteins. The mitoplast fraction, obtained by digitonin cell disruption, was pelleted at 14,000× *g* and resuspended in PBS [45].

4.9. Mitochondrial Respiratory Complex Activity Assay

Mitochondrial complexes activities were determined in mitochondrial membrane-enriched fractions from PC12 cells [45]. Aliquots of trypsinized cells were washed with ice-cold PBS, frozen in liquid nitrogen, and kept at −80 °C until use. To isolate the mitochondrial membrane-enriched fractions, cell pellets were thawed at 2–4 °C, suspended in 1 mL of 10 mM Tris–HCl (pH 7.5), supplemented with 1 mg/mL BSA, and exposed to ultrasound energy for 15 s at 0 °C. The ultrasound-treated cells were centrifuged (10 min at 600× *g*, 4 °C). The supernatant was obtained and centrifuged again (10 min at 14,000× *g*, 4 °C) to collect a mitochondrial pellet that was suspended in 0.1 mL of the respiratory medium. The activity of NADH:ubiquinone oxidoreductase (complex I) was measured in 40 mM potassium phosphate buffer, pH 7.4, 5 mM $MgCl_2$, in the presence of 3 mM KCN, 1 μg/mL antimycin, 200 μM decylubiquinone, using 50 μg of mitoplast proteins, by following the oxidation of 100 μM NADH at 340–425 nm ($\Delta\varepsilon = 6.81$ mM^{-1} cm^{-1}). The activity was corrected for the residual activity measured in the presence of 1 μg/mL rotenone. [46]. Succinate-cytochrome c oxidoreductase (complex II + III) activity was determined by following its reduction at 550–540 nm in the presence of cytochrome c ($\Delta\varepsilon = 19.1$ mM^{-1} cm^{-1}). The activity was also determined in the presence of antimycin A. The activity of Cytochrome c oxidase (complex IV) was established following the ferrocytochrome c oxidation at 550–540 nm ($\Delta\varepsilon = 19.1$ mM^{-1} cm^{-1}). Complex V activity (ATP hydrolase activity) was measured by an ATP-regenerating system. Frozen and thawed cells were suspended (at 0.1 mg protein/mL) in a buffer consisting of 375 mM sucrose, 75 mM KCl, 30 mM Tris–HCl pH 7.4, 3 mM $MgCl_2$, 2 mM PEP, 55 U/mL lactate dehydrogenase, 40 U/mL pyruvate kinase, 0.3 mM NADH. The reaction was started by the addition of 1 mM ATP and the oxidation of NADH was followed at 340 nm [45].

4.10. Protein Extraction and Western Blot Analysis

After treatment with DEHP (85 μM), alone or combined to HE (1 μM), for 24 h at 37 °C, PC12 cells (1×10^5) in 6-well plates were harvested, washed with PBS, and lysed in 100 μL lysis buffer

(Hepes 0.5 M containing 0.5% Nonidet-P40, 1 mM PMSF, 1 mg/mL aprotinin, 2 mg/mL leupeptin, pH 7.4), and incubated 20 min in ice before centrifugation. Protein concentrations were determined in cell lysates using Protein BioRad assay. Western blot was carried out as described in [39], proteins extracted for each sample, at equal concentration (50 μg) were boiled for 3 min in sample buffer (containing 40 mM Tris–HCl pH7.4, 2.5 % SDS, 5 % 2-mercaptoethanol, 5 % glycerol, 0.025 mg/mL of bromophenol blue), and then separated on a polyacrylamide mini gels precasting 4-20 % (codNB10420 NuSept Ltd. Australia). Separated proteins were transferred onto nitrocellulose membrane (BIO-RAD, Hercules, CA, USA) in transfer buffer containing (0.05 % SDS, 25 mM Tris, 192 mM glycine, and 20 % v/v methanol). The transfer of the proteins on the nitrocellulose membrane was confirmed by staining with Ponceau Red which was then removed by three washes in PBS (phosphate buffered saline) for 5 min each. Membranes were then incubated for 1 h at room temperature in 20 mM Tris pH 7.4, 150 mM NaCl, and Tween 20 (TBS-T) containing 2 % milk powder and incubated with appropriate primary anti-Hsp70 (SC-10789, Santa Cruz Biotech. Inc), anti-heme oxygenase-1 (HO-1) (SC-10789, Santa Cruz Biotech. Inc), anti-Thioredoxin (Trx) (Sc-13526, Santa Cruz Biotech. Inc.), anti-Sirt1 (SC-74465, Santa Cruz Biotech. Inc.), anti-Bax (SC-7480, Santa Cruz Biotech. Inc, anti-Bcl-2 (SC-7382, Santa Cruz Biotech. Inc), anti-p53 (SC-126, Santa Cruz Biotech. Inc), and anti-caspase-3 polyclonal (Santa Cruz Biotech. Inc.) overnight at 4 °C. The same membrane was incubated with a goat polyclonal antibody anti-beta-actin (SC-1615 Santa Cruz Biotech. Inc., Santa Cruz, CA, USA) to verify that the concentration of protein loaded in the gel was the same in each sample. Excess unbound antibodies were removed by three washes are with TBS-T for 5 min. After incubation with primary antibody, the membranes were washed three times for 5 min. in TBS-T and then incubated for 1 h at room temperature with the secondary polyclonal antibody conjugated with horseradish peroxidase (dilution1:500). The membranes were then washed three times with TBS-T for 5 min. Finally, the membranes were incubated for 3 min with Super Signal chemiluminescence detection system kit (Cod34080 Pierce Chemical Co, Rockford, IL, USA) to display the specific protein bands for each antibody. The immunoreactive bands were quantified by capturing the luminescence signal emitted from the membranes with the Gel Logic 2200 PRO (Bioscience) and analyzed with Molecular Imaging software for the complete analysis of regions of interest for measuring expression ratios.

4.11. Caspase-3 Activation Assay

The measure of caspase-3 activity was performed using a commercially available kit, according to the manufacturer's instructions (BD Pharmingen). At 50% confluence, PC12 cells were cultured in the presence of DEHP alone (85 μM), HE alone (1 μM), or DEHP with HE at 37 °C for 24 h. Cells were harvested and incubated in lysing buffer for 30 min and then incubated with 20 μM Ac-DEVD-AMC (a substrate for caspase-3-like proteases) at 37 °C for 1 h. The release of aminomethylcoumarin (AMC) was then measured on a Perkin-Elmer fluorimeter using an excitation/emission wave length of 380 nm/460 nm. The results were corrected for total protein content in the lysates using the bicinchoninic acid assay (Pierce, Rockford, IL, USA), and expressed as the percentage of activity in lysates from control cultures.

4.12. Statistical Analysis

Each experiment was done independently three times. Values are presented as the mean ± standard deviation (SD). The analysis parameters were tested for homogeneity of variance and normality, and they were found to be normally distributed. The data were therefore analyzed using a one-way analysis of variance (ANOVA) with a post hoc Tukey–Kramer test to identify significance between groups and their respective controls. In all cases, $p < 0.05$ (*), $p < 0.01$ (**), $p < 0.001$ (***) was considered statistically significant.

Author Contributions: I.A. and M.S. prepared the figures; A.Z., M.L.O., A.P., S.A.-E., and L.M. were involved in the drafting of this article; A.S. contributed to the methodology and to the revision of the manuscript; V.C. and A.T.S. modified and supervised the final version of the manuscript. All authors reviewed the final manuscript and gave approval for the presentation. I.A. and M.S., contributed equally to this work. All authors have read and agreed to the published version of the manuscript.

Funding: V.C. acknowledges support from Piano Ricerca Triennale-linea Intervento 2, University of Catania.

Acknowledgments: We acknowledge helpful discussions with Bill Ahern and Malcom Clark, from Mycology Research Laboratories Ltd., Luton, UK.

Conflicts of Interest: The authors declare no conflicts of interest.

References

1. Mariana, M.; Feiteiro, J.; Verde Cairrao, E. The effects of phthalates in the cardiovascular and reproductive systems. *Environ. Int.* **2016**, *94*, 758–776. [CrossRef] [PubMed]

2. Skinner, M.K. Endocrine disruptors epigenetic transgenerational inheritance. *Nat. Rev. Endocrinol.* **2015**, *12*, 68–70. [CrossRef] [PubMed]

3. Schmid, P.; Schlatter, C. Excretion and metabolism of di(2 ethylhexyl)phthalate in man. *Xenobiotica* **1985**, *15*, 251–256. [CrossRef] [PubMed]

4. Serrano, S.E.; Braun, J.; Trasande, L.; Dills, R.; Sathyanarayana, S. Phthalates and diet: A review of the food monitoring and epidemiology data. *Environ. Health A Glob. Access Sci. Source* **2014**, *13*. [CrossRef]

5. Aung, K.H.; Win-Shwe, T.T.; Kanaya, M.; Takano, H.; Tsukahara, S. Involvement of hemeoxygenase-1 in di(2-ethylhexyl) phthalate (DEHP)-induced apoptosis of Neuro-2a cells. *J. Toxicol. Sci.* **2014**, *39*, 217–229. [CrossRef] [PubMed]

6. Dhanya, C.R.; Indu, A.R.; Deepadevi, K.V.; Kurup, P. Inhibition of membrane Na(+)-K+ Atpase of the brain, liver and RBC in rats administered di(2-ethyl hexyl) phthalate (DEHP) a plasticizer used in polyvinyl chloride (PVC) blood storage bags. *Indian J. Exp. Biol.* **2003**, *41*, 814–820.

7. Moore, R.W.; Rudy, T.A.; Lin, T.M.; Ko, K.; Peterson, R.E. Abnormalities of sexual development in male rats with in utero and lactational exposure to the antiandrogenic plasticizer Di(2-ethylhexyl) phthalate. *Environ. Health Perspect.* **2001**, *109*, 229–237. [CrossRef]

8. Lin, H.; Yuan, K.; Li, L.; Liu, S.; Li, S.; Hu, G.; Lian, Q.Q.; Ge, R.S. In Utero Exposure to Diethylhexyl Phthalate Affects Rat Brain Development: A Behavioral and Genomic Approach. *Int. J. Environ. Res. Public Health* **2015**, *12*, 13696–13710. [CrossRef]

9. Komada, M.; Gendai, Y.; Kagawa, N.; Nagao, T. Prenatal exposure to di(2- ethylhexyl) phthalate impairs development of the mouse neoHEtex. *Toxicol. Lett.* **2016**, *259*, 69–79. [CrossRef]

10. Yan, M.H.; Wang, X.; Zhu, X. Mitochondrial defects and oxidative stress in Alzheimer's disease and Parkinson disease. *Free Radic. Biol. Med.* **2013**, *62*, 90–101. [CrossRef]

11. Jacobson, M.D.; Weil, M.; Raff, M.C. Programmed cell death in animal development. *Cell* **1997**, *88*, 347–354. [CrossRef]

12. Parsons, M.J.; Green, D.R. Mitochondria in cell death. *Essays Biochem.* **2010**, *47*, 99–114. [CrossRef] [PubMed]

13. Elsayed, E.A.; El Enshasy, H.; Wadaan, M.A.; Aziz, R. Mushrooms: A potential natural source of anti-inflammatory compounds for medical applications. *Mediat. Inflamm.* **2014**, 805841. [CrossRef] [PubMed]

14. El Enshasy, H.; Elsayed, E.A.; Aziz, R.; Wadaan, M.A. Mushrooms and truffles: Historical biofactories for complementary medicine in Africa and in the Middle East. *Evid. Based Complement. Alternat. Med.* **2013**, 620451. [CrossRef] [PubMed]

15. Xu, T.; Beelman, R.B.; Lambert, J.D. The cancer preventive effects of edible mushrooms. *Anticancer Agents Med. Chem.* **2012**, *12*, 1255–1263. [CrossRef]

16. Paterson, R.R.; Lima, N. Biomedical effects of mushrooms with emphasis on pure compounds. *Biomed. J.* **2014**, *37*, 357–368. [CrossRef]

17. Komura, D.L.; Ruthes, A.C.; Carbonero, E.R.; Gorin, P.A.; Iacomini, M. Water-soluble polysaccharides from Pleurotus ostreatus var. florida mycelial biomass. *Int. J. Biol. Macromol.* **2014**, *70*, 354–359. [CrossRef]

18. Wasser, S.P. Medicinal mushroom science: Current perspectives, advances, evidences, and challenges. *Biomed. J.* **2014**, *37*, 345–356. [CrossRef]

19. Lindequist, U.; Kim, H.W.; Tiralongo, E.; Van Griensven, L. Medicinal mushrooms. *Evid. Based Complement. Alternat. Med* **2014**, 806180. [CrossRef]

20. da Silva, A.F.; Sartori, D.; Macedo, F.C., Jr.; Ribeiro, L.R.; Fungaro, M.H.; Mantovani, M.S. Effects of beta-glucan extracted from Agaricus blazei on the expression of ERCC5, CASP9, and CYP1A1 genes and metabolic profile in HepG2 cells. *Hum. Exp. Toxicol.* **2013**, 32647–32654. [CrossRef]

21. Lai, C.L.; Lin, R.T.; Liou, L.M.; Liu, C.K. The role of event-related potentials in cognitive decline in Alzheimer's disease. *Clin. Neurophysiol.* **2010**, *121*, 194–199. [CrossRef] [PubMed]

22. Trovato Salinaro, A.; Pennisi, M.; Di Paola, R.; Scuto, M.; Crupi, R.; Teresa Cambria, M.T.; Ontario, M.L.; Tomasello, M.; Uva, M.; Maiolino, L.; et al. Neuroinflammation and neurohormesis in the pathogenesis of Alzheimer's disease and Alzheimer-linked pathologies: Modulation by nutritional mushrooms. *Immun. Ageing* **2018**, *15*, 8. [CrossRef] [PubMed]

23. Scuto, M.C.; Mancuso, C.; Tomasello, B.; Laura Ontario, M.; Cavallaro, A.; Frasca, F.; Maiolino, L.; Trovato Salinaro, A.; Calabrese, E.J.; Calabrese, V. Curcumin, Hormesis and the Nervous System. *Nutrients* **2019**, *11*, 2417. [CrossRef]

24. Calabrese, V.; Santoro, A.; Monti, D.; Crupi, R.; Di Paola, R.; Latteri, S.; Cuzzocrea, S.; Zappia, M.; Giordano, J.; Calabrese, E.J.; et al. Aging and Parkinson's Disease: Inflammaging, neuroinflammation and biological remodeling as key factors in pathogenesis. *Free Radic. Biol. Med.* **2018**, *115*, 80–91. [CrossRef] [PubMed]

25. Calabrese, E.J.; Dhawan, G.; Kapoor, R.; Iavicoli, I.; Calabrese, V. HORMESIS: A Fundamental Concept with Widespread Biological and Biomedical Applications. *Gerontology* **2016**, *62*, 530–535. [CrossRef] [PubMed]

26. Calabrese, E.J.; Dhawan, G.; Kapoor, R.; Iavicoli, I.; Calabrese, V. What is hormesis and its relevance to healthy aging and longevity? *Biogerontology* **2015**, *16*, 693–707. [CrossRef]

27. Cornelius, C.; Graziano, A.; Calabrese, E.J.; Calabrese, V. Hormesis and vitagenes in aging and longevity: Mitochondrial control and hormonal regulation. *Horm. Mol. Biol. Clin. Investig.* **2013**, *16*, 73–89. [CrossRef]

28. Calabrese, E.J.; Calabrese, V.; Giordano, J. The role of hormesis in the functional performance and protection of neural systems. *Brain Circ.* **2017**, *3*, 1–13. [CrossRef]

29. Calabrese, V.; Santoro, A.; Trovato Salinaro, A.; Modafferi, S.; Scuto, M.; Albouchi, F.; Monti, D.; Giordano, J.; Zappia, M.; Franceschi, C.; et al. Hormetic approaches to the treatment of Parkinson's disease: Perspectives and possibilities. *J. Neurosci. Res.* **2018**, *96*, 1641–1662. [CrossRef]

30. Calabrese, E.J.; Bhatia, T.N.; Calabrese, V.; Dhawan, G.; Giordano, J.; Hanekamp, Y.N.; Kapoor, R.; Kozumbo, W.J.; Leak, R.K. Cytotoxicity models of Huntington's disease and relevance of hormetic mechanisms: A critical assessment of experimental approaches and strategies. *Pharmacol. Res.* **2019**, *150*, 104371. [CrossRef]

31. Wang, D.; Calabrese, E.J.; Lian, B.; Lin, Z.; Calabrese, V. Hormesis as a mechanistic approach to understanding herbal treatments in traditional Chinese medicine. *Pharmacol. Ther.* **2018**, *184*, 42–50. [CrossRef] [PubMed]

32. Calabrese, V.; Giordano, J.; Crupi, R.; Di Paola, R.; Ruggieri, M.; Bianchini, R.; Ontario, M.L.; Cuzzocrea, S.; Calabrese, E.J. Hormesis, cellular stress response and neuroinflammation in schizophrenia: Early onset versus late onset state. *J. Neurosci. Res.* **2017**, *95*, 1182–1193. [CrossRef] [PubMed]

33. Pennisi, M.; Crupi, R.; Di Paola, R.; Ontario, M.L.; Bella, R.; Calabrese, E.J.; Crea, R.; Cuzzocrea, S.; Calabrese, V. Inflammasomes, hormesis, and antioxidants in neuroinflammation: Role of NRLP3 in Alzheimer disease. *J. Neurosci. Res.* **2017**, *95*, 1360–1372. [CrossRef] [PubMed]

34. Calabrese, V.; Giordano, J.; Signorile, A.; Laura Ontario, M.; Castorina, S.; De Pasquale, C.; Eckert, G.; Calabrese, E.J. Major pathogenic mechanisms in vascular dementia: Roles of cellular stress response and hormesis in neuroprotection. *J. Neurosci. Res.* **2016**, *94*, 1588–1603. [CrossRef]

35. Calabrese, V.; Giordano, J.; Ruggieri, M.; Berritta, D.; Trovato, A.; Ontario, M.L.; Bianchini, R.; Calabrese, E.J. Hormesis, cellular stress response, and redox homeostasis in autism spectrum disorders. *J. Neurosci. Res.* **2016**, *94*, 1488–1498. [CrossRef]

36. Dattilo, S.; Mancuso, C.; Koverech, G.; Di Mauro, P.; Ontario, M.L.; Petralia, C.C.; Petralia, A.; Maiolino, L.; Serra, A.; Calabrese, E.J.; et al. Heat shock proteins and hormesis in the diagnosis and treatment of neurodegenerative diseases. *Immun. Ageing* **2015**, *12*, 20. [CrossRef]

37. Trovato, A.; Siracusa, R.; Di Paola, R.; Scuto, M.; Fronte, V.; Koverech, G.; Luca, M.; Serra, A.; Toscano, M.A.; Petralia, A.; et al. Redox modulation of cellular stress response and lipoxin A4 expression by Coriolus versicolor in rat brain: Relevance to Alzheimer's disease pathogenesis. *Neurotoxicology* **2016**, *53*, 350–358. [CrossRef]

38. Scuto, M.; Di Mauro, P.; Ontario, M.L.; Amato, C.; Modafferi, S.; Ciavardelli, D.; Trovato Salinaro, A.; Maiolino, L.; Calabrese, V. Nutritional Mushroom Treatment in Meniere's Disease with Coriolus versicolor: A Rationale for Therapeutic Intervention in Neuroinflammation and Antineurodegeneration. *Int. J. Mol. Sci.* **2019**, *21*, 284. [CrossRef]

39. Trovato Salinaro, A.; Siracusa, R.; Di Paola, R.; Scuto, M.; Ontario, M.L.; Bua, O.; Serra, A. Redox modulation of cellular stress response and lipoxin A4 expression by Hericium Erinaceus in rat brain: Relevance to Alzheimer's disease pathogenesis. *Immun. Ageing* **2016**, *13*, 23. [CrossRef]

40. Hao, L.; Xie, Y.; Wu, G.; Cheng, A.; Liu, X.; Zheng, R.; Huo, H.; Zhang, J. Protective Effect of Hericium Erinaceus on Alcohol Induced Hepatotoxicity in Mice. *Evid. Based Complement. Alternat. Med.* **2015**, 418023. [CrossRef]

41. Kowaltowski, A.J.; de Souza-Pinto, N.C.; Castilho, R.F.; Vercesi, A.E. Mitochondria and reactive oxygen species. *Free Radic. Biol. Med.* **2009**, *47*, 333–343. [CrossRef] [PubMed]

42. Calabrese, V.; Cornelius, C.; Leso, V.; Trovato-Salinaro, A.; Ventimiglia, B.; Cavallaro, M.; Scuto, M.; Rizza, S.; Zanoli, L.; Neri, S.; et al. Oxidative stress, glutathione status, sirtuin and cellular stress response in type 2 diabetes. *Biochim. Biophys. Acta* **2012**, *1822*, 729–736. [CrossRef] [PubMed]

43. Zoratti, M.; Szabò, I. The mitochondrial permeability transition. *Biochim. Biophys. Acta* **1995**, *1241*, 139–176. [CrossRef]

44. Liao, B.; Zhou, C.; Liu, T.; Dai, Y.; Huang, H. A novel Hericium erinaceus polysaccharide: Structural characterization and prevention of H(2)O(2)-induced oxidative damage in GES-1 cells. *Int. J. Biol. Macromol.* **2019**, *20*. [CrossRef]

45. Signorile, A.; Micelli, L.; De Rasmo, D.; Santeramo, A.; Papa, F.; Ficarella, R.; Gattoni, G.; Scacco, S.; Papa, S. Regulation of the biogenesis of OXPHOS complexes in cell transition from replicating to quiescent state Involvement of PKA and effect of hydroxytyrosol. *Biochim. et Biophys. Acta* **2006**, 4539–4543. [CrossRef]

46. De Rasmo, D.; Micelli, L.; Santeramo, A.; Signorile, A.; Lattanzio, P.; Papa, S. cAMP regulates the functional activity, coupling efficiency and structural organization of mammalian FOF1 ATP synthase. *Biochim. Biophys. Acta* **2016**, *1857*, 350–358. [CrossRef]

47. Calabrese, V.; Cornelius, C.; Dinkova-Kostova, A.T.; Calabrese, E.J.; Mattson, M.P. Cellular stress responses, the hormesis paradigm and vitagenes: Novel targets for therapeutic intervention in neurodegenerative disorders. *Antioxid. Redox Signal.* **2010**, *13*, 1763–1811. [CrossRef]

48. Choi, S.I.; Lim, K.H.; Seong, B.L. Chaperoning roles of macromolecules interacting with proteins in vivo. *Int. J. Mol. Sci.* **2011**, *12*, 1979–1990. [CrossRef]

49. Mori, K.; Obara, Y.; Moriya, T.; Inatomi, S.; Nakahata, N. Effects of Hericium erinaceus on amyloid β(25-35) peptide-induced learning and memory deficits in mice. *Biomed. Res.* **2011**, *32*, 67–72. [CrossRef]

50. Catino, S.; Paciello, F.; Miceli, F.; Rolesi, R.; Troiani, D.; Calabrese, V.; Santangelo, R.; Mancuso, C. Ferulic Acid Regulates the Nrf2/Heme Oxygenase-1 System and Counteracts Trimethyltin-Induced Neuronal Damage in the Human Neuroblastoma Cell Line SH-SY5Y. *Front. Pharmacol.* **2016**, *6*, 305. [CrossRef]

51. Schuler, M.; Green, D.R. Mechanisms of P53-dependent apoptosis. *Biochem. Soc. Trans.* **2001**, *29*, 684–688. [CrossRef] [PubMed]

52. Badawy, A.A.-B. Kynurenine pathway of tryptophan metabolism: Regulatory and functional aspects. *Int. J. Tryptophan Res.* **2017**, *10*. [CrossRef] [PubMed]

53. Dantzer, R. Role of the kynurenine metabolism pathway in inflammation-induced depression: Preclinical approaches. *Curr. Top. Behav. Neurosci.* **2017**, *31*, 117–138. [PubMed]

54. Stone, T.W.; Mackay, G.M.; Forrest, C.M.; Clark, C.J.; Darlington, L.G. Tryptophan metabolites and brain disorders. *Clin. Chem. Lab. Med.* **2003**, *41*, 852–859. [CrossRef]

55. Sorgdrager, F.J.H.; Naudé, P.J.W.; Kema, I.P.; Nollen, E.A.; Deyn, P.P. Tryptophan Metabolism in Inflammaging: From Biomarker to Therapeutic Target. *Front. Immunol.* **2019**, *10*, 2565. [CrossRef]

56. Gostner, J.M.; Geisler, S.; Stonig, M.; Mair, L.; Sperner-Unterweger, B.; Fuchs, D. Tryptophan Metabolism and Related Pathways in Psychoneuroimmunology: The Impact of Nutrition and Lifestyle. *Neuropsychobiology* **2019**, *26*, 1–11. [CrossRef]

57. Ragusa, N.; Sfogliano, L.; Calabrese, V.; Rizza, V. Effects of multivitamin treatment on the activity of rat liver tryptophan pyrrolase during ethanol administration. *Acta Vitaminol. Enzymol.* **1981**, *3*, 199–204.

58. Calabrese, V.; Ragusa, N.; Rizza, V. Effect of pyrrolidone carboxylate (PCA) and pyridoxine on liver metabolism during chronic ethanol intake in rats. *Int. J. Tissue React.* **1995**, *17*, 15–20.

59. Gao, K.; Mu, C.L.; Farzi, A.; Zhu, W.Y. Tryptophan Metabolism: A Link Between the Gut Microbiota and Brain. *Adv. Nutr.* **2019**. [CrossRef]

60. Diling, C.; Chaoqun, Z.; Jian, Y.; Jian, L.; Jiyan, S.; Yizhen, X.; Guoxiao, L. Immunomodulatory Activities of a Fungal Protein Extracted from Hericium erinaceus through Regulating the Gut Microbiota. *Front. Immunol.* **2017**, *8*, 666. [CrossRef]

61. Ferreiro, E.; Pita, I.; Mota, S.; Valero, J.; Ferreira, N.; Fernandes, T.; Calabrase, V.; Fontes-Ribeiro, C.; Pereira, F.; Rego, A. Coriolus versicolor biomass increases dendritic arborization of newly-generated neurons in mouse hippocampal dentate gyrus. *Oncotarget* **2018**, *8*, 32929–32942. [CrossRef] [PubMed]

62. Wang, D.; Zhang, Y.; Yang, S.; Zhao, D.; Wang, M. A polysaccharide from cultured mycelium of Hericium erinaceus relieves ulcerative colitis by counteracting oxidative stress and improving mitochondrial function. *Int. J. Biol. Macromol.* **2019**, *125*, 572–579. [CrossRef] [PubMed]

63. Nagano, M.; Shimizu, K.; Kondo, R.; Hayashi, C.; Sato, D.; Kitagawa, K.; Ohnuki, K. Reduction of depression and anxiety by 4 weeks Hericium erinaceus intake. *Biomed. Res.* **2010**, *31*, 231–237. [CrossRef] [PubMed]

64. Bradford, M.M. A rapid and sensitive method for the quantification of microgram quantities of protein utilizing the principle of protein-dye binding. *Anal. Biochem.* **1976**, *72*, 248–254. [CrossRef]

65. Mosmann, T. Rapid colorimetric assay for cellular growth and survival: Application to proliferation and cytotoxicity assays. *J. Immunol. Methods* **1983**, *65*, 55–63. [CrossRef]

66. McLennan, H.R.; Degli Esposti, M. The contribution of mitochondrial respiratory complexes to the production of reactive oxygen species. *J. Bioenerg. Biomembr.* **2000**, *32*, 153–162. [CrossRef] [PubMed]

67. Debbasch, C.; Brignole, F.; Pisella, P.J.; Warnet, J.M.; Rat, P.; Baudouin, C. Quaternary ammoniums and other preservatives' contribution in oxidative stress and apoptosis on Chang conjunctival cells. *Investig. Ophthalmol. Vis. Sci.* **2001**, *42*, 642–652.

International Journal of
Molecular Sciences

Article

Healthspan Enhancement by Olive Polyphenols in *C. elegans* Wild Type and Parkinson's Models

Gabriele Di Rosa [1,†], Giovanni Brunetti [1,†], Maria Scuto [1], Angela Trovato Salinaro [1], Edward J. Calabrese [2], Roberto Crea [3], Christian Schmitz-Linneweber [4], Vittorio Calabrese [1,*] and Nadine Saul [4]

[1] Department of Biomedical and Biotechnological Sciences, University of Catania, 95125 Catania, Italy; dirosagabriele85@gmail.com (G.D.R.); q.burneti@gmail.com (G.B.); mary-amir@hotmail.it (M.S.); trovato@unict.it (A.T.S.)

[2] Department of Environmental Health Sciences, Morrill I, N344, University of Massachusetts, Amherst, MA 01003, USA; edwardc@schoolph.umass.edu

[3] Oliphenol LLC., 26225 Eden Landing Road, Unit C, Hayward, CA 94545, USA; robertocrea48@gmail.com

[4] Faculty of Life Sciences, Institute of Biology, Molecular Genetics Group, Humboldt University of Berlin, Philippstr. 13, House 22, 10115 Berlin, Germany; christian.schmitz-linneweber@rz.hu-berlin.de (C.S.-L.); nadine.saul@gmx.de (N.S.)

* Correspondence: calabres@unict.it

† These authors contributed equally to this work.

Received: 11 May 2020; Accepted: 26 May 2020; Published: 29 May 2020

Abstract: Parkinson's disease (PD) is the second most prevalent late-age onset neurodegenerative disorder, affecting 1% of the population after the age of about 60 years old and 4% of those over 80 years old, causing motor impairments and cognitive dysfunction. Increasing evidence indicates that Mediterranean diet (MD) exerts beneficial effects in maintaining health, especially during ageing and by the prevention of neurodegenerative disorders. In this regard, olive oil and its biophenolic constituents like hydroxytyrosol (HT) have received growing attention in the past years. Thus, in the current study we test the health-promoting effects of two hydroxytyrosol preparations, pure HT and Hidrox® (HD), which is hydroxytyrosol in its "natural" environment, in the established invertebrate model organism *Caenorhabditis elegans*. HD exposure led to much stronger beneficial locomotion effects in wild type worms compared to HT in the same concentration. Consistent to this finding, in OW13 worms, a PD-model characterized by α-synuclein expression in muscles, HD exhibited a significant higher effect on α-synuclein accumulation and swim performance than HT, an effect partly confirmed also in swim assays with the UA44 strain, which features α-synuclein expression in DA-neurons. Interestingly, beneficial effects of HD and HT treatment with similar strength were detected in the lifespan and autofluorescence of wild-type nematodes, in the neuronal health of UA44 worms as well as in the locomotion of rotenone-induced PD-model. Thus, the hypothesis that HD features higher healthspan-promoting abilities than HT was at least partly confirmed. Our study demonstrates that HD polyphenolic extract treatment has the potential to partly prevent or even treat ageing-related neurodegenerative diseases and ageing itself. Future investigations including mammalian models and human clinical trials are needed to uncover the full potential of these olive compounds.

Keywords: *C. elegans*; polyphenols; olive oil; healthspan; lifespan; ageing; Parkinson's disease

---

1. Introduction

Emerging research has recently focused on increasing the life expectancy of humans which is, however, accompanied by a progressively greater prevalence of neurodegenerative disorders, notably

Parkinson's disease (PD). PD is the second most prevalent late-age onset neurodegenerative disorder affecting 1% of the population after the age of about 60 years old and 4% of those over 80 years old, causing motor impairments, cognitive dysfunction, sleep difficulties, autonomic dysfunctions, and pain [1]. PD is characterized by the progressive loss of dopaminergic neurons in the substantia nigra pars compacta (*SNpc*) of the midbrain area [2]. At the cellular level, the neuropathological hallmarks of PD include intra-cytoplasmic inclusions that contain α-synuclein aggregates, a primary component of intraneuronal Lewy bodies and Lewy neurites in vulnerable neurons of the brain [3]. The loss of dopaminergic neurons results in major motor impairments including resting tremor, muscle rigidity, bradykinesia and postural instability. Since PD affects neurons in the central and peripheral nervous systems, patients typically also exhibit multiple non-motor symptoms including anxiety, depression, memory loss, perturbed proteostasis, mitochondrial dysfunction, oxidative stress, dysregulation of redox homeostasis as well as neurotoxicity [4]. Although the etiology of PD is currently unknown, genetic and environmental triggers are two major factors that play a role in the development of the disease, with the environment accounting for over two-thirds of all cases [5]. Recently, longitudinal studies have identified at least 23 loci and 19 disease-causing genes for familial parkinsonism associated with the progression of non-motor symptoms in PD patients [6,7]. Moreover, several studies have suggested that mitochondrial complex I deficiency in different brain regions is associated with impairment of energy metabolism and neuronal death in PD [8]. It has been postulated that neuroinflammatory processes might play a crucial role in the pathogenesis of PD. The proteomic approach revealed accumulation of neurotoxic misfolded α-synuclein aggregates inducing microglial activation associated to loss of dopaminergic neurons in nigrostriatal system underlying PD pathogenesis [9]. Although treatments are available to alleviate motor symptoms, currently, there are no preventive therapies that can target and lessen PD progression. Recently, epidemiological and clinical studies have supported the idea that Mediterranean diet (MD) is strongly associated with lifespan extension as well as with healthy aging process by reducing the progression of age-related pathologies [10]. The beneficial effects of natural polyphenols and derivatives comprise multi-target activities including the anti-amyloid aggregation, antioxidant, antimicrobial, antihypertensive, hypoglycemic, antiproliferative and vasodilator effects, as well as redox homeostasis activities through a direct modulation of enzymes and proteins involved in stress response pathways [11–14]. Particular attention has been paid recently by our laboratory to the effects of natural olive polyphenols such as hydroxytyrosol (HT) and oleuropein aglycone (OLE), [10,15] known to possess healthspan benefits against α-synuclein aggregation into intracellular Lewy bodies, as found in PD neurons of the mesencephalic substantia nigra [16]. In addition, in vitro and in vivo studies have shown that HT exerts various protective effects, particularly, strong anti-oxidant and radical scavenging activities [17,18]. With regard to mechanism of polyphenol action, the biological concept of hormesis has emerged as a significant dose response model in the field of neuroprotection elicited by low dose of olive polyphenols [19]. Notably, increasing evidence suggests that mild stressors such as HT may offer beneficial effects in a hormetic-like manner by activating Nrf2-stress response pathway and enhancing brain resilience, neuroplasticity as well as lifespan in vitro and in experimental PD models [20,21]. Moreover, HT activates the Nrf2–antioxidant response element (ARE) pathway, leading to the activation of phase II detoxifying enzymes and the protection of dopaminergic neurons exposed to hydrogen peroxide or to 6-hydroxydopamine [22,23]. This is consistent with the idea that neurohormesis may have anti-aging effects due to induction of adaptive pathways triggered to cope with a mild neuronal stress and open novel potential therapeutic strategies for clinical interventions against the onset and/or progression of PD in humans. In these ways, neurohormetic polyphenols might protect neurons against injury and disease by stimulating the production of antioxidant enzymes, neurotrophic factors, protein chaperones and other proteins that help cells to withstand stress [24,25]. Interestingly, our recent in vivo study with olive polyphenols has demonstrated that HT and OLE exert neuroprotective effects, an improved overall healthspan and, in part, longevity in *Caenorhabditis elegans* (*C. elegans*) models of PD and wild type [15].

The nematode *C. elegans* is a multicellular model organism that offers several advantages for investigating both aging and neurodegenerative disorders [26]. The biological features of *C. elegans* are multiple and including short life cycle (i.e., about 3.5 days from egg to adult) and a lifespan of only about 20 days, transparent body, conserved gene network and neurological pathways [27,28]. Moreover, comparative proteomics indicates that for 83% of the *C. elegans* proteome human homologous can be identified [29]. Comparative genomic analysis also shows that nearly 53% of the human protein-coding genome has recognizable worm orthologues [30]. Furthermore, there is a tight connection between lifespan extension and resistance to diverse environments [31]. In this regard, several studies indicate that different stressors acting in hormetic-like manner extend lifespan in *C. elegans* [32], and suggest that hormetic effects could be exploited to prevent the onset of neurodegenerative diseases [20]. Most importantly, its complete genomic sequence is available. Therefore, *C. elegans* as a model organism is widely applied for screening natural bioactive compounds [33]. Several polyphenols effectively increase healthspan and lifespan as well as mitochondrial function in *C. elegans* [34–38]. Notably, *C. elegans* represents an excellent model to study the neuroprotective effects of olive polyphenols. In this context, recent research has demonstrated that extracts from olive leaves efficiently scavenged free radicals in vitro and significantly increased the expression of antioxidant enzymes extending lifespan and increased stress resistance in *C. elegans* [39,40].

In the current study we focus on the health-promoting effects of two hydroxytyrosol preparations, pure hydroxytyrosol and Hidrox®. Hidrox®/ Olivenol Plus™ (HD) is a patented freeze-dried hydroxytyrosol-rich formulation obtained from the acidic hydrolysis of olive vegetation water (OVW or olive juice) and where hydroxytyrosol (40–45% at the total water-soluble olive polyphenols) is maintained in its "natural" environment [41]. Olive juice (aqueous fraction) represents 50% of the weight of the olive fruit and is normally discarded as wastewater. Several findings have reported that HD displays different activity than pure or synthetic HT [40]. International in vitro and in vivo studies showed the health benefits and efficacy of HD as anti-inflammatory, anti-oxidant, anti-scavenger as well as anti-aggregating compound, particularly in PD [42].

We hypothesized that the polyphenolic treatments, Hidrox® and pure hydroxytyrosol have the capacity to increase the mean lifespan of *C. elegans* in the presence and absence of thermal stress. Furthermore, it is assumed that they are able to counteract the age-related deterioration of general health parameters, which were assessed by determining the swim performance as a measure of overall body fitness as well as the autofluorescence as one of the most popular ageing biomarkers [43]. Moreover, numerous in vitro studies were already successfully performed to verify the anti-PD effects of olive ingredients [14], however, in vivo studies were rarely conducted. Therefore, by using one chemically-induced and two transgenic PD models of *C. elegans*, the polyphenolic treatments were tested for their anti-PD effects in vivo. Besides the swim performance, neuronal degeneration as well as α-synuclein accumulation were taken into account to assess the anti-PD potential. Finally, it was hypothesized that HD is even more effective in preventing PD- and ageing-related symptoms than HT, which was tested by a direct comparison of the action of HT and HD.

2. Results

2.1. HD and HT Enhanced the Health and Lifespan of Wild Type Nematodes

As stress resistance is one of the key features characterizing the health status of an organism [44], heat stress resistance was determined in the presence and absence of HD in different concentrations. This test was also used to find the optimal concentration for further investigations. The results of HT treatment are shown in addition to enable the direct comparison between pure and mixed polyphenol treatments.

We observed the survival of worms starting from the 3rd day of adulthood, which was the day of heat stress exposure, until all worms died. The mean survival, which refers to the time between that stress exposure day until the end of the test, was increased by about 22% (from 2.23 days in the control group to 2.71 days in the treated group) during 250 µg/mL HT treatment (Figure 1A), whereas 250 µg/mL HD extended the survival by even 40% (from 2.96 in the control to 4.13 days) (Figure 1B). Moreover, the maximum survival, defined as the time point when 90% of the population was dead, increased by 63% during 250 µg/mL HD treatment ($p \leq 0.001$ with Fisher's Exact Test) and only by 14% during 250 µg/mL HT treatment ($p \leq 0.05$ with Fisher's Exact Test) compared to the respective control. 250 µg/mL was the most effective concentration for both treatments in terms of mean and maximum survival after stress exposure (Table S1), thus, this concentration was selected for the following experiments.

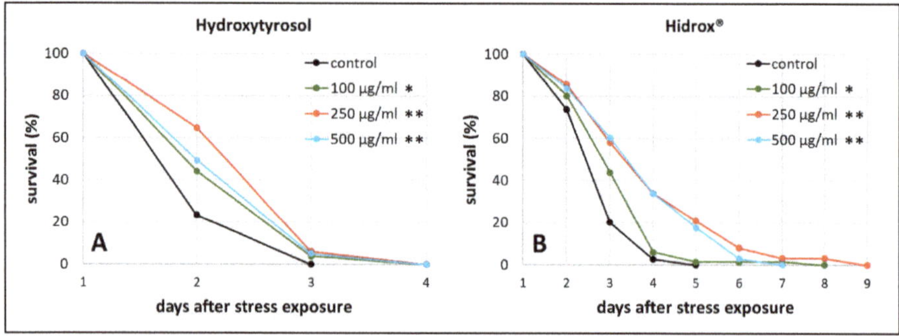

Figure 1. Heat stress resistance in the presence of HT and HD. At the third day of adulthood (day 1) wild type nematodes were exposed to heat stress at 37 °C for 3 h prior monitoring survival. The survival is plotted as the percentage of the initial population in the control group as well as in the HT (**A**) and HD (**B**) treated groups. Three biological replicates were combined with a total of ≥52 nematodes per treatment. Statistical significance was calculated by log-rank test. Differences compared to control were considered significant at $p < 0.05$ (*) and $p < 0.001$ (**).

Besides stress resistance, locomotion is one of the most important features that reflects the general fitness and health status of nematodes [45,46] and shows a constant decline during the ageing process [47]. Therefore, the swimming behavior was monitored in three different age classes with and without polyphenol treatment. Three age-dependent motion-parameters were chosen, that is wave initiation rate (often referred to as thrashing speed), activity index, and body wave number. The wave initiation rate is the number of body waves per minute, which indicates the movement-speed, whereas the activity index adds up the number of pixels that are covered by the nematode during the time spent for two strokes as an indicator for the vigorousness of bending over time. Furthermore, the body wave number, which is low in healthy and young worms, determines the waviness of the body at each time point. The data obtained in the current study verify the age-dependence of all selected swim parameters (Figure 2A–C). As previously described in Restif et al. [47] and Ibáñez-Ventoso et al. [48], the wave initiation rate (Figure 2A) and the activity index (Figure 2C) decreased with age, whereas the body wave number (Figure 2B) increased during ageing.

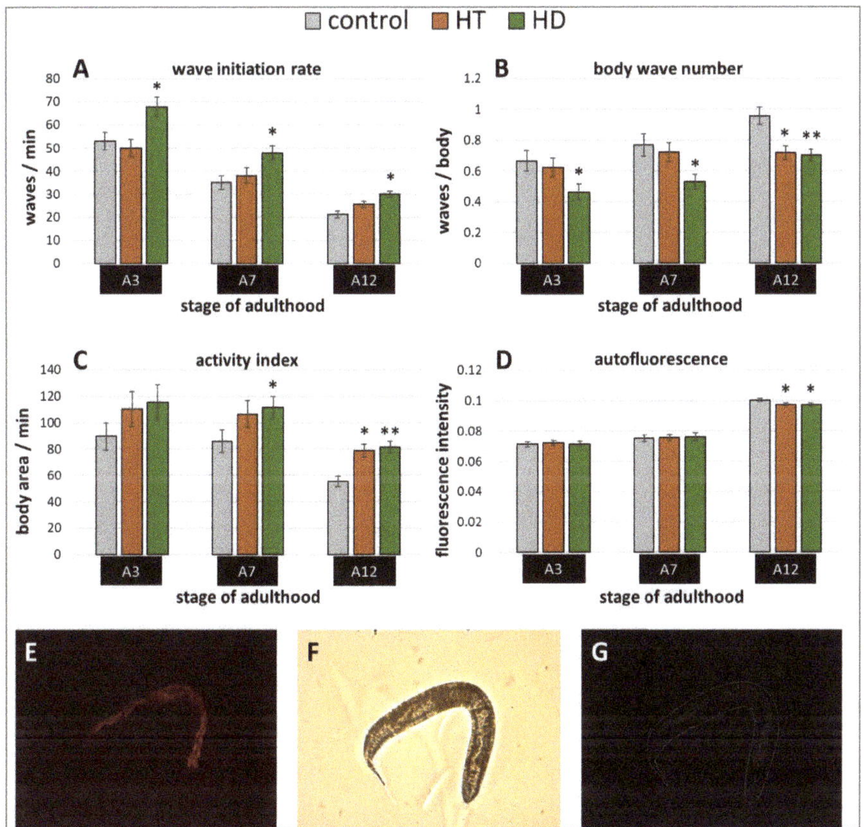

Figure 2. Healthspan benefits in wild type nematodes treated with HD and HT. The analysis of locomotion after polyphenol treatments comprises three parameters: the wave initiation rate (**A**), the body wave number (**B**) and the activity index (**C**). In two independent repeats, a total of ≥63 nematodes were analysed per treatment and age. In addition, the autofluorescence was monitored in two biological repeats with a total of ≥37 nematodes per treatment and age (**D**). Data are represented as mean ± SEM and statistical differences compared to control were considered significant at $p < 0.05$ (*) and $p < 0.001$ (**). A3, A7, A12: 3rd, 7th and 12th day of adulthood. Finally, an example for the typical appearance of red autofluorescence is shown (**E**) with the respective shot in bright field (**F**) as well as the merged and processed picture for the analysis in CellProfiler (**G**).

Interestingly, the movement speed was not influenced by 250 µg/mL HT in any age group, but 250 µg/mL HD could provoke an increase of 28%, 36%, and 42% in the wave initiation rate at the 3rd, 7th, and 12th day of adulthood, respectively (Figure 2A). Furthermore, HD was also able to enhance healthspan by decreasing the body wave number by at least 27% in all three age groups (Figure 2B) and by increasing the activity index by 30% and 48% at the 7th and 12th day of adulthood (Figure 2C), whereas HT did only positively influence these parameters at the 12th day of adulthood.

In addition, a well-known biomarker was investigated to analyse the beneficial effects of 250 µg/mL HD on the healthspan of *C. elegans*. The amount of autofluorescent material, sometimes referred to as lipofuscin or "age pigment", increases in *C. elegans* during ageing [49]. It was shown that red autofluorescence, which is mainly located in the intestine, reflects the ageing and health status of nematodes in the most reliable way [43]. Therefore, red fluorescence was measured in *C. elegans* during ageing, whereas the total intensity was calculated per worm body as illustrated by Figure 2E–G. Both treatments, HT and HD, were able to decrease the accumulation of the fluorescent material at the 12th day of adulthood by 3%, an effect which was absent in younger worms (Figure 2D).

Finally, the influence of HT and HD on the mortality rate was measured in wild type nematodes under standard laboratory conditions. The treatment with 250 mg/mL HT and HD resulted in an increase of mean lifespan by 14% (Figure 3A) and 12% (Figure 3B), respectively. The maximum lifespan, however, was only slightly (without significance according to the Fisher's Exact Test) increased by 7% after HD-treatment and the biggest increase was visible in the median lifespan (16%). A similar pattern was observed in the HT-treated group.

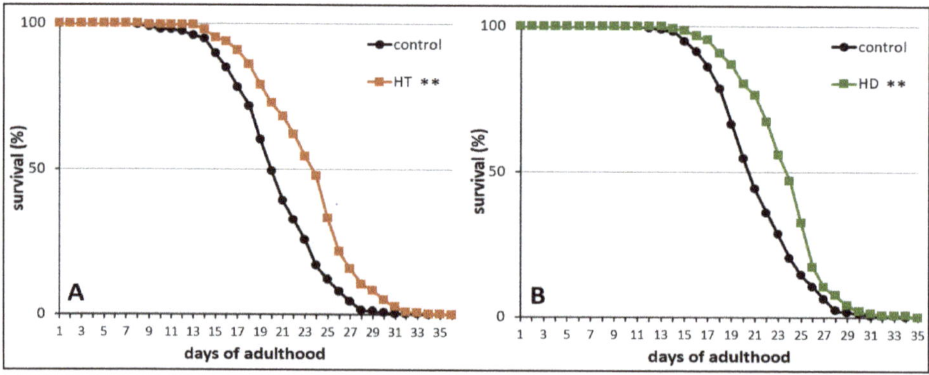

Figure 3. Life prolonging effects of HT and HD in wild type. The survival curves of control and polyphenol treated nematodes are shown. Survival is expressed as a percentage of the initial population per day. The curves represent three independent trials with a total of 250 and 286 nematodes in the control and HT treated nematodes, respectively (**A**) and two independent trials with a total of 172 and 144 nematodes in the control and HD treated nematodes, respectively (**B**). Statistical significance was calculated by log-rank test; differences compared to control were considered significant at $p < 0.001$ (**).

2.2. Rotenone-Induced Parkinsonian Models in C. elegans Profit from HT and HD

Exposure to the pesticide rotenone induces the parkinsonian syndrome in wild type *C. elegans*, which manifests in impaired fitness and movement [50]. A swim assay (illustrated in Figure 4D,E) was performed to determine whether HD and HT are able to reduce the rotenone induced symptoms. 10 μM rotenone strongly impaired the movement capacities at the 3rd and 7th day of adulthood in all tested parameters (Figure 4A–C) when compared to untreated nematodes at the same age (Figure 2A–C). Interestingly, both polyphenol treatments showed similar beneficial effects on nematodes suffering from rotenone at both tested ages. The body wave number (illustrated in Figure 4F) was decreased by up to 30% and 24% after HD- and HT-treatment, respectively (Figure 4B). The thrashing speed was increased after HT-treatment by 68% and 56% at the 3rd and 7th day of adulthood, respectively and by 119% and 55% at the 3rd and 7th day of adulthood, respectively, after HD- treatment (Figure 4A). But the strongest effects were observed in the activity index with an increase of 142% (A3) and 116% (A7) after HT-treatment and 209% (A3) and 58% (A7) after HD-treatment (Figure 4C).

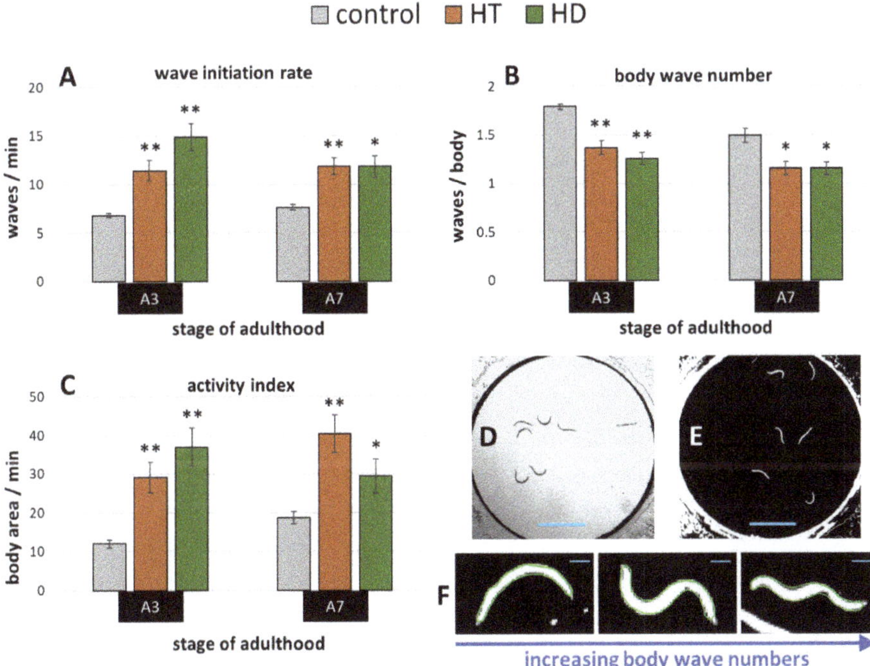

Figure 4. Effect of HD and HT on rotenone induced locomotion impairment. 10 μM rotenone administration with and without simultaneous HD or HT treatment started at the fourth larval stage. The wave initiation rate (**A**), the body wave number (**B**) and the activity index (**C**) are shown at the third (A3) and seventh (A7) day of adulthood. Data are pooled from two biological repeats with ≥44 worms per treatment and age. The bars represent the mean ± SEM and differences compared to control were considered significant at $p < 0.05$ (*) and $p < 0.001$ (**). The appearance of single frames in swim analysis are shown before (**D**) and after (**E**) image processing. In addition, single nematodes recognized by CeleST (as indicated by the green outline) are shown with increasing number of body waves (**F**). The blue scale bars represent 2.5 mm (**D,E**) and 200 μm (**F**), respectively.

2.3. Benefits by HT and HD in α-Synuclein-Induced Parkinsonian Models

The transgenic *C. elegans* synucleinopathies-model 'OW13' features α-synuclein expression in the body wall muscle cells driven by the muscle specific *unc-54*-promoter. The resulting movement deficits were already described by Van Ham et al. [51] and were also visible in the current study. The wave initiation rate, for instance, deteriorated by more than 50% when comparing untreated wild type animals (Figure 2A) with OW13 nematodes (Figure 5A) at the 3rd day of adulthood. Both polyphenols were able to mitigate the α-synuclein-induced locomotion impairments, whereas HD showed slightly higher capacities in aged (A7) nematodes (Figure 5A–C): At the 7th day of adulthood, the wave initiation rate (Figure 5A), the body wave number (Figure 5B), and the activity index (Figure 5C) were improved by HD-treatment by 96%, 42%, and 70%, respectively, whereas HT led to an enhancement by 47%, 25 %, and 34%, respectively.

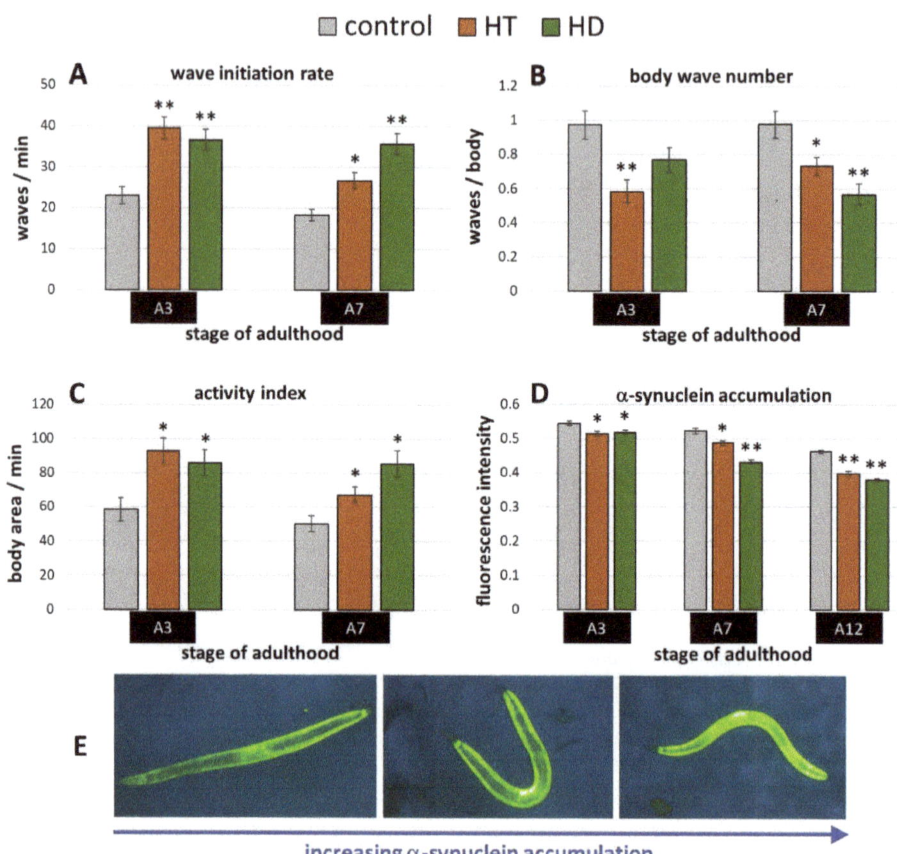

Figure 5. Benefits from HD & HT treatment in the OW13 strain. The nematode strain OW13 is characterized by α-synuclein expression in the body wall muscle cells. After polyphenol treatment, the wave initiation rate (**A**), the body wave number (**B**) and the activity index (**C**) were determined in two independent trials with ≥51 nematodes per treatment and age. Furthermore, the α-synuclein accumulation in muscle cells was quantified in two trials with ≥35 nematodes per treatment and age by fluorescence microscopy using a yellow filter (**D**). Data are presented as mean ± SEM. Differences compared to control were considered significant at $p < 0.05$ (*) and $p < 0.001$ (**). A3, A7, A12: 3rd, 7th and 12th day of adulthood. Finally, examples for the appearance of α-synuclein accumulation with increasing fluorescence intensities are shown (**E**).

The accumulation of α-synuclein in the muscle cells can be directly observed and quantified in the OW13-strain (Figure 5E). This is possible due to the transparent nature of *C. elegans* and due to the linkage of the yellow fluorescent protein (YFP) to the synthesized α-synuclein. Thus, fluorescence microscopy enabled the detection of potential α-synuclein-inhibiting abilities of the tested polyphenol treatments. Indeed, both treatments led to a reduction of accumulated YFP (Figure 5D), which is a direct indication for the decrease of the α-synuclein amount. The enhancement by the polyphenols is quite similar at A3 (5–6%), whereas HD showed clearly stronger effects at A7 and A12 with a decrease of 17% and 18%, respectively, compared to HT-treated nematodes with a decrease of 7% and 14% (Figure 5D). The overall reduction of the fluorescence intensities with age is based on the aging-dependent decline of *unc-54* expression [52].

Furthermore, the Parkinson's model 'UA44' was used to investigate the anti-Parkinsonian capacities of the polyphenol treatments. This strain is characterized by the expression of α-synuclein in dopaminergic neurons, which does not lead to distinct movement deficits [53] but to accelerated neurodegeneration [54]. Interestingly, only the wave initiation rate could profit from HT and HD treatment in this model (Figure 6A): HT increased the rate by 11% (A3) and 26% (A7) and HD by 45% (A3) and 28% (A7). No enhancement could be observed in the body wave number (Figure 6B) or the activity index (Figure 6C) by either polyphenol treatment.

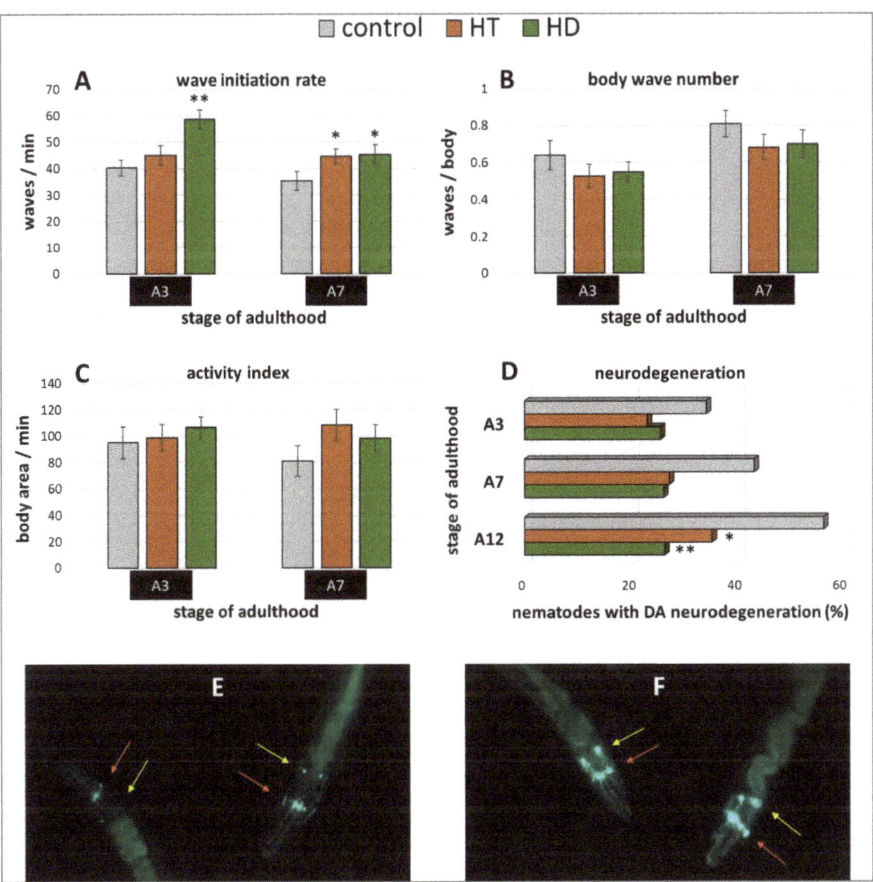

Figure 6. Benefits from HD & HT treatment in the UA44 strain. The nematode strain UA44 is characterized by α-synuclein as well as GFP expression in dopaminergic neurons. After polyphenol treatment, the wave initiation rate (**A**), the body wave number (**B**) and the activity index (**C**) were determined in two independent trials with ≥48 nematodes per treatment and age. Furthermore, the neuronal viability was analysed in three trials with ≥39 nematodes per treatment and age by fluorescence microscopy using a green filter. Shown are the percentages of nematodes with degenerated dopaminergic anterior neurons with and without polyphenolic treatment (**D**). Data are presented as mean ± SEM. Differences compared to control were considered significant at $p < 0.05$ (*) and $p < 0.001$ (**). A3, A7, A12: 3rd, 7th and 12th day of adulthood. Two nematodes with neurodegeneration characterized by missing or weak fluorescence in the DA neurons (**E**) as well as two nematodes with intact neuronal appearance (**F**) are shown. The DA neurons are sub-classified as four CEP neurons (red arrows), which are superimposed in most pictures and two ADE neurons (yellow arrows).

In the UA44 strain, the green fluorescent protein (GFP) is linked to the dopamine transporter in dopaminergic neurons, thus, the vitality of the six anterior and two posterior dopaminergic (DA) neurons can be observed with a fluorescent microscope. The α-synuclein-induced damage of the nerve cells manifests as lowered or missing fluorescence in single neurons, whereas the classification of intact and degenerated anterior DA neurons were performed as described in Harrington et al. [55] and as illustrated in Figure 6E,F. The increase of degenerated neurons with age could be completely blocked by 250 µg/mL HD (Figure 6D). In all age classes, the quantity of degenerated anterior DA neurons constantly amounts to about 26% in the HD-treated group, whereas the quantity of degenerated neurons in the control group increased by 22% from 34% (A3) to 56% (A12). HT-treated nematodes also featured a neuroprotective effect, however, an increase of neurodegeneration with age is still visible.

2.4. Summary and Comparison of HD- and HT-Action in All Bioassays

To compare the efficiencies of the action of HT and HD in all bioassays, the percentage changes relative to the respective control without polyphenol treatment were calculated and illustrated (Figure 7), whereas saturated and light-coloured bars represent significant and non-significant changes to the control, respectively. Furthermore, the measurements in the HT- and HD-treated groups were statistically compared with each other and labelled with * ($p < 0.05$) or ** ($p < 0.001$). Both polyphenol treatments are very efficient in enhancing all swim parameters in the rotenone PD-model and no significant differences could be detected between the HT- and HD-treated groups. However, in the wild type, the advantage of HD compared to HT is visible in two of the three locomotion characteristics. Since the measurements of lifespan and heat stress resistance were performed separately and with different control-groups, no direct comparison was possible in these cases. The effect of HD and HT was quite similar in the UA44 transgenic strain; only the wave initiation rate of young UA44-nematodes differed in HT- and HD-treated nematodes with a significant advantage in the HD-treated group. Moreover, the advantage of HD compared to HT is clearly visible in older OW13 nematodes, where α-synuclein accumulation, wave initiation rate and activity index featured a more pronounced benefit from the HD-treatment. None of the measurements in any strain exhibited an advantage of HT compared to HD.

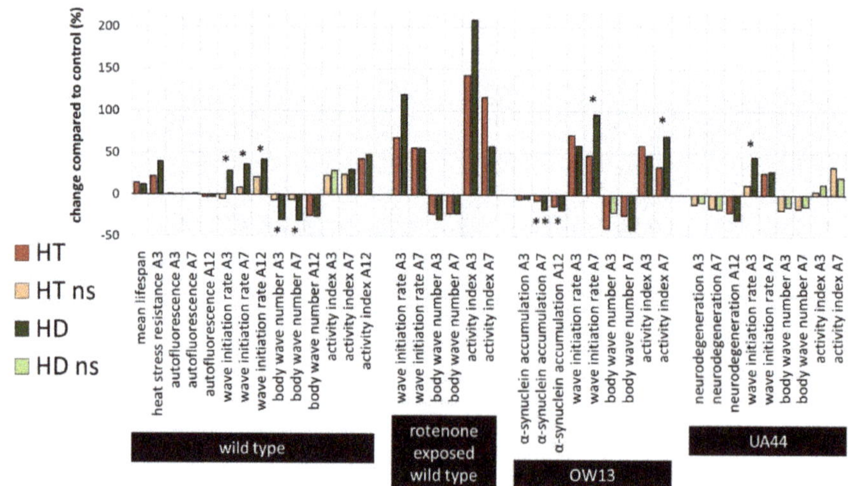

Figure 7. Summarized percentage changes after polyphenol treatment in all bioassays. The graph shows the percentage changes in all measured parameters compared to the respective control.

Dark red (HT) and dark green (HD) bars represent significant ($p < 0.05$ or $p < 0.001$) and light green or red bars non-significant (ns) differences to the control. In addition, the differences between the HT- and HD-treated groups were statistically determined and the differences between the treated groups were considered significant at $p < 0.05$ (*) and $p < 0.001$ (**). Due to the distinct performance of the assays with temporal distance and separate control-groups, no direct comparison of HT and HD was conducted for the lifespan and heat stress resistance assy. A3, A7, A12: 3rd, 7th and 12th day of adulthood.

3. Discussion

3.1. HD and HT Enhance Health and Longevity of Wild type C. elegans

Many studies reported the beneficial effects of polyphenols and flavonoids to improve health and extend lifespan in *C. elegans* [56,57]. Furthermore, healthy aging can be enhanced by several broad factors such as hormesis, autophagy and calorie restriction that increase stress resistance and longevity in *C. elegans* [58,59]. Healthspan is hard to define since it comprises the core features of health, that is physiological, cognitive, physical and reproductive function as well as a lack of disease [44]. To determine and compare the effects of treatments in terms of healthspan, it is necessary to operationalize features of health and thus, enable an objective way to measure healthspan. In this regard, the identification of predictive biomarkers and molecular pathways of health are mandatory to finally suggest applicable interventions, such as nutrition and exercise [44]. In the present study, we have demonstrated that the polyphenolic preparations HD and HT at a concentration of 250 µg/mL prolong lifespan and improve healthspan, determined via several physiological and functional parameters such as stress resistance, age pigment accumulation and swim behaviour in old wild type worms. The most powerful predictor of longevity and healthspan in old worms seems to be movement [60,61]. Similar to humans, the ability of *C. elegans* to move diminishes with aging [62]. Interestingly, HD also exerted optimal performances regarding the impact on locomotion in older wild type nematodes, which can be interpreted as an anti-ageing effect, while HT treatment did not show such evident improvements.

To treat *C. elegans* as naturally as possible, a live bacterial food was used throughout their life in this study. However, this protocol could also create some problems, which should be mentioned here. By adding these compounds to the living feeding bacteria, compound-bacteria interactions cannot be excluded. It is known that several plant polyphenols possess antibacterial activities [63]. These potential antimicrobial abilities could result in the inhibition of bacterial proliferation, which in turn would reduce the harmful intestinal accumulation of bacteria during ageing and prolong healthspan in *C. elegans* [64–66]. However, a recent study by Medina—Martínez et al. [67] showed, that at least 400 µg/mL HT were necessary to produce a perceptible growth deceleration of different *E. coli* strains. Furthermore, also olive leaf extract was shown to be only weakly active against *E. coli* bacteria [68]. On the other hand, it also cannot be excluded that compound-bacteria interactions lead to the degradation of the test compound by the bacteria. At least for hydroxytyrosol, this possibility can be neglected according to the study by Medina—Martínez et al. [67] who showed that degradation by *E. coli* was present to only a small extent. Nevertheless, in future studies the metabolic profile of *C. elegans* after the digestion of HT and HD in combination with alive *E. coli* should be checked for potential degradation products.

3.2. Hormesis Is Involved in Neuroprotective Effects from HD

Mild stress-induced hormesis represents a promising strategy to improve longevity and healthy aging. A recent paper demonstrated that hormesis leads to ageing-deceleration and highlighted its beneficial effects on lifespan, overall healthspan and especially stress resistance by activation of *daf-16*, *sod-3*, *ctl-1*, *hsp-16.2* and *sir-2.1* longevity genes in *C. elegans* [69]. Other recent studies also indicated the anti-aging effect of hormesis and considered it as an overcompensation stress response to the disruption in homeostasis via HSF-1 and SKN-1/Nrf2 signalling pathways [70,71]. The transcription factor SKN-1, the *C. elegans* orthologue of mammalian Nrf2 protein, is a well-known longevity factor that plays an essential role in oxidative stress response. It has also been reported that activation of SKN-1 induces the suppression of DAF-16, another well-known longevity gene, leading to detrimental effects on stress resistance and lifespan in *C. elegans*. In the same study, it has been demonstrated that oleic acid induces protective actions by regulating DAF-16 to promote lifespan extension and health in ageing worms [72]. In addition, a recent study reported that a chaperone complex mediates longevity response between HSF-1 and DAF-16/FOXO by reducing insulin/IGF-1 signaling to increase the lifespan in *C. elegans* [73]. The 90-kDa heat-shock protein (HSP-90) is an essential, evolutionarily conserved eukaryotic molecular chaperone [74]. Consistent with this concept, a recent study showed that DAF-21/HSP-90 is required for *C. elegans* longevity and provides a functional crosstalk between the proteostasis and nutrient signaling networks by ensuring DAF-16/FOXO isoform A activity [75]. Interestingly, the sirtuin family, named after mammalian Sirtuin 1 (SIRT1), features a high number of sirtuin-orthologs in several organisms, such as SIR-2.1 in *C. elegans*. Recently, it has been demonstrated that HSP-90/Hsp90 modulates lifespan via SIR-2.1/SIRT1 in *C. elegans* and in mammalian cells [76].

The Brunetti et al. [15] findings were supportive of a hormetic dose response for hydroxytyrosol with the optimal concentration at 250 μg/mL, the same concentration employed in the present study within a similar preconditioning experimental system. These findings are consistent with a substantial body of research that shows that preconditioning experiments which employed an adequate number of conditioning doses typically demonstrate a hormetic dose response [77,78], which is characterized by a low dose stimulation with the maximum response typically about 30–60% greater than the control response, similar to what was reported here. Thus, a hormetic background mechanisms for the action of HD seems plausible. Hidrox® has an excellent safety profile, with no adverse effects even at a very high dose [41]. Consistent with this notion, in vitro studies have evaluated HT as a non-genotoxic and non-mutagenic compound with NOAEL (No Observed Adverse Effects Level) classification, suitable for long-term consumption [79]. Thus, the safety profile of Hidrox® makes it an excellent food supplement.

3.3. Implications in PD Pathogenesis

Delaying aging, e.g., as seen after HD treatment, is a neuroprotective mechanism that may provide potential prevention in worm models of PD [80]. Parkinson's disease (PD) is a neurodegenerative disorder characterized by a severe depletion in number of dopaminergic cells of the *substantia nigra* (SN). As an effect of this reduction in dopaminergic neurons, a significant fall in brain dopamine (DA) levels occurs [81]. Although several hypotheses have been raised, including (i) defective DNA repair mechanisms, (ii) specific genetic defects, (iii) mitochondrial dysfunction, or (iv) toxic compounds in the environment, none of these, alone, completely explains the cascade of events responsible for the onset and progression of PD [81–84]. A large body of evidence demonstrates that free radicals play a key role in the pathogenesis of PD. In fact, there is a 10-fold increase in hydroperoxide levels in SN in PD, and dopaminergic neurons produce hydrogen peroxide either enzymatically, through the activity of mono-amine oxidase-A (MAO-A), or non-enzymatically via the intracellular autoxidation of DA [81,85]. Once formed, hydrogen peroxide by reacting with the reduced form of transition metals, such as Fe (II) and Cu(I), gives rise to the powerful oxidant hydroxyl radical and oxidative damage to nigral membrane lipids, proteins, and DNA ensues [86]. Reduced glutathione (GSH) significantly contributes to the detoxification of hydroxyl radical; in fact it reacts with the free thiol group of GSH which is oxidized to GSSG [86,87]. Unfortunately, SN has very low levels of GSH compared with other brain areas and this contributes to the triggering of PD pathogenesis by free radicals [88].

A number of worm models of PD have been generated through either exposing worms to a neurotoxic chemical such as 1-methyl-4-phenylpyridium ions (MPP+) or *6-hydroxydopamine* (6-OHDA) or by reproducing the genetic defect present in the inherited form of PD [53]. In particular, many recent data suggest a key role played by phenolic components of extra virgin olive oil (EVOO) in counteracting protein misfolding and proteotoxicity, with a particular emphasis on the mechanisms leading to the onset and progression of PD. Notably, HT efficiently neutralizes free radicals and protects biomolecules from ROS-induced oxidative damage. In this regard, HT activates the Nrf2–antioxidant response element (ARE) pathway, leading to the activation of phase II detoxifying enzymes and the protection of dopaminergic neurons exposed to hydrogen peroxide or to 6-hydroxydopamine [89–91]. These protective enzymes include NADPH quinone oxidoreductase-1, heme oxygenase-1, glutathione S-transferase, and the modifier subunit of glutamate cysteine ligase, which catalyzes the first step in the synthesis of GSH [10]. It is noteworthy that HT, a product of dopamine metabolism, is present in the brain [92]. Specifically, monoamine oxidase (MAO) catalyzes oxidative deamination of dopamine in a neurotoxic metabolite, DOPAL (3,4-dihydroxyphenylaldehyde) in dopaminergic neurons [93]. The latter can be oxidized by aldehyde dehydrogenase (ALDH) to DOPAC (3,4-dihydroxyphenylacetic acid), the major metabolite of dopamine in the brain or may be reduced to HT by alcohol dehydrogenase (ADH). At the same time, DOPAC reductase can transform DOPAC into HT [94]. DOPAL is a highly reactive metabolite, suggesting that it might be a neurotoxic dopamine metabolite with a role in the pathogenesis of PD.

Several studies reported the neurotoxicity of DOPAL in vivo and strongly suggest its role in PD pathogenesis [95].

The health effects of olive polyphenols, particularly HD and HT on the ageing process in old worms (i.e., increased locomotion and reduced intestinal autofluorescence) have been reported here. Our study reinforces the hypothesis that HD is protective against PD pathogenesis. This is in agreement with other studies showing that the olive fruit derivatives oleuropein and HT, as well as other polyphenols, such as the green tea derivative epigallocatechin 3-gallate (EGCG) inhibit DA-related toxicity and protect against environmental or genetic factors that induce DA neuron degeneration by the modulation of Nrf2-Keap1 and PGC-1α anti-oxidative signaling pathways in vitro and in vivo [96–98]. In addition, another recent study suggested that tyrosol from EVOO is effective in reducing α-synuclein inclusions, resulting in a lower toxicity and extended lifespan of treated transgenic nematodes [99].

Notably, α-synuclein is an aggregation-prone neuronal protein whose cellular function is not well known. As *C. elegans* has no orthologue of this protein, worm models have been generated by overexpression of wild-type or mutant forms of human α-synuclein in different tissues (i.e., either body-wall muscle, pan-neuronal, or only in the dopaminergic neurons). In most cases, overexpression leads to locomotion defects or the degeneration of dopaminergic neurons [26].

Interestingly, both polyphenols mitigated, age-dependently, the build-up of human α-synuclein in the body wall muscle cells of a transgenic *C. elegans* model (strain OW13) and improved swim performance. Our results are relevant to PD pathogenesis, due to the central role of mitochondria in metabolism, ROS regulation, and proteostasis [100]. The extent to which these pathways, including the mitochondrial unfolded protein response (UPR) and mitophagy, are active may predict severity and progression of these disorders, as well as sensitivity to chemical stressors. This holds true when considering, in a PD-like context, that transgenic nematodes express the Lewy body constituent protein α-synuclein. In fact, recent studies suggested that co-expression of α-synuclein and ATFS-1-associated dysregulation of the UPR synergistically potentiate dopaminergic neurotoxicity [101]. Moreover, other evidences have demonstrated that, in *C. elegans*, the inducible transcription factor SKN-1, directly controls UPR signaling and controls the transcription factor genes of XBP-1 and ATF-6 [102].

3.4. HD Is More Effective in Health Maintenance than Pure HT

In the present paper, we compared the health effects of Hidrox® and hydroxytyrosol as neuroprotective agents in *C. elegans* wild type and PD models. An important consideration that has emerged from this study relates to the different strength of biological activities delivered by hydroxytyrosol in its "natural" environment versus hydroxytyrosol in its purified, synthetic form. Although HT and HD were formulated to the same concentrations of 250 µg/mL, the two formulations used in the experiments contain dramatically differing hydroxytyrosol concentrations. HD as a raw formulation, indeed, contains a much smaller amount of hydroxytyrosol. 250 µg of HD correspond to 30 µg total polyphenols and only approximately 15 µg hydroxytyrosol. Thus, in comparison to the HT formulation (100% hydroxytyrosol), HD has 1/17th of the hydroxytyrosol in HT, but provides similar or even higher activity in all the assays so far described.

The difference between a "natural" formulation of olive polyphenols and purified fractions of olive polyphenols is, however, not surprising. Several studies published in the literature [14,21,42] have confirmed that hydroxytyrosol, stripped from its natural environment and/or cofactors which can be in minute concentrations in the juice of olives, provided a much less and limited range of activities in vitro and in vivo. Therefore, HD activity cannot just be accounted by and attributed solely to hydroxytyrosol. While this study confirms that "natural" HD has a superior range of activities than its purified or synthetic hydroxytyrosol counterpart, further studies are needed to support this conclusion.

Due to its composition (see chromatogram in Figure S1), it is plausible that other polyphenols than HT alone, might be involved in the beneficial action of HD. HD, in fact, contains compounds such as oleuropein, oleuropein aglycone, tyrosol, and gallic acid, to only name a few [41,103]. Previous studies already showed their abilities to improve the life- and healthspan of the nematodes [15,35,40,98], thus, it could be argued that the beneficial action of HD is due to one of those polyphenolic ingredients independent of HT. However, oleuropein, for instance, being the second most abundant polyphenol in HD, is only present in trace amounts. Only about 1% of HD consists of oleuropein so that the final concentration of 250 µg/mL HD delivers only about 2.5 µg/mL oleuropein. In our previous study, we could show that even 30 µg/mL HT or oleuropein aglycone were not sufficient to provide healthspan benefits [15]. Thus, it is unlikely that any single polyphenol included in HD is responsible for the observed effects. More likely, the polyphenols act in concert, whereas HT as the most abundant polyphenol in HD plays a central role. Additional experiments with chemically defined mixtures of polyphenols and other HD ingredients must be performed in future to find the responsible elements for healthspan enhancement by HD.

Our findings are in agreement with other studies indicating a higher cytoprotective activity of HD than that of purified HT. Consistent with this line of evidence, in vitro studies have measured the damage produced to the cell membrane by potent oxidant agents like H_2O_2 and 15-HPETE in bovine heart endothelial cells, and revealed that HD was extremely active in preventing membrane damage even at concentrations in micromolar range (10^{-6}), while pure HT obtained by HPLC separation was not protective against oxidant insult and damage [104]. On the contrary, high concentrations of purified HT produced a pro-oxidation effect and increased the cytotoxicity of the oxidants [103]. Another in vitro study confirmed that HD had higher antioxidant and anti-inflammatory activity than the same amount of pure HT, as demonstrated following the oxidation of mitochondrial membrane lipids by free radicals applying Electron Spin Resonance (ESR) spectroscopy (using Superoxide, HO radical and NO radicals as toxic agents) [104]. Furthermore, by measuring the effect of HT and HD on atherosclerosis lesions in an animal model, it was shown that the phenolic-enriched olive product (HT), out of its original matrix, could not only be not beneficial but actually harmful. The results suggest that the formulation of functional foods may require maintaining the natural environment in which these molecules are found [105]. In humans, HD administrated orally (1 mL of Olivenol, 2.5 mg HT) significantly increased plasma antioxidant activity and, in addition, bioavailability studies have established that, after ingestion, the absorption of HT from olive oil is dose-dependent, with low adsorption bioavailability after administration of a high amount of pure HT [106].

4. Materials and Methods

4.1. C. elegans Maintenance

The wild type *C. elegans* strain N2 (var. Bristol), the transgenic *C. elegans* strain OW13 (grk-1(ok1239); pkIs2386 [unc-54p::α-synuclein::YFP + unc-119(+)]) as well as the *Escherichia coli* feeding strain OP50 were obtained from the Caenorhabditis Genetics Centre (CGC, University of Minnesota, Minneapolis, MN, USA). The *C. elegans* strain UA44 (baIn11[Pdat-1::α-synuclein, Pdat-1::GFP]) was kindly provided by the Caldwell laboratory (University of Alabama, Tuscaloosa, AL, USA) [107]. All nematodes were maintained on standard nematode growth medium (NGM) agar plates at 22 °C, seeded with OP50 according to Brenner [108]. Prior to all tests, synchronized *C. elegans* populations were obtained by dissolving young adults in a 3% sodium hypochlorite solution according to Stiernagle [109]. The obtained eggs hatched in M9 buffer overnight, and were transferred to new NGM plates the following day. About 48 h later, L4 larvae were transferred to treatment and control plates. In order to inhibit the reproduction, 100 µM 5-fluoro-2-deoxyuridine (FUdR) was added to each plate [110].

4.2. Polyphenol and Rotenone Treatment

The treatment plates were prepared with the polyphenol hydroxytyrosol (HT; Sigma-Aldrich, St. Louis, MO, USA or with the aqueous olive pulp extract "HIDROX®" (HD). The HD extract was provided by Oliphenol LLC (Hayward, CA, USA), and total polyphenolic content of the HD extract was declared as 12% [111]. Among the polyphenolics, the major constituent of HD is hydroxytyrosol (40–50%), while other polyphenols present include oleuropein (5–10%), tyrosol (0.3%), and about 20% of other polyphenols including oleuropein aglycone and gallic acid [41]. HD is a freeze-dried powder prepared from the acidic hydrolysis (citric acid 1%) of the aqueous fraction of olives extracted from the defatted olive pulp, a byproduct during the processing of olives (*Olea europaea* L.) for olive oil extraction [41]. Hidrox® is titrated on the total content of olive polyphenols (12%).

HT and HD were dissolved in bidistilled water at 60 mg/mL and the solutions were stored at −20 °C. HT or HD, respectively, were added to the bacteria and agar at a final concentration of 100–500 µg/mL.

To trigger the Parkinsonian phenotype, wild type nematodes were exposed to rotenone. A stock solution of 0.5 mg/mL rotenone was prepared in DMSO and added to a final concentration of 10 µM to the control and polyphenol plates. After distribution with a spatula and drying for 24 h in the dark, OP50 (including 10 µM rotenone and the respective polyphenol) was spread on the plates. L4 nematodes were transferred to the rotenone plates until they were used for bioassays.

4.3. Lifespan and Heat Stress Assay

Synchronized wild type nematodes were observed with a stereo microscope and scored for their survival by gently touching them with a platinum wire. The animals were counted daily from the first day of adulthood until all died. Contaminated plates as well as vanished nematodes were censored. The heat stress test was performed according to the lifespan protocol, however, nematodes were stressed by heat (37 °C) for 3 h at the third day of adulthood and counting of dead and alive worms started 1 day after stress exposure.

4.4. Fluorescence Microscopy Analysis

For the fluorescence observation, several nematodes were placed on a 2% agar pad on a microscope slide and anesthetized with 4 µL NaN$_3$ (1M). The images were taken with the aid of the Axioskop fluorescence microscope (Carl Zeiss, Oberkochen, Germany) and filter sets from the Zeiss 4880 series.

To determine the autofluorescence in wild type nematodes, the images were captured at 100× magnification with a red filter set (TRITC, 545/30 nm ex, 610/70 nm em). In addition, bright field images were used to define the shape of each worm. The CellProfiler Software [112,113] was used to determine the mean red fluorescence intensity per total worm body.

The OW13 transgenic strain features yellow fluorescent protein linked to α-synuclein in the body wall muscle cells. Therefore, the nematodes were captured by using a yellow barrier filter with 100× magnification. The images were processed using the CellProfiler software and the yellow fluorescence intensity emitted per total worm body was calculated.

The UA44 transgenic strain features GFP linked to the dopamine transporter in the six dopaminergic neurons of the head and two in the tail as well as α-synuclein expression in dopaminergic neurons. To analyse the vitality of the neurons, the green barrier filter was used with a 200× magnification.

The number of detectable anterior neurons were finally counted and assayed for patterns of degeneration, as described previously from Harrington et al. [55].

4.5. Swim Behaviour Assay

Wells with a depth of 0.5 mm and Ø 10 mm were created with two self-adhesive marking films for microscope slides and filled with M9 buffer. 5 to 10 nematodes were transferred per well, covered by a cover slip, and recorded for 60 s with a connected camera. Each video was converted into single frames which were analysed with the CeleST software as described by Restif et al. [47] and Ibáñez-Ventoso et al. [48]. The wave initiation rate, the body wave number and the activity index were evaluated as representative parameters of locomotive behaviour.

4.6. Statistical Analysis

All experiments were independently conducted at least two times. The Online Application for Survival Analysis (OASIS 2; https://sbi.postech.ac.kr/oasis2/) [114] was used for comparing survival differences between two conditions. Fluorescence intensities as well as swim behaviour parameters were calculated as mean ± SEM and statistical significance was calculated by two tailed t-test using GraphPad (https://www.graphpad.com/quickcalcs/). Chi-square test was used to compare the number of worms with intact and degenerated neurons in the UA44 strain.

5. Conclusions

Both polyphenol treatments—pure hydroxytyrosol and the natural preparation 'Hidrox®'—were able to similarly improve the lifespan, stress resistance as well as age pigment accumulation in wild type nematodes. Furthermore, the beneficial locomotion effects of HD and HT are quite equally strong in the rotenone-stressed PD-model of *C. elegans*. However, the abilities of HD and HT also provide some differences: HD exposure led to much stronger beneficial locomotion effects in wild type worms compared to HT. Moreover, also the PD-model characterized by α-synuclein expression in muscles (strain OW13) did benefit significantly more from HD than HT. Only in the UA44 strain, which features α-synuclein expression in DA-neurons, the beneficial effects of HD and HT are rather weak with only one minor advantage of HD over HT. Thus, the hypothesis that HD features higher healthspan-promoting abilities than HT was only partly confirmed. Further studies are needed to uncover the molecular background mechanisms which led to this distribution of effects. Nevertheless, both polyphenolic treatments have the potential to partly prevent or even treat ageing-related neurodegenerative diseases and ageing itself. Future investigations including mammalian models and human clinical trials are needed to uncover the full potential of these olive ingredients.

Supplementary Materials: Supplementary materials can be found at http://www.mdpi.com/1422-0067/21/11/3893/s1.

Author Contributions: Conceptualization, V.C. and N.S.; methodology, N.S.; validation, N.S., V.C.; formal analysis, G.B., G.D.R.; investigation, G.B., G.D.R.; resources, V.C., M.S.; writing—original draft preparation, G.B., A.T.S., G.D.R. and N.S.; writing—review and editing, V.C., R.C., C.S.-L., M.S., E.J.C.; supervision, N.S., C.S.-L., and V.C.; funding acquisition, N.S. All authors have read and agreed to the published version of the manuscript.

Funding: This project has received funding from the European Union's Horizon 2020 research and innovation programme under Grant agreement No 633589 (Aging with Elegans). This publication reflects only the authors' views and the Commission is not responsible for any use that may be made of the information it contains.

Acknowledgments: We thank the Caenorhabditis Genetics Centre (CGC), which is funded by NIH Office of Research Infrastructure Programs (P40 OD010440) and the Caldwell laboratory, University of Alabama for the supply of the Caenorhabditis elegans strains. Furthermore, we thank Thea Böttcher and Shumon Chakrabarti for their support in the lab. Not least, a special thanks goes to Christian E.W. Steinberg who enabled this project and cooperation.

Conflicts of Interest: One of the authors, Roberto Crea, is the inventor of Hidrox® which is currently sold by Oliphenol LLC, California. The other authors declare no conflict of interest.

Abbreviations

ADH: Alcohol dehydrogenase; ALDH: Aldehyde dehydrogenase; ARE: Antioxidant response element; C. *elegans*: Caenorhabditis elegans; CGC: Caenorhabditis Genetics Centre; DA: Dopaminergic neurons; DOPAC: 3,4-dihydroxyphenylacetic acid; DOPAL: 3,4-dihydroxyphenylaldehyde; EGCG: Epigallocatechin 3-gallate; ESR: Electron Spin Resonance; EVOO: Extra virgin olive oil; GFP: Green fluorescent protein; GSH: Reduced glutathione: GSSG: Oxidized glutathione; HD: Hidrox®/Olivenol Plus™; HT: Hydroxytyrosol; HSP-90: Heat-shock protein-90; MAO-A: Mono-amine oxidase-A; MD: Mediterranean diet; MPP+: 1-methyl-4-phenylpyridium ions; NGM: Nematode growth medium; 6-OHDA: 6-hydroxydopamine; OLE: Oleuropein aglycone; OVW: Olive vegetation water; PD: Parkinson's disease; SN: Substantia nigra; SNpc: Substantia nigra pars compacta: SIRT1: Sirtuin 1; UPR: Unfolded protein response; YFP: Yellow fluorescent protein

References

1. Beitz, J.M. Parkinson's disease: A review. *Front. Biosci.* **2014**, *6*, 65–74. [CrossRef] [PubMed]
2. Bellucci, A.; Mercuri, N.B.; Venneri, A.; Faustini, G.; Longhena, F.; Pizzi, M.; Missale, C.; Spano, P. Review: Parkinson's disease: From synaptic loss to connectome dysfunction. *Neuropathol. Appl. Neurobiol.* **2016**, *42*, 77–94. [PubMed]
3. Dickson, D.W. Neuropathology of Parkinson disease. *Park. Relat. Disord.* **2018**, *46* (Suppl. 1), S30–S33.
4. Grover, S.; Somaiya, M.; Kumar, S.; Avasthi, A. Psychiatric aspects of Parkinson's disease. *J. Neurosci. Rural. Pr.* **2015**, *6*, 65–76. [CrossRef] [PubMed]
5. Trinh, J.; Farrer, M. Advances in the genetics of Parkinson disease. *Nat. Rev. Neurol.* **2013**, *9*, 445–454. [CrossRef]
6. Deng, H.; Wang, P.; Jankovic, J. The genetics of Parkinson disease. *Ageing Res. Rev.* **2018**, *42*, 72–85. [CrossRef]
7. Deng, X.; Xiao, B.; Allen, J.C.; Ng, E.; Foo, J.N.; Lo, Y.L.; Tan, L.C.S.; Tan, E.K. Parkinson's disease GWAS-linked Park16 carriers show greater motor progression. *J. Med. Genet.* **2019**, *56*, 765–768.
8. Gatt, A.P.; Duncan, O.F.; Attems, J.; Francis, P.T.; Ballard, C.G.; Bateman, J.M. Dementia in Parkinson's disease is associated with enhanced mitochondrial complex I deficiency. *Mov. Disord.* **2016**, *31*, 352–359.
9. Sarkar, S.; Dammer, E.B.; Malovic, E.; Olsen, A.L.; Raza, S.A.; Gao, T.; Xiao, H.; Oliver, D.L.; Duong, D.; Joers, V.; et al. Molecular Signatures of Neuroinflammation Induced by αSynuclein Aggregates in Microglial Cells. *Front. Immunol.* **2020**, *11*, 33. [CrossRef]
10. Leri, M.; Scuto, M.; Ontario, M.L.; Calabrese, V.; Calabrese, E.J.; Bucciantini, M.; Stefani, M. Healthy Effects of Plant Polyphenols: Molecular Mechanisms. *Int. J. Mol. Sci.* **2020**, *21*, 1250. [CrossRef]
11. Scuto, M.; Di Mauro, P.; Ontario, M.L.; Amato, C.; Modafferi, S.; Ciavardelli, D.; Trovato Salinaro, A.; Maiolino, L.; Calabrese, V. Nutritional Mushroom Treatment in Meniere's Disease with Coriolus versicolor: A Rationale for Therapeutic Intervention in Neuroinflammation and Antineurodegeneration. *Int. J. Mol. Sci.* **2019**, *21*, 284. [CrossRef] [PubMed]
12. Scuto, M.C.; Mancuso, C.; Tomasello, B.; Laura Ontario, M.; Cavallaro, A.; Frasca, F.; Maiolino, L.; Trovato Salinaro, A.; Calabrese, E.J.; Calabrese, V. Curcumin, Hormesis and the Nervous System. *Nutrients* **2019**, *11*, 2417. [CrossRef] [PubMed]
13. Trovato, A.; Siracusa, R.; Di Paola, R.; Scuto, M.; Ontario, M.L.; Bua, O.; Di Mauro, P.; Toscano, M.A.; Petralia, C.C.T.; Maiolino, L.; et al. Redox modulation of cellular stress response and lipoxin A4 expression by Hericium Erinaceus in rat brain: Relevance to Alzheimer's disease pathogenesis. *Immun. Ageing* **2016**, *13*, 23.
14. Angeloni, C.; Malaguti, M.; Barbalace, M.C.; Hrelia, S. Bioactivity of olive oil phenols in neuroprotection. *Int. J. Mol. Sci.* **2017**, *18*, 2230. [CrossRef] [PubMed]
15. Brunetti, G.; Di Rosa, G.; Scuto, M.; Leri, M.; Stefani, M.; Schmitz-Linneweber, C.; Calabrese, V.; Saul, N. Healthspan Maintenance and Prevention of Parkinson's-like Phenotypes with Hydroxytyrosol and Oleuropein Aglycone in *C. elegans*. *Int. J. Mol. Sci.* **2020**, *21*, 2588. [CrossRef]
16. Palazzi, L.; Leri, M.; Cesaro, S.; Stefani, M.; Bucciantini, M.; Polverino de Laureto, P. Insight into the molecular mechanism underlying the inhibition of α-synuclein aggregation by hydroxytyrosol. *Biochem. Pharmacol.* **2020**, *173*, 113722. [CrossRef]
17. Visioli, F.; Bellomo, G.; Galli, C. Free radical-scavenging properties of olive oil polyphenols. *Biochem. Biophys. Res. Commun.* **1998**, *247*, 60–64. [CrossRef]
18. Rietjens, S.J.; Bast, A.; Haenen, G.R. New insights into controversies on the antioxidant potential of the olive oil antioxidant hydroxytyrosol. *J. Agric. Food Chem.* **2007**, *55*, 7609–7614.

19. Menendez, J.A.; Joven, J.; Aragonès, G.; Barrajón-Catalán, E.; Beltrán-Debón, R.; Borrás-Linares, I.; Camps, J.; Corominas-Faja, B.; Cufí, S.; Fernández-Arroyo, S.; et al. Xenohormetic and anti-aging activity of secoiridoid polyphenols present in extra virgin olive oil: A new family of gerosuppressant agents. *Cell Cycle* **2013**, *12*, 555–578.

20. Calabrese, V.; Santoro, A.; Trovato Salinaro, A.; Modafferi, S.; Scuto, M.; Albouchi, F.; Monti, D.; Giordano, J.; Zappia, M.; Franceschi, C.; et al. Hormetic approaches to the treatment of Parkinson's disease: Perspectives and possibilities. *J. Neurosci. Res.* **2018**, *96*, 1641–1662.

21. Romana-Souza, B.; Saguie, B.O.; Pereira de Almeida Nogueira, N.; Paes, M.; Dos Santos Valença, S.; Atella, G.C.; Monte-Alto-Costa, A. Oleic acid and hydroxytyrosol present in olive oil promote ROS and inflammatory response in normal cultures of murine dermal fibroblasts through the NF-κB and NRF2 pathways. *Food Res. Int.* **2020**, *131*, 108984. [PubMed]

22. Yu, G.; Deng, A.; Tang, W.; Ma, J.; Yuan, C.; Ma, J. Hydroxytyrosol induces phase II detoxifying enzyme expression and effectively protects dopaminergic cells against dopamine- and 6-hydroxydopamine induced cytotoxicity. *Neurochem. Int.* **2016**, *96*, 113–120. [PubMed]

23. Funakohi-Tago, M.; Sakata, T.; Fujiwara, S.; Sakakura, A.; Sugai, T.; Tago, K.; Tamura, H. Hydroxytyrosol butyrate inhibits 6-OHDA-induced apoptosis through activation of the Nrf2/HO-1 axis in SH-SY5Y cells. *Eur. J. Pharmacol.* **2018**, *834*, 246–256. [PubMed]

24. Trovato Salinaro, A.; Pennisi, M.; Di Paola, R.; Scuto, M.; Crupi, R.; Cambria, M.T.; Ontario, M.L.; Tomasello, M.; Uva, M.; Maiolino, L.; et al. Neuroinflammation and neurohormesis in the pathogenesis of Alzheimer's disease and Alzheimer-linked pathologies: Modulation by nutritional mushrooms. *Immun. Ageing* **2018**, *15*, 8. [CrossRef]

25. Gorzynik-Debicka, M.; Przychodzen, P.; Cappello, F.; Kuban-Jankowska, A.; Marino Gammazza, A.; Knap, N.; Wozniak, M.; Gorska-Ponikowska, M. Potential Health Benefits of Olive Oil and Plant Polyphenols. *Int. J. Mol. Sci.* **2018**, *19*, 686.

26. Maulik, M.; Mitra, S.; Bult-Ito, A.; Taylor, B.E.; Vayndorf, E.M. Behavioral phenotyping and pathological indicators of Parkinson's disease in *C. elegans* models. *Front. Genet.* **2017**, *8*, 77. [CrossRef]

27. Johnson, T.E. Advantages and disadvantages of *Caenorhabditis elegans* for aging research. *Exp Gerontol.* **2003**, *38*, 1329–1332. [CrossRef]

28. Chen, X.; Barclay, J.W.; Burgoyne, R.D.; Morgan, A. Using, *C. elegans* to discover therapeutic compounds for ageing-associated neurodegenerative diseases. *Chem. Cent. J.* **2015**, *9*, 65. [CrossRef]

29. Lai, C.H.; Chou, C.Y.; Ch'ang, L.Y.; Liu, C.S.; Lin, W. Identification of novel human genes evolutionarily conserved in *Caenorhabditis elegans* by comparative proteomics. *Genome Res.* **2000**, *10*, 703–713. [CrossRef]

30. Kim, W.; Underwood, R.S.; Greenwald, I.; Shaye, D.D. OrthoList 2: A New Comparative Genomic Analysis of Human and *Caenorhabditis elegans* Genes. *Genetics* **2018**, *210*, 445–461. [CrossRef]

31. Lithgow, G.J.; Walker, G.A. Stress resistance as a determinate of *C. elegans* lifespan. *Mech. Ageing Dev.* **2002**, *123*, 765–771. [CrossRef]

32. Pietsch, K.; Saul NChakrabarti, S.; Stürzenbaum, S.R.; Menzel RSteinberg, C.E. Hormetins, antioxidants and prooxidants: Defining quercetin-caffeic acid- and rosmarinic acid-mediated life extension in *C. elegans*. *Biogerontology* **2011**, *12*, 329–347. [CrossRef] [PubMed]

33. Shen, P.; Yue, Y.; Zheng, J.; Park, Y. *Caenorhabditis elegans*: A convenient in vivo model for assessing the impact of food bioactive components on obesity, aging, and alzheimer's disease. *Annu. Rev. Food Sci. Technol.* **2017**, *9*, 1–22. [CrossRef] [PubMed]

34. Papaevgeniou, N.; Chondrogianni, N. Anti-aging and Anti-aggregation Properties of Polyphenolic Compounds in *C. elegans*. *Curr. Pharm. Des.* **2018**, *24*, 2107–2120. [CrossRef]

35. Saul, N.; Pietsch, K.; Stürzenbaum, S.R.; Menzel, R.; Steinberg, C.E. Diversity of polyphenol action in *Caenorhabditis elegans*: Between toxicity and longevity. *J. Nat. Prod.* **2011**, *74*, 1713–1720. [CrossRef] [PubMed]

36. Dilberger, B.; Passon, M.; Asseburg, H.; Silaidos, C.V.; Schmitt, F.; Schmiedl, T.; Schieber, A.; Eckert, G.P. Polyphenols and Metabolites Enhance Survival in Rodents and Nematodes-Impact of Mitochondria. *Nutrients* **2019**, *11*, 1886. [CrossRef] [PubMed]

37. Wang, J.; Deng, N.; Wang, H.; Li, T.; Chen, L.; Zheng, B.; Liu, R.H. Effects of Orange Extracts on Longevity, Healthspan, and Stress Resistance in *Caenorhabditis elegans*. *Molecules* **2020**, *25*, 351. [CrossRef]

38. Lin, C.; Zhang, X.; Su, Z.; Xiao, J.; Lv, M.; Cao, Y.; Chen, Y. Carnosol Improved Lifespan and Healthspan by Promoting Antioxidant Capacity in *Caenorhabditis elegans*. *Oxid. Med. Cell. Longev.* **2019**, *2019*, 5958043. [CrossRef]

39. Luo, S.; Jiang, X.; Jia, L.; Tan, C.; Li, M.; Yang, Q.; Du, Y.; Ding, C. In Vivo and in Vitro Antioxidant Activities of Methanol Extracts from Olive Leaves on *Caenorhabditis elegans*. *Molecules* **2019**, *24*, 704. [CrossRef]

40. Garcia-Moreno, J.C.; Porta de la Riva, M.; Martínez-Lara, E.; Siles, E.; Cañuelo, A. Tyrosol, a simple phenol from EVOO, targets multiple pathogenic mechanisms of neurodegeneration in a *C. elegans* model of Parkinson's disease. *Neurobiol. Aging* **2019**, *82*, 60–68. [CrossRef]

41. Soni, M.G.; Burdock, G.A.; Christian, M.S.; Bitler, C.M.; Crea, R. Safety assessment of aqueous olive pulp extract as an antioxidant or antimicrobial agent in foods. *Food Chem. Toxicol.* **2006**, *44*, 903–915. [CrossRef]

42. Calabrese, V.; Crea, R. Potential prevention and treatment of Neurodegenerative diseases: Olive polyphenols and hydroxytyrosol. *Eur. J. Neurodegener. Dis.* **2016**, *5*, 81–108.

43. Pincus, Z.; Mazer, T.C.; Slack, F.J. Autofluorescence as a measure of senescence in *C. elegans*: Look to red, not blue or green. *Aging* **2016**, *8*, 889.

44. Fuellen, G.; Jansen, L.; Cohen, A.A.; Luyten, W.; Gogol, M.; Simm, A.; Saul, N.; Cirulli, F.; Berry, A.; Antal, P.; et al. Health and Aging: Unifying Concepts, Scores, Biomarkers and Pathways. *Aging Dis.* **2019**, *10*, 883–900. [CrossRef]

45. Hahm, J.-H.; Kim, S.; DiLoreto, R.; Shi, C.; Lee, S.-J.V.; Murphy, C.T.; Nam, H.G.C. *C. elegans* maximum velocity correlates with healthspan and is maintained in worms with an insulin receptor mutation. *Nat. Commun.* **2015**, *6*, 8919. [CrossRef]

46. Li, G.; Gong, J.; Liu, J.; Liu, J.; Li, H.; Hsu, A.-L.; Liu, J.; Xu, X.Z.S. Genetic and pharmacological interventions in the aging motor nervous system slow motor aging and extend life span in *C. elegans*. *Sci. Adv.* **2019**, *5*, eaau5041.

47. Restif, C.; Ibáñez-Ventoso, C.; Vora, M.M.; Guo, S.; Metaxas, D.; Driscoll, M. CeleST: Computer vision software for quantitative analysis of *C. elegans* swim behavior reveals novel features of locomotion. *PLoS Comput. Biol.* **2014**, *10*, e1003702.

48. Ibáñez-Ventoso, C.; Herrera, C.; Chen, E.; Motto, D.; Driscoll, M. Automated Analysis of *C. elegans* Swim Behavior Using CeleST Software. *JoVE* **2016**, *118*, e54359. [CrossRef]

49. Pincus, Z.; Slack, F.J. Developmental biomarkers of aging in *Caenorhabditis elegans*. *Dev. Dyn.* **2010**, *239*, 1306–1314.

50. Sherer, T.B.; Betarbet, R.; Testa, C.M.; Seo, B.B.; Richardson, J.R.; Kim, J.H.; Miller, G.W.; Yagi, T.; Matsuno-Yagi, A.; Greenamyre, J.T. Mechanism of toxicity in rotenone models of Parkinson's disease. *J. Neurosci.* **2003**, *23*, 10756–10764. [CrossRef]

51. Van Ham, T.J.; Thijssen, K.L.; Breitling, R.; Hofstra, R.M.; Plasterk, R.H.; Nollen, E.A. *C. elegans* model identifies genetic modifiers of α-synuclein inclusion formation during aging. *PLoS Genet.* **2008**, *4*, e1000027. [CrossRef] [PubMed]

52. Adamla, F.; Ignatova, Z. Somatic expression of unc-54 and vha-6 mRNAs declines but not pan-neuronal rgef-1 and unc-119 expression in aging *Caenorhabditis elegans*. *Sci. Rep.* **2015**, *5*, 10692. [PubMed]

53. Lakso, M.; Vartiainen, S.; Moilanen, A.M.; Sirviö, J.; Thomas, J.H.; Nass, R.; Blakely, R.D.; Wong, G. Dopaminergic neuronal loss and motor deficits in *Caenorhabditis elegans* overexpressing human α-synuclein. *J. Neurochem.* **2003**, *86*, 165–172. [CrossRef]

54. Gaeta, A.L.; Caldwell, K.A.; Caldwell, G.A. Found in Translation: The Utility of *C. elegans* Alpha-Synuclein Models of Parkinson's Disease. *Brain Sci.* **2019**, *9*, 73. [CrossRef]

55. Harrington, A.J.; Knight, A.L.; Caldwell, G.A.; Caldwell, K.A. *Caenorhabditis elegans* as a model system for identifying effectors of α-synuclein misfolding and dopaminergic cell death associated with Parkinson's disease. *Methods* **2011**, *53*, 220–225.

56. Ayuda-Durán, B.; González-Manzano, S.; Miranda-Vizuete, A.; Sánchez-Hernández, E.; RRomero, M.; Dueñas, M.; Santos-Buelga, C.; González-Paramás, A.M. Exploring Target Genes Involved in the Effect of Quercetin on the Response to Oxidative Stress in *Caenorhabditis elegans*. *Antioxidants* **2019**, *8*, 585. [CrossRef]

57. Zheng, S.Q.; Huang, X.B.; Xing, T.K.; Ding, A.J.; Wu, G.S.; Luo, H.R. Chlorogenic Acid Extends the Lifespan of *Caenorhabditis elegans* via Insulin/IGF-1 Signaling Pathway. *J. Gerontol. A Biol. Sci. Med. Sci.* **2017**, *72*, 464–472.

58. Kumsta, C.; Hansen, M. Hormetic heat shock and HSF-1 overexpression improve *C. elegans* survival and proteostasis by inducing autophagy. *Autophagy* **2017**, *13*, 1076–1077. [CrossRef]

59. Carocho, M.; Ferreira, I.C.F.R.; Morales, P.; Soković, M. Antioxidants and Prooxidants: Effects on Health and Aging 2018. *Oxid. Med. Cell. Longev.* **2019**, *2019*, 7971613. [CrossRef]

60. Bansal, A.; Zhu, L.H.J.; Yen, K.; Tissenbaum, H.A. Uncoupling lifespan and healthspan in *Caenorhabditis elegans* longevity mutants. *Proc. Natl. Acad. Sci. USA* **2015**, *112*, E277–E286.

61. Zhang, W.B.; Sinha, D.B.; Pittman, W.E.; Hvatum, E.; Stroustrup, N.; Pincus, Z. Extended Twilight among Isogenic *C. elegans* Causes a Disproportionate Scaling between Lifespan and Health. *Cell Syst.* **2016**, *3*, 333–345.e4.

62. Marck, A.; Berthelot, G.; Foulonneau, V.; Marc, A.; Antero-Jacquemin, J.; Noirez, P.; Anne MBronikowski Morgan, T.J.; Garland, T.; Carter, P.A.; Hersen, P.; et al. Age-related changes in locomotor performance reveal a similar pattern for *Caenorhabditis elegans*, *Mus domesticus*, *Canis familiaris*, *Equus caballus*, and *Homo sapiens*. *J. Gerontol—Ser. A Biol. Sci. Med. Sci.* **2017**, *72*, 455–463.

63. Coppo, E.; Marchese, A. Antibacterial activity of polyphenols. *Curr. Pharm. Biotechnol.* **2014**, *15*, 380–390.

64. Cabreiro, F.; Gems, D. Worms need microbes too: Microbiota, health and aging in *Caenorhabditis elegans*. *EMBO Mol. Med.* **2013**, *5*, 1300–1310.

65. Garigan, D.; Hsu, A.-L.; Fraser, A.G.; Kamath, R.S.; Ahringer, J.; Kenyon, C. Genetic analysis of tissue aging in *Caenorhabditis elegans*: A role for heat-shock factor and bacterial proliferation. *Genetics* **2002**, *161*, 1101–1112.

66. Portal-Celhay, C.; Bradley, E.R.; Blaser, M.J. Control of intestinal bacterial proliferation in regulation of lifespan in *Caenorhabditis elegans*. *BMC Microbiol.* **2012**, *12*, 49. [CrossRef]

67. Medina-Martínez, M.S.; Truchado, P.; Castro-Ibáñez, I.; Allende, A. Antimicrobial activity of hydroxytyrosol: A current controversy. *Biosci. Biotechnol. Biochem.* **2016**, *80*, 801–810. [CrossRef]

68. Sudjana, A.N.; D'Orazio, C.; Ryan, V.; Rasool, N.; Ng, J.; Islam, N.; Riley, T.V.; Hammer, K.A. Antimicrobial activity of commercial *Olea europaea* (olive) leaf extract. *Int. J. Antimicrob. Agents* **2009**, *33*, 461–463. [CrossRef]

69. Sun, T.; Wu, H.; Cong, M.; Zhan, J.; Li, F. Meta-analytic evidence for the anti-aging effect of hormesis on *Caenorhabditis elegans*. *Aging* **2020**, *12*, 2723–2746. [CrossRef]

70. Schmeisser, S.; Schmeisser, K.; Weimer, S.; Groth, M.; Priebe, S.; Fazius, E.; Kuhlow, D.; Pick, D.; Einax, J.W.; Guthke, R.; et al. Mitochondrial hormesis links low-dose arsenite exposure to lifespan extension. *Aging Cell* **2013**, *12*, 508–517. [CrossRef]

71. Govindan, S.; Amirthalingam, M.; Duraisamy, K.; Govindhan, T.; Sundararaj, N.; Palanisamy, S. Phytochemicals-induced hormesis protects *Caenorhabditis elegans* against α-synuclein protein aggregation and stress through modulating HSF-1 and SKN-1/Nrf2 signaling pathways. *Biomed. Pharmacother.* **2018**, *102*, 812–822. [CrossRef]

72. Deng, J.; Dai, Y.; Tang, H.; Pang, S. SKN-1 Is a Negative Regulator of DAF-16 and Somatic Stress Resistance in *Caenorhabditis elegans*. *G3 (Bethesda)* **2020**, *10*, 1707–1712. [CrossRef]

73. Son, H.G.; Seo, K.; Seo, M.; Park, S.; Ham, S.; An, S.W.A.; Choi, E.S.; Lee, Y.; Baek, H.; Kim, E.; et al. Prefoldin 6 mediates longevity response from heat shock factor 1 to FOXO in *C. elegans*. *Genes Dev.* **2018**, *32*, 1562–1575. [CrossRef]

74. Taipale, M.; Jarosz, D.F.; Lindquist, S. HSP90 at the hub of protein homeostasis: Emerging mechanistic insights. *Nat. Rev. Mol. Cell Biol.* **2010**, *11*, 515–528. [CrossRef]

75. Somogyvári, M.; Gecse, E.; Sőti, C. DAF-21/Hsp90 is required for *C. elegans* longevity by ensuring DAF-16/FOXO isoform A function. *Sci. Rep.* **2018**, *8*, 12048. [CrossRef]

76. Nguyen, M.T.; Somogyvári, M.; Sőti, C. Hsp90 Stabilizes SIRT1 Orthologs in Mammalian Cells and *C. elegans*. *Int. J. Mol. Sci.* **2018**, *19*, 3661. [CrossRef]

77. Calabrese, E.J. Preconditioning is hormesis part I: Documentation, dose-response features and mechanistic foundations. *Pharmacol. Res.* **2016**, *110*, 242–264. [CrossRef]

78. Calabrese, E.J. Preconditioning is hormesis part II: How the conditioning dose mediates protection: Dose optimization within temporal and mechanistic frameworks. *Pharmacol. Res.* **2016**, *110*, 265–275. [CrossRef]

79. Auñon-Calles, D.; Giordano, E.; Bohnenberger, S.; Visioli, F. Hydroxytyrosol is not genotoxic in vitro. *Pharmacol. Res.* **2013**, *74*, 87–93. [CrossRef]

80. Cooper, J.F.; Dues, D.J.; Spielbauer, K.K.; Machiela, E.; Senchuk, M.M.; Van Raamsdonk, J.M. Delaying aging is neuroprotective in Parkinson's disease: A genetic analysis in *C. elegans* models. *NPJ Parkinsons Dis.* **2015**, *1*, 15022. [CrossRef]

81. Hald, A.; Lotharius, J. Oxidative stress and inflammation in Parkinson's disease: Is there a causal link? *Exp. Neurol.* **2005**, *193*, 279–290. [CrossRef]

82. Calabrese, V.; Santoro, A.; Monti, D.; Crupi, R.; Di Paola, R.; Latteri, S.; Cuzzocrea, S.; Zappia, M.; Giordano, J.; Calabrese, E.J.; et al. Aging and Parkinson's Disease: Inflammaging, neuroinflammation and biological remodeling as key factors in pathogenesis. *Free Radic. Biol. Med.* **2018**, *115*, 80–91. [CrossRef]

83. Calabrese, V.; Cornelius, C.; Rizzarelli, E.; Owen, J.B.; Dinkova-Kostova, A.T.; Butterfield, D.A. Nitric oxide in cell survival: A janus molecule. *Antioxid Redox Signal.* **2009**, *11*, 2717–2739. [CrossRef]

84. Poon, H.F.; Frasier, M.; Shreve, N.; Calabrese, V.; Wolozin, B.; Butterfield, D.A. Mitochondrial associated metabolic proteins are selectively oxidized in A30P alpha-synuclein transgenic mice—A model of familial Parkinson's disease. *Neurobiol. Dis.* **2005**, *18*, 492–498. [CrossRef]

85. Dexter, D.T.; Nanayakkara, I.; Goss-Sampson, M.A.; Muller, D.P.; Harding, A.E.; Marsden, C.D.; Jenner, P. Nigral dopaminergic cell loss in vitamin E deficient rats. *Neuroreport* **1994**, *5*, 1773–1776. [CrossRef]

86. Calabrese, V.; Cornelius, C.; Leso, V.; Trovato-Salinaro, A.; Ventimiglia, B.; Cavallaro, M.; Scuto, M.; Rizza, S.; Zanoli, L.; Neri, S.; et al. Oxidative stress, glutathione status, sirtuin and cellular stress response in type 2 diabetes. *Biochim. Biophys. Acta* **2012**, *1822*, 729–736. [CrossRef]

87. Calabrese, V.; Testa, G.; Ravagna, A.; Bates, T.E.; Stella, A.M. HSP70 induction in the brain following ethanol administration in the rat: Regulation by glutathione redox state. *Biochem. Biophys. Res. Commun.* **2000**, *269*, 397–400. [CrossRef]

88. Perry, T.L.; Godin, D.V.; Hansen, S. Parkinson's disease: A disorder due to nigral glutathione deficiency? *Neurosci. Lett.* **1982**, *33*, 305–310. [CrossRef]

89. Goldstein, D.S.; Jinsmaa, Y.; Sullivan, P.; Holmes, C.; Kopin, I.J.; Sharabi, Y. 3,4-Dihydroxyphenylethanol (Hydroxytyrosol) mitigates the increase in spontaneous oxidation of dopamine during monoamine oxidase inhibition in pc12 cells. *Neurochem. Res.* **2016**, *41*, 2173–2178. [CrossRef]

90. Gallardo-Fernández, M.; Hornedo-Ortega, R.; Cerezo, A.B.; Troncoso, A.M.; García-Parrilla, M.C. Melatonin, protocatechuic acid and hydroxytyrosol effects on vitagenes system against alpha-synuclein toxicity. *Food Chem. Toxicol.* **2019**, *134*, 110817. [CrossRef]

91. Lambert de Malezieu, M.; Courtel, P.; Sleno, L.; Abasq, M.L.; Ramassamy, C. Synergistic properties of bioavailable phenolic compounds from olive oil: Electron transfer and neuroprotective properties. *Nutr. Neurosci.* **2019**, *9*, 1–14.

92. De la Torre, R.; Covas, M.I.; Pujadas, M.A.; Fito, M.; Farre, M. Is dopamine behind the health benefits of red wine? *Eur. J. Nutr.* **2006**, *45*, 307–310. [CrossRef]

93. Rodríguez-Morató, J.; Xicota, L.; Fito, M.; Farre, M.; Dierssen, M.; de la Torre, R. Potential role of olive oil phenolic compounds in the prevention of neurodegenerative diseases. *Molecules* **2015**, *20*, 4655–4680. [CrossRef]

94. Xu, C.L.; Sim, M.K. Reduction of dihydroxyphenylacetic acid by a novel enzyme in the rat brain. *Biochem. Pharmacol.* **1995**, *50*, 1333–1337. [CrossRef]

95. Burke, W.J.; Li, S.W.; Williams, E.A.; Nonneman, R.; Zahm, D.S. 3,4-Dihydroxyphenylacetaldehyde is the toxic dopamine metabolite in vivo: Implications for Parkinson's disease pathogenesis. *Brain Res.* **2003**, *989*, 205–213. [CrossRef]

96. Mohammad-Beigi, H.; Aliakbari, F.; Sahin, C.; Lomax, C.; Tawfike, A.; Schafer, N.P.; Amiri-Nowdijeh, A.; Eskandari, H.; Møller, I.M.; Hosseini-Mazinani, M.; et al. Oleuropein derivatives from olive fruit extracts reduce α-synuclein fibrillation and oligomer toxicity. *J. Biol. Chem.* **2019**, *294*, 4215–4232.

97. Perez-Barron, G.A.; Montes, S.; Rubio-Osornio, M.; Avila-Acevedo, J.G.; Garcia-Jimenez, S.; Rios, L.C.; Monroy-Noyola, A. Hydroxytyrosol inhibits MAO isoforms and prevents neurotoxicity inducible by MPP+ in vivo. *Front. Biosci.* **2020**, *12*, 25–37.

98. Zhou, Z.D.; Xie, S.P.; Saw, W.T.; Ho, P.G.H.; Wang, H.; Lei, Z.; Yi, Z.; Tan, E.K. The Therapeutic Implications of Tea Polyphenols against Dopamine (DA) Neuron Degeneration in Parkinson's Disease (PD). *Cells* **2019**, *8*, 911. [CrossRef] [PubMed]

99. Cañuelo, A.; Esteban, F.J.; Peragón, J. Gene expression profiling to investigate tyrosol-induced lifespan extension in *Caenorhabditis elegans*. *Eur. J. Nutr.* **2016**, *55*, 639–650. [CrossRef]

100. Sironi, L.; Restelli, L.M.; Tolnay, M.; Neutzner, A.; Frank, S. Dysregulated Interorganellar Crosstalk of Mitochondria in the Pathogenesis of Parkinson's Disease. *Cells* **2020**, *9*, 233. [CrossRef]

101. Martinez, B.A.; Petersen, D.A.; Gaeta, A.L.; Stanley, S.P.; Caldwell, G.A.; Caldwell, K.A. Dysregulation of the Mitochondrial Unfolded Protein Response Induces Non-Apoptotic Dopaminergic Neurodegeneration in *C. elegans* Models of Parkinson's Disease. *J. Neurosci.* **2017**, *37*, 11085–11100. [CrossRef] [PubMed]

102. Glover-Cutter, K.M.; Lin, S.; Blackwell, T.K. Integration of the unfolded protein and oxidative stress responses through SKN-1/Nrf. *PLoS Genet.* **2013**, *9*, e1003701. [CrossRef] [PubMed]
103. Richard, N.; Arnold, S.; Hoeller, U.; Kilpert, C.; Wertz, K.; Schwager, J. Hydroxytyrosol is the major anti-inflammatory compound in aqueous olive extracts and impairs cytokine and chemokine production in macrophages. *Planta Med.* **2011**, *77*, 1890–1897. [CrossRef] [PubMed]
104. Lins, P.G.; Marina Piccoli Pugine, S.; Scatolini, A.M.; de Melo, M.P. In vitro antioxidant activity of olive leaf extract (*Olea europaea* L.) and its protective effect on oxidative damage in human erythrocytes. *Heliyon* **2018**, *4*, e00805. [CrossRef]
105. Acín, S.; Navarro, M.A.; Arbonés-Mainar, J.M.; Guillén, N.; Sarría, A.J.; Carnicer, R.; Surra, J.C.; Orman, I.; Segovia, J.C.; Torre Rde, L.; et al. Hydroxytyrosol administration enhances atherosclerotic lesion development in apo E deficient mice. *J. Biochem.* **2006**, *140*, 383–391. [CrossRef]
106. Visioli, F.; Wolfram, R.; Richard, D.; Abdullah, M.I.; Crea, R. Olive phenolic increase glutathione levels in healthy volunteers. *J. Agric. Food Chem.* **2009**, *57*, 1793–1796. [CrossRef]
107. Cao, S.; Gelwix, C.C.; Caldwell, K.A.; Caldwell, G.A. Torsin-mediated protection from cellular stress in the dopaminergic neurons of *Caenorhabditis elegans*. *J. Neurosci.* **2005**, *25*, 3801–3812.
108. Brenner, S. The genetics of *Caenorhabditis elegans*. *Genetics* **1974**, *77*, 71–94.
109. Stiernagle, T. Maintenance of *C. elegans*. *C. elegans* **1999**, *2*, 51–67.
110. Mitchell, D.H.; Stiles, J.W.; Santelli, J.; Sanadi, D.R. Synchronous growth and aging of *Caenorhabditis elegans* in the presence of fluorodeoxyuridine. *J. Gerontol.* **1979**, *34*, 28–36. [CrossRef]
111. Crea, R.; Liu, S.; Zhu, H.; Yang, Y.; Pontoniere, P. Validation of neuroprotective action of a commercially available formulation of olive polyphenols in a zebra-fish model vis-a-vis pure hydroxytyrosol. *J. Agric. Sci. Technol.* **2017**, *1*, 22–26.
112. Hamilton, N. Quantification and its applications in fluorescent microscopy imaging. *Traffic* **2009**, *10*, 951–961. [CrossRef] [PubMed]
113. McQuin, C.; Goodman, A.; Chernyshev, V.; Kamentsky, L.; Cimini, B.A.; Karhohs, K.W.; Doan, M.; Ding, L.; Rafelski, S.M.; Thirstrup, D. CellProfiler 3.0: Next-generation image processing for biology. *PLoS Biol.* **2018**, *16*, e2005970.
114. Han, S.K.; Lee, D.; Lee, H.; Kim, D.; Son, H.G.; Yang, J.-S.; Lee, S.-J.V.; Kim, S. OASIS 2: Online application for survival analysis 2 with features for the analysis of maximal lifespan and healthspan in aging research. *Oncotarget* **2016**, *7*, 56147. [CrossRef]

International Journal of
Molecular Sciences

Article

Healthspan Maintenance and Prevention of Parkinson's-like Phenotypes with Hydroxytyrosol and Oleuropein Aglycone in *C. elegans*

Giovanni Brunetti [1,†], **Gabriele Di Rosa** [1,†], **Maria Scuto** [1], **Manuela Leri** [2,3], **Massimo Stefani** [2], **Christian Schmitz-Linneweber** [4], **Vittorio Calabrese** [1,*] and **Nadine Saul** [4,*]

[1]   Department of Biomedical and Biotechnological Sciences, University of Catania, 95125 Catania, Italy; q.burneti@gmail.com (G.B.); dirosagabriele85@gmail.com (G.D.R.); mary-amir@hotmail.it (M.S.)
[2]   Department of Experimental and Clinical Biomedical Sciences "Mario Serio", University of Florence, Viale Morgagni 50, 50134 Florence, Italy; manuela.leri@unifi.it (M.L.); massimo.stefani@unifi.it (M.S.)
[3]   Department of Neuroscience, Psychology, Area of Medicine and Health of the Child of the University of Florence, Viale Pieraccini, 6 - 50139 Florence, Italy
[4]   Humboldt University of Berlin, Faculty of Life Sciences, Institute of Biology, Molecular Genetics Group, Philippstr. 13, House 22, 10115 Berlin, Germany; christian.schmitz-linneweber@rz.hu-berlin.de
*   Correspondence: calabres@unict.it (V.C.); nadine.saul@gmx.de (N.S.)
†   These authors contributed equally to this work.

Received: 13 March 2020; Accepted: 7 April 2020; Published: 8 April 2020

Abstract: Numerous studies highlighted the beneficial effects of the Mediterranean diet (MD) in maintaining health, especially during ageing. Even neurodegeneration, which is part of the natural ageing process, as well as the foundation of ageing-related neurodegenerative disorders like Alzheimer's and Parkinson's disease (PD), was successfully targeted by MD. In this regard, olive oil and its polyphenolic constituents have received increasing attention in the last years. Thus, this study focuses on two main olive oil polyphenols, hydroxytyrosol (HT) and oleuropein aglycone (OLE), and their effects on ageing symptoms with special attention to PD. In order to avoid long-lasting, expensive, and ethically controversial experiments, the established invertebrate model organism *Caenorhabditis elegans* was used to test HT and OLE treatments. Interestingly, both polyphenols were able to increase the survival after heat stress, but only HT could prolong the lifespan in unstressed conditions. Furthermore, in aged worms, HT and OLE caused improvements of locomotive behavior and the attenuation of autofluorescence as a marker for ageing. In addition, by using three different *C. elegans* PD models, HT and OLE were shown i) to enhance locomotion in worms suffering from α-synuclein-expression in muscles or rotenone exposure, ii) to reduce α-synuclein accumulation in muscles cells, and iii) to prevent neurodegeneration in α-synuclein-containing dopaminergic neurons. Hormesis, antioxidative capacities and an activity-boost of the proteasome & phase II detoxifying enzymes are discussed as potential underlying causes for these beneficial effects. Further biological and medical trials are indicated to assess the full potential of HT and OLE and to uncover their mode of action.

Keywords: *C. elegans*; polyphenols; olive oil; healthspan; lifespan; ageing; Parkinson's disease

---

1. Introduction

Neurodegenerative diseases are becoming increasingly prevalent in the ageing populations of industrialized nations, going hand in hand with the increase in life expectancy. These disorders, which include Alzheimer's disease and Parkinson's disease, share a common feature: the accumulation of misfolded proteins in pathological inclusions [1].

Parkinson's disease (PD) is a chronic, age-related and adult-onset neurodegenerative disorder characterized by the loss of dopaminergic neurons in an area of the midbrain called substantia nigra (SN) along with intraneuronal inclusions known as Lewy bodies, which contain amyloid aggregates of misfolded α-synuclein [2–4]. PD is considered today as the most common movement disorder that affects 1–2 per 1000 of the population and since the prevalence is increasing with age, PD affects 1% of the population above 60 years [5]. There are dozens of PD-related symptoms and signs but the most typical are motor deficits including tremors, muscle rigidity, bradykinesia, and impaired gait. Among non-motor symptoms, the most common are olfactory dysfunction, cognitive impairment, psychiatric symptoms, and autonomic dysfunction [6]. PD is a multifactorial disorder and the majority of PD cases are sporadic with unknown aetiology possibly caused by an association of genetic and environmental risk factors. At least 23 loci and 19 disease-causing genes for PD have been identified and designated as PD-causing genes [7].

Several hypotheses have been proposed regarding the cause of loss of dopaminergic neurons in PD, whereas oxidative stress, in particular, is strongly associated with the development of PD [8,9]. Other studies have shown defects in the mitochondrial complex-I of neurons, which lead to impaired energy metabolism and cell death [10]. Furthermore, the proteolytic hypothesis describes nigral neuron loss in PD as a result of toxic accumulation of aggregates of misfolded proteins, notably α-synuclein, resulting in neuro-inflammation [11]. Dopaminergic neurons of substantia nigra pars compacta appear particularly vulnerable to the harmful effects of α-synuclein aggregates [12]. Since ageing is a major risk for PD, it has been hypothesized that PD could be, at least in part, a type of segmental ageing, in which the viability of dopaminergic (DA) neurons is impaired by so far unknown localized and accelerated ageing mechanisms [13].

Neurodegenerative disorders are associated with high morbidity and mortality, and few effective options are available for their treatment [1]. Thus, many studies have been conducted focusing on natural compounds present in food as important molecules against neurodegenerative diseases such as PD [14–17]. Several lines of evidence support the beneficial effect of the Mediterranean diet (MD) in preventing neurodegeneration, possibly due to its richness in polyphenols [18,19]. Natural polyphenols exert numerous biological activities, like antioxidant, anti-inflammatory, antiviral, antibacterial, antiproliferative, and anticarcinogenic capacities (reviewed in Stevanovic, et al. [20]), as well as cellular redox state modulation activities through direct action on enzymes, proteins and receptors [21,22]. In addition, in patients affected by osteoarthritis or cardiovascular diseases, beneficial epigenetic chromatin modifications were also caused by polyphenols [23–25].

One possible mode of action of natural polyphenols is the hormesis effect. The biological processes underlying hormetic dose–response, recently, focused attention in the field of neuroprotection, which was mainly elucidated through the exploitation of bioactive polyphenols against the main age-related diseases, particularly in PD [13]. In this light, low levels of exogenous and endogenous stressors have been reported to display hormetic characteristics that induce neurophysiological mechanisms of maintenance and repair, including heat shock, the application of pro-oxidants, as well as the application of polyphenols from plants [26]. Recently, it has been postulated that the MD exerts healthy effects through hormetic mechanisms, as specific olive oil polyphenols (e.g., oleuropein and hydroxytyrosol) likely counteract the effects of neuro-inflammatory stimuli by acting as modulators of stress responsive mechanisms, which result in adaptive stress resistance [27]. Moreover, in vivo studies suggest that a diet rich in phytochemicals may enhance neuroplasticity and stress resistance to neuro-inflammation, mitigating or preventing neurodegenerative changes in the brain that are typical in a number of age-related disorders, including PD [28–30].

It has been hypothesized that extra virgin olive oil polyphenols could be among the main determinants of the beneficial effect of the MD [31–33]. Extra virgin olive oil (EVOO) contains approximately 36 phenolic compounds [21], which represent the main group of antioxidants found in virgin olive oil. The main phenolic subclasses present in olive oil are phenolic alcohols, phenolic acids, flavonoids, lignans, and secoiridoids [34], whereas the latter represent the largest quantity in the EVOO. The main secoiridoids in olive oil are oleuropein aglycone and ligstroside, which undergo transformation to hydroxytyrosol or tyrosol, respectively through two enzymes (beta glucosidase and esterase) in the gastrointestinal tract [35].

In the present paper, we focus on health effects of hydroxytyrosol (HT; Figure 1B) and oleuropein aglycone (OLE; Figure 1A) as neuroprotective agents against PD in the light of recent studies indicating that OLE and HT can be beneficial against PD by stabilizing the monomeric state of α-synuclein, thus, favouring the growth of aggregates devoid of toxicity [36,37]. Furthermore, previous studies have shown that these molecules are strongly protective against neurodegeneration in different transgenic models of Aβ deposition [38–40]. Moreover, we have to consider that hydroxytyrosol is a by-product of dopamine oxidative metabolism [41] and in the last years the gold standard therapy against PD has relied on restoring the optimum level of dopamine [42].

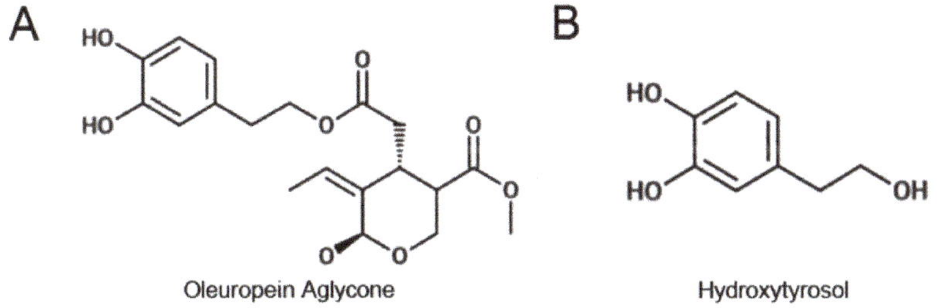

Figure 1. Chemical structure of (**A**) oleuropein aglycone and (**B**) hydroxytyrosol.

Tests with mammalian models are very powerful, but are expensive, long-lasting and cause ethical concerns. Therefore, the widely used model organism *Caenorhabditis elegans* (*C. elegans*) was applied for this study. Despite its simple structure, *C. elegans* features several tissues and organs in alignment to higher animals, like muscles, a nervous system, an epidermis, a gastrointestinal tract, and gonads [43]. Furthermore, about 50% of the human protein-coding genome has recognizable worm orthologs [44]. Last but not least, neurological pathways are highly conserved between invertebrates and mammals, and numerous neurodegenerative disease-related transgenic and mutant strains are available in the nematodes [45]. In addition to several neuro-protective substance screenings [45], numerous polyphenols were also successfully tested for their general health and lifespan benefits in *C. elegans* [46–48], making this nematode an optimal model to study the neuroprotective effects of olive polyphenols.

It was hypothesized that the polyphenol OLE and its main metabolite HT increase the mean lifespan of *C. elegans* in the presence and in the absence of stress conditions. Furthermore, it is assumed that they are able to counteract the age-related decline of general health parameters, which were assessed by determining the swim behaviour as a measure of overall body fitness as well as the intestinal autofluorescence, being one of the most popular ageing biomarkers [49]. Furthermore, numerous cell culture studies were already successfully performed to verify the anti-PD effects of olive ingredients, as summarized in Angeloni, Malaguti, Barbalace, and Hrelia [22], however, in vivo studies are hardly present. Therefore, by using a chemically induced and two transgenic PD models of *C. elegans*, the polyphenolic treatments were tested for their anti-PD effects in vivo. Although *C. elegans* is not able to develop PD, the PD models feature characteristic attributes related to PD. Besides the swim performance, neuronal degeneration as well as α-synuclein accumulation were taken into account to assess the anti-PD potential.

2. Results

2.1. Oleuropein Aglycone and Hydroxytyrosol Extended the Survival of Wild-type C. elegans *after Heat Stress*

To find the optimal concentration for this study, the treatments with polyphenols were initially tested at different concentrations in a heat stress-resistance test. Due to the relatively fast and simple execution, heat stress resistance is frequently measured to screen treatments for potential health and lifespan benefits [50–52]. Since stress resistance abilities are strongly correlated to ageing, ageing-associated diseases and lifespan [53–56], this test was also used as a pre-test in the current study. The monitored survival after stress exposure revealed an improved stress resistance in OLE-treated nematodes (Figure 2A). Two-hundred-and-fifty µg/mL and 500 µg/mL OLE provoked statistically significant changes in all biological repeats performed. The mean lifespan after heat stress increased by 15% and 22% in the OLE 250 µg/mL and OLE 500 µg/mL treated group, respectively. However, the survival differences between OLE 250 µg/mL and OLE 500 µg/mL treated nematodes were not significant. Furthermore, no significant survival benefits were observed with 30 and 100 µg/mL compared to control. Due to the highest percentage benefit, 500 µg/mL OLE was chosen for subsequent experiments.

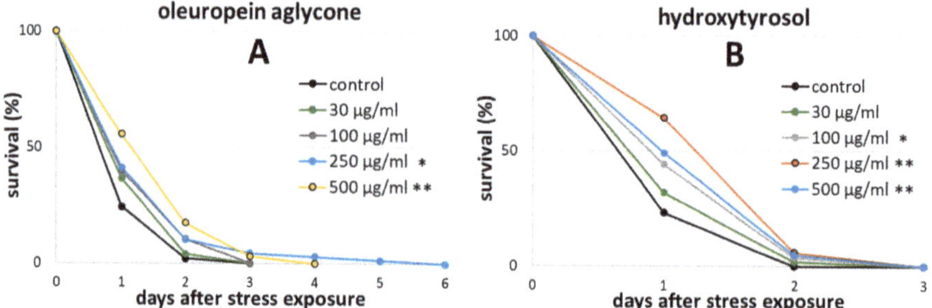

Figure 2. Heat stress survival in presence of different OLE and HT concentrations. Survival is expressed as a percentage of the initial population combined with three biological replications. At the third day of adulthood (day 0), wild type nematodes were exposed to heat stress at 37 °C for 3 h prior monitoring survival. (**A**) Survival curves during OLE treatment with n (control) = 70, n (30 µg/mL) = 74, n (100 µg/mL) = 44, n (250 µg/mL) = 68, and n (500 µg/mL) = 64; (**B**) Survival curves during HT treatment with n (control) = 90, n (30 µg/mL) = 50, n (100 µg/mL) = 52, n (250 µg/mL) = 67, and n (500 µg/mL) = 79. Statistical significance was calculated by log-rank test including Bonferroni correction. Differences compared to control were considered significant at $p < 0.05$ (*) and $p < 0.001$ (**). n: number of tested nematodes.

Similar results were achieved by using HT (Figure 2B). The mean lifespan after heat stress was increased by about 11% by treatment with 100 µg/mL HT, by 22% with 250 µg/mL HT and by 14% with 500 µg/mL HT. There was no significant difference between the survival curves of these treatment groups among each other. Again, 30 µg/mL was not sufficient to improve the survival. Based on these results, 250 µg/mL HT was applied for further experiments. None of the tested concentrations and compounds exerted a harmful effect on the survival of the nematodes after stress exposure.

2.2. Hydroxytyrosol Prolonged the Lifespan of Wild Type C. elegans

Several olive polyphenols and preparations have proven to be effective in extending the lifespan in *C. elegans* [57,58]. However, the olive oil polyphenols investigated in this study have not been tested so far in this sense.

Surprisingly, OLE treatment did not result in any significant lifespan enhancement (Figure 3A): The mean lifespan of wild type nematodes was only hardly noticeably increased by 2.7%, which is probably the result of a minor, not significant, increase in the median lifespan from 22.55 days to 23.31 days (Table 1). However, the treatment with HT led to an increase of mean lifespan by 14.1% (Figure 3B). This life prolongation was not only visible in terms of mean and median lifespan, but was also reflected in terms of minimum and maximum lifespan (the time point, when 25% or 90%, respectively, of the individuals are dead) as well as in the time point when final death occurred (Table 1). No obvious side effects, such as extrusion of internal organs through the vulva or morphological alterations in movement in the polyphenols-treated groups compared to the controls were seen during lifelong observation.

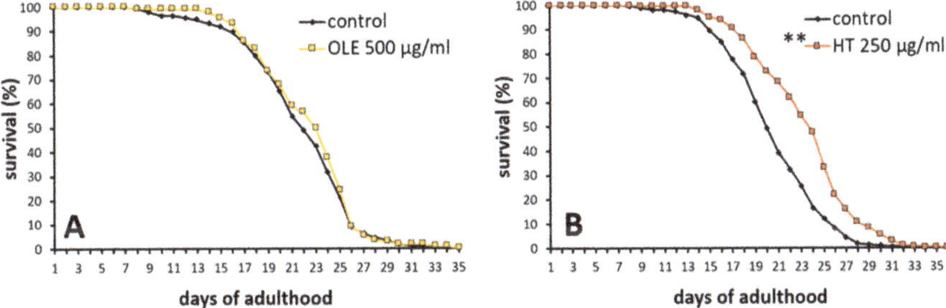

Figure 3. Effect of OLE and HT on lifespan in *C. elegans*. The survival curves of controls and polyphenol-treated nematodes are shown. Survival is expressed as a percentage of the initial population per day. (**A**) The curve represents two independent experiments (*n* control = 184, *n* OLE = 111); (**B**) Representative survival curve of three independent experiments with control and HT-treated worms (*n* control = 250, *n* HT = 286). Statistical significance was calculated by log-rank test; differences compared to control were considered significant at $p < 0.05$ (*) and $p < 0.001$ (**). *n*: number of tested nematodes.

Table 1. Lifespan characteristics during OLE and HT treatment.

Treatment	*n*	Mean Lifespan (days)	SEM	Days until Deaths of Population Reached				
				25%	50%	75%	90%	100%
control (DMSO)	228	22.15	0.32	18.73	22.55	25.00	26.09	34.00
Oleuropein	159	22.75	0.37	18.89	23.31	25.17	26.10	35.00
control (water)	305	20.61	0.26	17.50	19.94	23.07	25.51	34.00
Hydroxytyrosol	335	23.51**	0.26	19.64**	23.75**	25.76**	28.36**	36.00

Differences compared to control were considered significant at $p < 0.05$ (*) and $p < 0.001$ (**). *p*-value determination was realized with log-rank test for the mean lifespan and Mann–Whitney U test for specific time points.

2.3. Polyphenols Improved Age Pigment Accumulation and Locomotive Behaviour in Wild Type Nematodes

The intestinal autofluorescence, one of the most prominent ageing and health biomarker in *C. elegans* [49], was surveyed to determine the overall health status. Observations with a fluorescence microscope fitted with a red filter set allowed to detect the accumulation of the "age pigment" in different age classes. As expected, the red fluorescent intensity increased with age (Figure 4), whereas the increase was only weakly pronounced at the 7th day of adulthood. Both polyphenolic compounds were able to reduce age-related gain in autofluorescence. The quantity of fluorescent pigments was slightly, yet significantly, diminished at the 12th day, but not at the 3rd or 7th day, of adulthood (Figure 4).

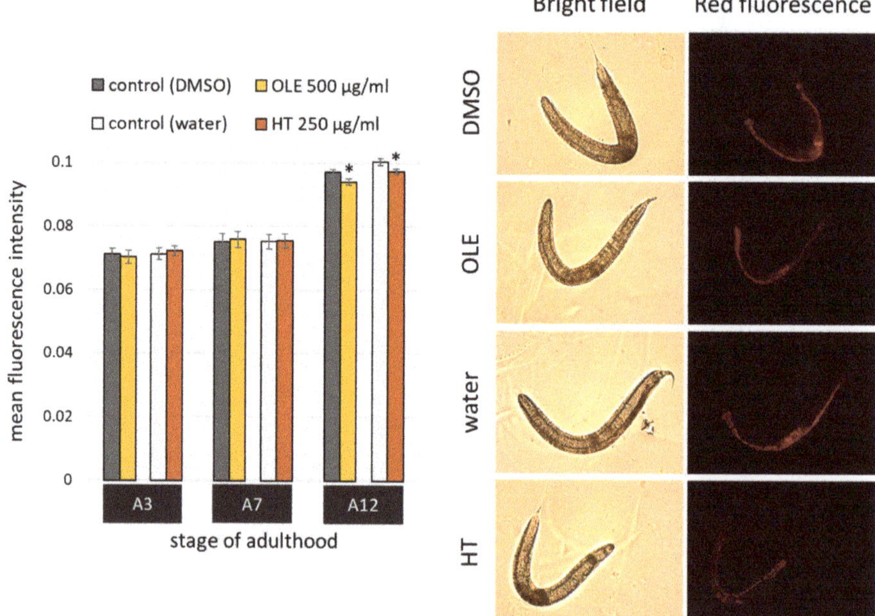

Figure 4. Age pigment quantification after OLE and HT treatment in *C. elegans*. Nematodes were observed by fluorescence microscopy in red spectrum at day 3, 7, and 12 of adulthood in two biological repeats. The bar charts (left) show the mean red fluorescence intensity of OLE (*n* A3 = 34; *n* A7 = 38; *n* A12 = 88) and DMSO (*n* A3 = 39; *n* A7 = 43; *n* A12 =84)-treated nematodes as well as HT (*n* A3 = 38; *n* A7 = 45; *n* A12 = 94)-treated nematodes and their respective water control (*n* A3 = 38; *n* A7 =44; *n* A12 = 95). Data are represented as mean ± SEM, and statistical differences compared to control were considered significant at $p < 0.05$ (*). *n*: number of tested nematodes; A3, A7, A12: 3rd, 7th and 12th day of adulthood. In addition, example pictures (right) representing bright field and red fluorescence shots at the 12th day of adulthood in the control (DMSO and water) and polyphenol-treated (OLE and HT) groups are shown (all scale bars = 200 µm).

Since ageing is marked by physical decline, sarcopenia is considered a valuable parameter of health status in organisms of metazoans, including *C. elegans* [59,60]. Therefore, the ability of both polyphenols to boost the health of nematodes was additionally assessed with a swim assay. We measured the thrashing rate, the body wave number and the activity index to determine the physical performance at different ages. The thrashing rate is the number of body thrashes per minute as an indicator for the speed of movement whereas the activity index sums up the number of pixels that are covered by the body during the time needed for two strokes as an indicator for the vigorousness of bending [61]. Furthermore, the body wave number, a feature that increases with age, determines the waviness of the body at each time point. Indeed, the vigorousness and speed of movement of untreated worms declined with age, as indicated by the differences in both endpoints between the 3rd and 7th or 12th day of adulthood, respectively (Figure 5A,C). Moreover, the body wave number increased with age (Figure 5B), in agreement with the results from Restif et al. [61], thus verifying the correct performance of the test.

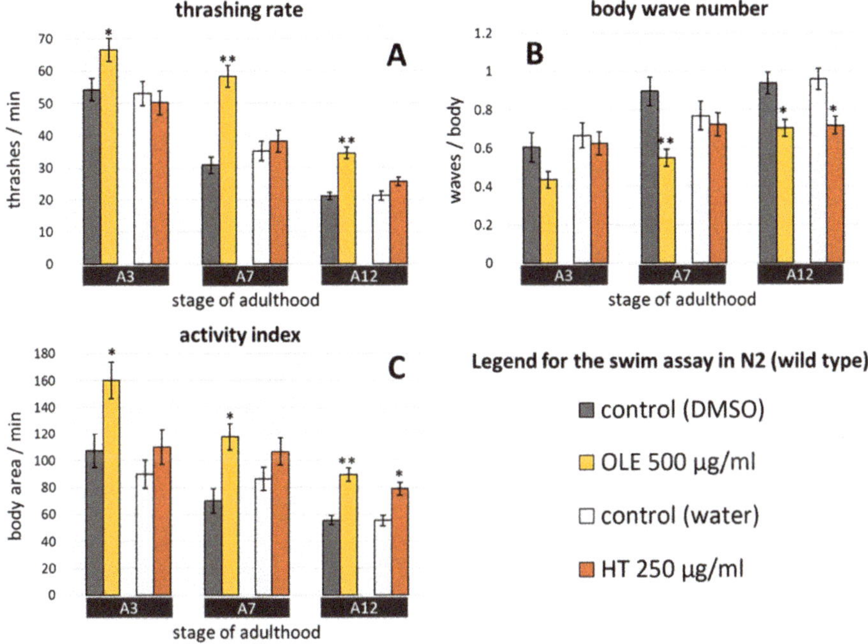

Figure 5. Swim behaviour characteristics in wild type *C. elegans* treated with OLE and HT. Locomotory performances were determined at day 3, 7, and 12 of adulthood in two independent repeats. The determination of locomotion differences comprises three parameters: (**A**) the thrashing rate, (**B**) the body wave number, and (**C**) the activity index. The number of analysed animals accounts for: DMSO control = 67 (A3), 70 (A7) and 117 (A12); OLE = 63 (A3), 70 (A7) and 106 (A12); water control = 76 (A3), 70 (A7) and 111 (A12); HT = 70 (A3), 70 (A7) and 104 (A12). Data are presented as mean ± SEM and differences compared to control were considered significant at $p < 0.05$ (*) and $p < 0.001$ (**). A3, A7, A12: day 3, 7, 12 of adulthood.

OLE treatment resulted in a remarkable increase of the number of thrashes per minute (Figure 5A) and of the activity index (Figure 5C) displayed by worms at all three tested stages, whereas the body wave number was decreased at the 7th and 12th day of adulthood (Figure 5B). The percentage increase was about 23% (A3), 89% (A7) and 64% (A12) and about 49% (A3), 69% (A7) and 61% (A12) for thrashing rate and activity index, respectively. The decrease of the body wave number reached its maximum at the 7th day of adulthood with a reduction of 39%. Surprisingly, HT was not able to enhance the thrashing rate of the nematodes at any adult-day (Figure 5A). However, at the 12th day of adulthood, an increase of 43% was detected by analysing the covered pixel per body and minute (Figure 5C). In addition, a decrease of 25% in the body wave number was found at A12 as well (Figure 5B). No polyphenol led to a reduction of motor performance in treated worms.

2.4. C. elegans *Parkinsonian Models Profit from Olive Oil Polyphenol Treatments*

Exposure to the pesticide rotenone or the transgenic expression of human α-synuclein induces the Parkinsonian-like syndrome in C. *elegans*, which manifests in impaired movement [62,63]. In order to assess whether the polyphenols are able to reduce this symptom, the swimming analysis was performed with nematodes suffering from Parkinson's-like symptoms induced by exposure to rotenone of wild type worms. Furthermore, transgenic Parkinson's models expressing human α-synuclein in muscle cells (strain OW13) and dopaminergic neurons (strain UA44), respectively, were treated with polyphenols as well.

Treatment with 10 µM rotenone led to dramatically decreased movement abilities. Given that untreated nematodes exhibit more than 50 thrashes per minute at A3, the thrashing rate was reduced by more than 80% to less than 10 thrashes per minute in rotenone-exposed C. *elegans* (Figure 6A). Similar proportions could be monitored for the activity index of young control nematodes with or without rotenone treatment (Figure 6C). Furthermore, the body wave number was more than doubled due to exposure to rotenone compared to untreated nematodes (Figure 6B). Interestingly and in contrast to rotenone-untreated nematodes, the locomotion abilities after exposure to rotenone did not decline after the 3rd day of adulthood (Figure 6). OLE was able to partly inhibit the rotenone-induced movement decline in both tested ages and all swim traits by more than doubling the measured values (Figure 6A,C) or by reducing them by at least 42% (Figure 6B). Quite strong effects were also visible by using HT in all tested ages and endpoints: HT increased the thrashing rate by at least 56% and the activity index by a minimum of 116%. The body wave number was decreased by at least 23%.

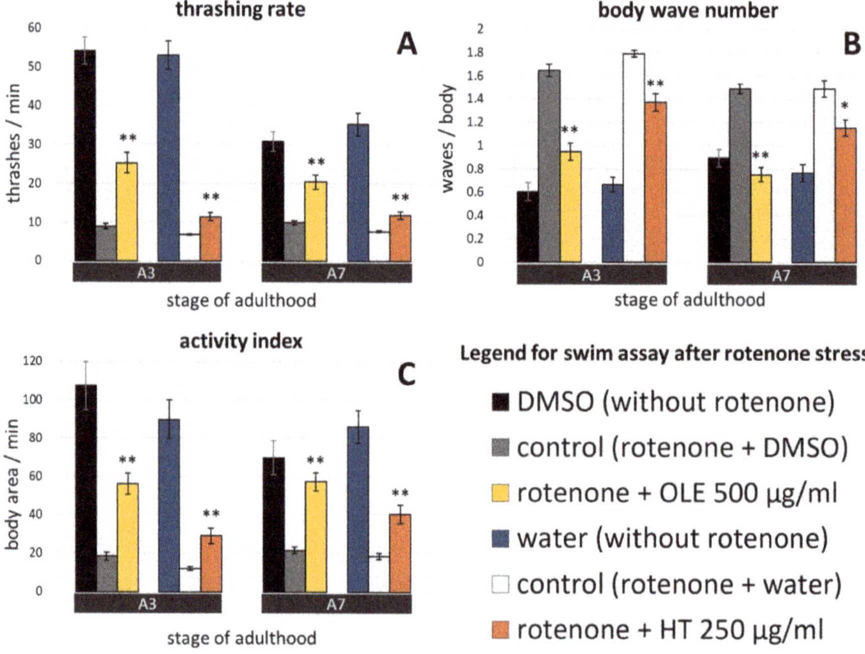

Figure 6. Effect of OLE and HT on rotenone-induced locomotion deficits. The administration of 10 µM rotenone, starting at the fourth larval stage, led to Parkinsonian-like phenotype exhibiting movement impairments at the 3rd and 7th day of adulthood. (**A**) The thrashing rate, (**B**) the body wave number, and (**C**) the activity index are shown with and without simultaneous OLE or HT administration. The number of tested nematodes was: DMSO control = 58 (A3) and 69 (A7); OLE = 60 (A3) and 63 (A7); water control = 62 (A3) and 46 (A7); HT = 44 (A3) and 61 (A7). A3, A7: day 3 and 7 of adulthood. Data are pooled from two biological repeats and presented as mean ± SEM, and differences compared to control were considered significant at $p < 0.05$ (*) and $p < 0.001$ (**). To enable direct comparisons, data from nematodes without rotenone and polyphenol exposures (see Figure 5) are shown in addition.

To check the anti-Parkinson's effect of olive oil polyphenols, the motor activity was also assessed in the transgenic C. *elegans* OW13 and UA44 strains, both models of synucleinopathies. The presence of α-synuclein in the body wall muscle cells (strain OW13) leads to movement deficits, as previously described by Van Ham, et al. [64]. Indeed, the thrashing rate clearly deteriorated from about 50 thrashes/min in young untreated wild type adults to about 35 and 23 thrashes per minute, respectively, in young untreated OW13 worms (Figure 7A). This effect, albeit weaker, was also seen, in part, in older nematodes as well as for the activity index (Figure 7C). The body wave number was very stable between the 3rd and 7th day of adulthood (Figure 7B). Interestingly, control DMSO treatment already led to a mild beneficial effect for all three swim parameters in the OW13 strain.

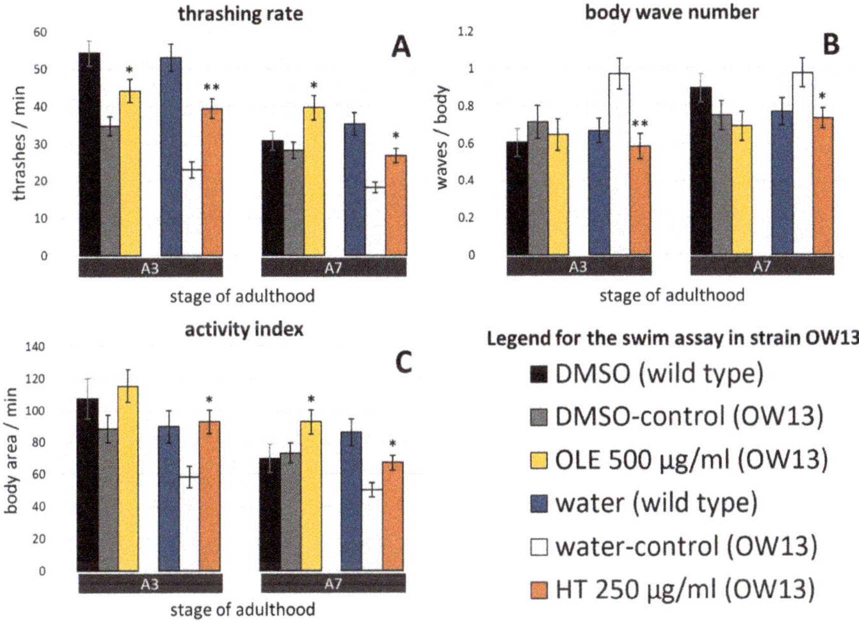

Figure 7. OLE and HT effects on the swim performance in the OW13 strain. (**A**) The thrashing rate, (**B**) the body wave number, and (**C**) the activity index were determined at day 3 and day 7 of adulthood with and without OLE or HT treatment. Here, the nematode strain OW13, characterized by α-synuclein in the body wall muscle cells, was used. The number of tested nematodes was: DMSO control = 65 (A3) and 71 (A7); OLE = 61 (A3) and 60 (A7); water control = 51 (A3) and 71 (A7); HT = 61 (A3) and 66 (A7). A3, A7: day 3 and 7 of adulthood. Data are collected in two independent trials and are presented as mean ± SEM. Differences compared to control were considered significant at $p < 0.05$ (*) and $p < 0.001$ (**). To enable direct comparisons, data from wild type nematodes without polyphenol exposures (see Figure 5) are shown in addition.

Treatment with either polyphenol improved several swim performance features in this strain over control treatments (Figure 7A–C). OLE administration provoked a 27% increase of thrashes per minute at day 3 and a 40% increase at day 7 (Figure 7A), whereas the increase of the activity index (27%) was detected only at day 7 (Figure 7C) and no significant change was seen in the body wave number (Figure 7B). HT displayed its advantageous effects in both tested age groups and in all swim parameters (Figure 7A–C), whereas HT remarkably increased the thrashing rate by 71% in A3.

Interestingly, the strain UA44, which is characterized by the presence of α-synuclein in dopaminergic neurons, did only show a negligible decline in locomotion from day 3 to day 7 of adulthood (Figure 8). Furthermore, the difference in swim performance between young wild type and young UA44 nematodes was hardly recognizable, a finding that agrees with data from a previous study, showing that α-synuclein controlled by a dopaminergic promotor did not disturb the thrashing speed [65]. Surprisingly, treatment with either polyphenol resulted only in a limited improvement in swim behaviour of the UA44 strain. OLE treatment increased the activity index in young nematodes (Figure 8C) but resulted only in minor and non-significant changes of the thrashing rate and body wave number for UA44 (Figure 8A,B). On the other hand, HT supplementation showed beneficial effects only on the number of thrashes per minute in older worms (Figure 8A), but not on the magnitude of movement or the waviness (Figure 8B,C).

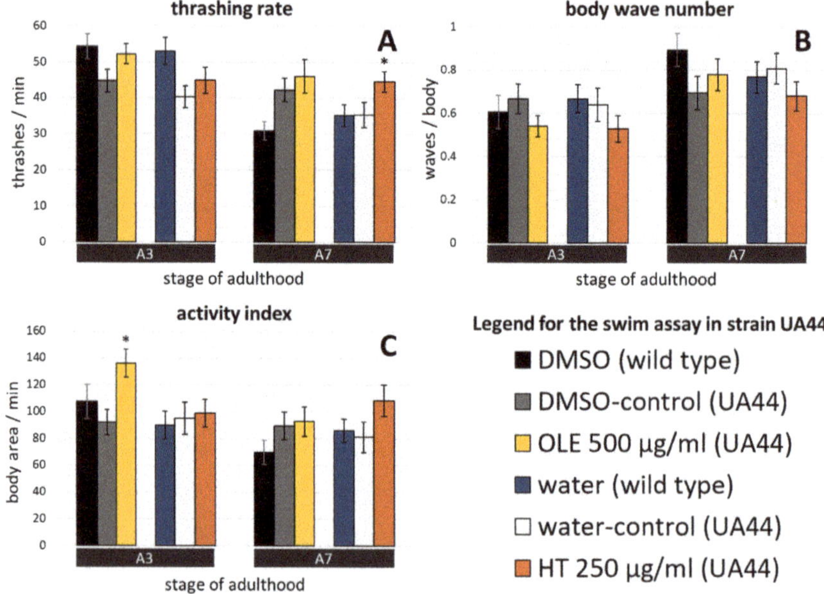

Figure 8. Effect of OLE and HT on the swim performance in the *C. elegans* UA44 strain. (**A**) The number of thrashes per minute, (**B**) the number of waves running through the body, and (**C**) the area covered by the body per minute were observed in presence and absence of OLE and HT at day 3 and 7 of adulthood in two biological repeats. The UA44 strain used in this study is characterized by α-synuclein in dopaminergic neurons. The bar charts represent the following number of individuals: DMSO control (*n* A3 = 63, *n* A7 = 50), OLE (*n* A3 = 72; *n* A7 = 55), water control (*n* A3 = 57, *n* A7 = 55) and HT (*n* A3 = 48, *n* A7 = 65). *n*: number of tested nematodes. A3, A7: day 3, 7 of adulthood. Data are represented as mean ± SEM and differences compared to control were considered significant at $p < 0.05$ (*). To enable direct comparisons, data from wild type nematodes without polyphenol exposures (see Figure 5) are shown in addition.

To summarize, HT and OLE treatments resulted in enhanced swim performance in nematodes suffering from rotenone exposure or α-synuclein expression in muscle cells (strain OW13). However, the strain UA44, characterized by α-synuclein expression in dopaminergic neurons, only weakly profited from the polyphenol treatments.

2.5. α-synuclein Induced Damages and α-synuclein Accumulation was Targeted by Olive Oil Polyphenols In Vivo

The advantage of the *C. elegans* OW13 strain is the yellow fluorescent labelling of synthesized α-synuclein driven by the muscle specific *unc-54*-promoter. Therefore, the potential polyphenolic inhibition of the pathological α-synuclein accumulation in muscles can be observed via fluorescence microscopy. Treatment of this model with either polyphenol resulted in a progressive reduction of α-synuclein accumulation in the body wall muscle cells compared to the control groups: The reduction of α-synuclein accumulation was about 5% at day 3 and 8% at day 7 and 12 of adulthood in OLE-treated groups (Figure 9). Even more pronounced effects were monitored by using HT, with a reduction of α-synuclein accumulation by 6% at day 3, 7% at day 7, and 14% at day 12 of adulthood. Overall, the fluorescence intensity declined with age, a finding in line with the ageing-dependent decline of *unc-54* expression observed by Adamla and Ignatova [66].

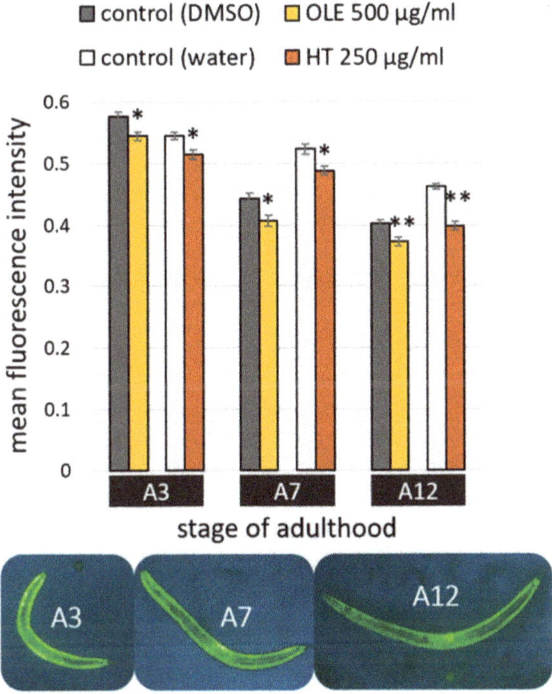

Figure 9. Changed α-synuclein accumulation in muscle cells of the OW13 strain after OLE and HT treatment. Nematodes from the OW13 strain were subjected to polyphenolic treatments starting from L4 and analysed by fluorescence microscopy using a yellow filter at day 3, 7, and 12 of adulthood. The bars represent the mean fluorescent intensity ± SEM from two biological repeats and the number of tested nematodes were: DMSO control (*n* A3 = 32, *n* A7 = 48, *n* A12 = 38), OLE (*n* A3 = 28; *n* A7 = 43, *n* A12 = 30), water control (*n* A3 = 47, *n* A7 = 75, *n* A12 = 66) and HT (*n* A3 = 35, *n* A7 = 39, *n* A12 = 44). A3, A7, A12: day 3, 7, 12 of adulthood. Differences compared to control were considered significant at $p < 0.05$ (*) and $p < 0.001$ (**). In addition, three example pictures from untreated OW13 nematodes visualising the age-dependent fluorescent change are shown (all scale bars = 200 µm).

In addition to the OW13 strain, the UA44 transgenic animals were also monitored under fluorescent conditions. The α-synuclein accumulation in dopaminergic neurons causes neurodegeneration in the UA44 transgenic strain [63]. Moreover, GFP linked to the dopamine transporter in dopaminergic nerve cells allows to visualize the quantity and quality of the six anterior and two posterior dopaminergic neurons after α-synuclein-induced damage [67].

Counting and description of each dopaminergic neuron was performed as proposed by Harrington, et al. [68]. We counted six anterior dopaminergic neurons (four CEP neurons and two ADE neurons) and two posterior DA neurons (PDE neurons) [69]. Every type of alteration, such as the loss of uniformity of the neuronal body and the reduction of the fluorescence up to neuronal shutdown, were noted in order to classify the neurons as degenerated (as shown in Figure 10B) or intact (Figure 10C). The fraction of worms with damaged dopaminergic neurons was growing with age (Figure 10A), however, HT was able to minimize neuronal damages especially in older nematodes (Figure 10A). A smaller and non-significant neuroprotective effect on dopaminergic neurons was also obtained with OLE treatment (Figure 10A).

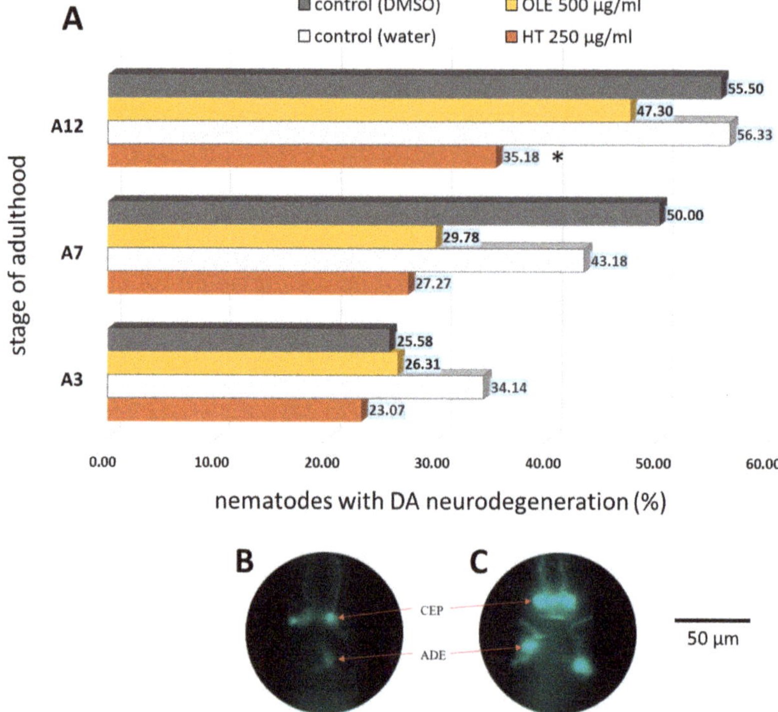

Figure 10. The impact of OLE and HT on dopaminergic neurodegeneration caused by α-synuclein in the *C. elegans* UA44 strain. (**A**) Shown are the percentages of nematodes with degenerated dopaminergic anterior neurons with and without polyphenolic treatment. The determination of the type and frequency of aberrations of dopaminergic neuronal viability was performed at day 3, 7, and 12 of adulthood. The number of tested nematodes in three biological repeats was: DMSO control = 43 (A3), 31 (A7) and 45 (A12); OLE = 38 (A3), 47 (A7) and 39 (A12); water control = 41 (A3), 44 (A7) and 71 (A12); HT = 39 (A3), 44 (A7) and 54 (A12). A3, A7, A12: day 3, 7, 12 of adulthood. Data were analysed using chi-square test with *$p < 0.05$ and **$p < 0.001$. In addition, an example of α-synuclein-induced degeneration in the anterior DA neurons of the *C. elegans* UA44 strain, expressing both P*dat-1*::GFP + P*dat-1*::α-syn is shown (**B,C**). The DA neurons are sub-classified as four CEP neurons, which are superimposed in most pictures, and two ADE neurons. (**B**) Degeneration of CEP and ADE; (**C**) intact DA neurons.

3. Discussion

3.1. HT and OLE Boost the Health, but not Necessarily the Lifespan, of Wild Type C. elegans

Lifespan analyses in *C. elegans* treated with plant polyphenols and other natural compounds were often reported with positive outcome (reviewed in Collins, et al. [70] and Pallauf, et al. [71]). The mechanisms behind this beneficial action are discussed in different directions, ranging from antioxidant, pro-oxidant, hormetic, direct pathway targeting, or caloric restriction mimetic effects, to recall only a few [72–76]. In the present study, HT was shown to improve the lifespan of *C. elegans* as well as the accumulation of age-pigment and swim behaviour in old worms. Thus, health and lifespan were targeted in parallel as expected. Interestingly, OLE also exerted very good performances regarding the impact on locomotion, stress resistance and age-pigment accumulation, yet in the absence of a life extending effect. However, since longer life does not automatically indicate healthier life [77,78], the suggestion that vice versa improved health is not a guarantee for a longer life, is not unreasonable.

Healthspan is hard to define and a lot of parameters need to be considered, but, simplifying, it can be described as the period of life in which the individual is functionally independent and free from serious diseases [79,80]. Uncoupling of the correlation between lifespan and healthspan was discussed in detail for *daf-2 C. elegans* knockout mutants, in which the lifespan-extending inhibition of insulin signalling did [60,81] or did not [78] increase healthspan in parallel. However, this discussion is not restricted to worms, but also takes place for flies, mice and humans [77,82,83]. Thus, it remains unclear how and to what extent healthspan and lifespan correlate. Furthermore, the different outcomes in the labs as exemplified by *daf-2* underline that healthspan-related features should be tested in a standardized way; not only to maintain comparability between results from different labs, but also to fully characterize the potential relationship between lifespan and healthspan in an objective manner.

Nevertheless, here the question arises, why OLE only affected healthspan but not lifespan? The stability of OLE in aqueous and ethanol solutions was shown to be better than other polyphenols [84,85] and during UV-induced degradation, the life-extending hydroxytyrosol is one of the main end products of its metabolism. Thus, an elevated level of OLE degradation during long-lasting lifespan analysis is not a sufficient explanation, also because such an effect should be much stronger for other, less stable polyphenols like quercetin (reviewed in Wang, et al. [86]), which it is not [87,88]. Another explanation is provided by the mode of action of the green tea ingredient epigallocatechin 3-gallate (EGCG). Brown et al. [89] and Zhang et al. [90] reported that, in spite of several health benefits in EGCG-treated worms, including prolonged survival under stress, no survival advantages were monitored during standard culture conditions. The antioxidant capacity of EGCG was emphasized as the main biochemical mechanisms responsible for the improvement of diverse health attributes. Due to the known strong antioxidant characteristics of OLE [91], this could also be true for OLE; accordingly, the missing lifespan prolongation by EGCG and OLE is not unexpected considering that Pun et al. [92] observed that antioxidant actions do not lead to longevity in *C. elegans*. Based on this consideration, the action of HT needs to be reconsidered. It must be concluded that the lifespan extension by HT is probably independent of its antioxidant power.

Finally, it needs to be mentioned that all tests were performed in the presence of 5-fluoro-2-deoxyuridine (FUdR). Since FUdR was shown to have (mainly positive) influences on the stress resistance and lifespan in *C. elegans* [93–95], it cannot be excluded that this could lead to false-negative results. This could be an alternative explanation for the missing life-prolonging effects in the OLE-treated group.

3.2. Anti-Parkinson's Syndrome Effects: Evidence from Three Different PD Models of C. elegans

Cooper, et al. [96] described that decelerating ageing may provide a possible treatment for PD. The beneficial effects of OLE and HT on the age-related intestinal autofluorescence and on the locomotion in old worms are strong indicators that the ageing process itself is reduced by these polyphenols. Thus, the anti-PD action of OLE and HT was not entirely surprising. Indeed, both polyphenols showed strong and convincing anti-PD activity. However, which is the underlying mechanism?

DA neurons are suffering from oxidative stress in the rotenone PD model because this pesticide generates reactive oxygen species [62]. Therefore, the neuroprotective role of the analysed polyphenols in the *C. elegans* rotenone model, characterized by defects in swimming behaviour, may be related to their antioxidant activity. Indeed, the treatment and prevention of PD with antioxidants was discussed and tested repeatedly [14,97,98], but the results were sobering, suggesting that antioxidant treatments might not be the key to combat PD [99].

The protective properties of plant polyphenols on DA neurons could also be associated not only with the structure of HT, also a product of dopamine metabolism, but also to its ability to induce phase II detoxifying enzymes. These include NADPH quinone oxidoreductase-1, heme oxygenase-1, glutathione S-transferase, and the modifier subunit of glutamate cysteine ligase which catalyses the first step in the synthesis of GSH [22]. Indeed, a *Drosophila* PD model was used to show that boosted phase II enzyme activity reduces α-synuclein-mediated neuronal loss [100].

Interestingly, both polyphenols reduced, age-dependently, the build-up of human α-synuclein in the body wall muscle cells of a transgenic *C. elegans* model (strain OW13) and improved swim performance. However, it is not clear whether the polyphenols are able to eliminate accumulated α-synuclein or prevent its accumulation and which mechanistic process is responsible for their action. One further idea regarding their mode of action is delivered by Angeloni et al. [22] who showed that polyphenols are able to modulate the proteostatic machinery, both at the proteasome complex and at the autophagic flux. More specifically, one study evaluated the effect of OLE on the proteasome complex of human embryonic fibroblasts, showing an improvement of the proteasome activity [101]. Since the impairment of the ubiquitin-proteasome complex is deeply involved in the pathogenesis of neurodegenerative diseases [102], the induction of proteasome activity might be a possible reason for the anti-PD effects of polyphenols. Other studies with OLE and HT reported a remarkable activation of the autophagic flux in a murine model of Aβ deposition, with a significant reduction of plaque load following activation of microglia [38,39]. These data also support the anti-PD activity of these polyphenols following reduction of α-synuclein aggregates in the affected brain areas.

To validate the neuroprotective action of polyphenols on dopaminergic neurons, experiments were performed also in another *C. elegans* model of PD, the strain UA44, where α-synuclein induces qualitative and quantitative damages of the six anterior and two posterior dopaminergic neurons. Our data showed that HT was able to minimize neuronal damage especially in older nematodes.

As reported above, HT is also endogenous to the brain as a catabolite of neurotransmitter breakdown. The neurotoxic action by dopamine and its intermediate metabolites is described as an autotoxic mechanism that contributes to the selective loss of dopaminergic neurons in PD [3]. HT, also known as DOPET (3,4-dihydroxyphenylethanol), is produced from dopamine by dopamine oxidative metabolism [103] in order to reduce the levels of the neurotoxic intermediate product 3,4-dihydroxyphenyl-acetaldehyde (DOPAL) in dopaminergic neurons [104]. Whether this metabolic pathway is connected to the observed beneficial action of exogenous HT remains an open question. A smaller and non-significant neuroprotective effect on dopaminergic neurons also resulted from OLE treatment. In future studies, the question needs to be answered, whether the longevity effect and the beneficial effects in the UA44 strain following HT treatment are based on the same mechanisms, which are not present or weaker in OLE-treated worms.

Tyrosol differs from HT only by one hydroxyl group; it was also shown to be a potent health- and lifespan-boosting substance in *C. elegans* [58]. However, in contrast to HT, tyrosol could not exert any preventive effects in kidney cells subjected to oxidative stress [105]. The beneficial action of HT compared to that of tyrosol could be explained by the higher scavenging and antioxidant activity of HT due to the additional hydroxyl group [106]. Nevertheless, this increased antioxidant power seems not to be the only mode of action as suggested by the convincing antioxidant-study in *C. elegans* from Pun et al. [92], by additional studies, questioning the power of antioxidants, reviewed in [107] as well as by the mitohormesis concept, which underlines the importance of ROS [108,109].

Another possible background mechanism for the observed effects could be hormesis, already considered in the case of the effect of tyrosol in *C. elegans* [58]. Several studies indicate that different stressors extend lifespan in *C. elegans* in a hormetic-like manner [110–112] and suggest that hormetic effects could be exploited to prevent the onset of various diseases [13,113,114], including neurodegenerative disorders, and to slow down the ageing process [115,116].

In addition, a recent study demonstrated that hormetic dietary phytochemicals might improve health and extend lifespan through mild elevation of ROS, which activate a number of stress adaptive genes in *C. elegans* via HSF-1 and SKN-1/Nrf2 signalling pathways [117]. Govindan and colleagues also note that the hormetic stress by phytochemicals suppresses the late age onset of misfolding and aggregation of proteins such as α-synuclein in PD. The close link between stress and ageing suggests that interventions harnessing the hormetic mechanisms may extend lifespan or delay age-associated functional decline. Taken together, these data indicate that low concentrations of natural polyphenols such as OLE and HT generate a moderate functional stress that extends healthspan in wild type and experimental models of PD. This is consistent with the idea that "neurohormesis" may have anti-ageing effects thanks to the induction of adaptive pathways triggered to cope with a mild neuronal stress, opening novel potential therapeutic strategies for clinical interventions against the onset and/or progression of PD in humans.

4. Materials and Methods

4.1. C. elegans Strains and Culture Conditions

The wild type *C. elegans* strain N2 (Var. Bristol), the transgenic *C. elegans* strain OW13 (*grk-1*(ok1239), pkIs2386 [*unc-54*p::α-synuclein::YFP + *unc-119*(+)]), as well as the *E. coli* OP50 feeding strain were obtained from the Caenorhabditis Genetics Centre, University of Minnesota (Minneapolis, MN, USA). The *C. elegans* strain UA44 (baIn11[P*dat-1*::α-synuclein, P*dat-1*::GFP]) was kindly provided by the Caldwell laboratory, University of Alabama (Tuscaloosa, AL, USA) [118]. *C. elegans* wild type and transgenic strains were grown on standard nematode growth medium (NGM) at 22 °C, seeded with *E. coli* OP50 and maintained following standard protocols as described previously [119]. Prior to all tests, synchronous L1 larvae were obtained via "egg prep" by strongly shaking at least 2000 young adults for about 4 min in 10 mL bleaching solution (0.5 mL NaOH (10 M), 2.5 mL sodium hypochlorite-solution (12%), and 7 mL bidest water). After washing with M9 buffer at least three times, the resulting egg pellet was slightly shaken overnight in 4 mL M9 buffer. The following day, the hatched L1 larvae were distributed to NGM plates seeded with OP50. Forty-eight hours later, L4 larvae were transferred to treatment plates as described below. 5-fluoro-2-deoxyuridine (FUdR; Tokyo Chemical Industry, Eschborn, Germany) was used to inhibit fertilization [120] and was dropped onto each plate with a final concentration of 100 µM (according to agar volume).

4.2. Polyphenol and Rotenone Treatment

The treatment plates were prepared with the polyphenols oleuropein aglycone (Extrasynthese) and hydroxytyrosol (Sigma-Aldrich, St. Louis, MO, USA). Glycated oleuropein was de-glycosylated by treatment with almond β-glucosidase (EC 3.2.1.21, Fluka, Sigma Aldrich, St. Louis, MO, USA) as previously described [121], with minor modifications. Briefly, 10 mM oleuropein in 0.1 M phosphate buffer (pH 7.0) was incubated overnight with β-glucosidase (8.9 I.U.) at RT. Then, the reaction mixture was centrifuged, the precipitate re-suspended in 50% (*v/v*) dimethyl sulfoxide (DMSO) and the solution kept frozen. Complete deglycosylation was assessed by assaying the released glucose. HT powder was dissolved in bidest water at 60 mg/mL and the solution stored at −20 °C. OLE and HT were added to the bacteria and agar at a final concentration of 30 µg/mL, 100 µg/mL, 250 µg/mL, or 500 µg/mL, respectively. A final concentration of 0.05% DMSO (for OLE assays) or equal amounts of bidest water (for HT assays) were applied in all treatment and control plates as well as feeding bacteria.

To trigger the Parkinsonian-related phenotype, wild type nematodes were treated with rotenone (Sigma-Aldrich, St. Louis, MO, USA). A stock solution of 0.5 mg/mL rotenone was prepared in DMSO and added with a final concentration of 10 µM to the control and polyphenol plates. After distribution with a spatula and drying for 24 h in the dark, OP50, including 10 µM rotenone and the respective polyphenol, was spread on the plates. L4 nematodes were transferred to the rotenone plates until they were used for bioassays.

4.3. Lifespan and Heat Stress Assay

Synchronized wild-type nematodes were observed and scored for their survival throughout their lifespan. The animals were counted daily from the first day of adulthood until all died. To assess the viability of the nematodes, they were first gently touched with a platinum wire at the tail and the head. If no movements were recognizable, the pharyngeal pumping was observed. The worms were considered dead when none of these movements were recognizable. The nematodes at the edge of the plate or in the depth of the agar were considered lost and excluded from the count.

The heat stress test was performed according to the lifespan protocol with the difference that nematodes at the third day of adulthood were stressed for 3 h at 37 °C and counting of dead and alive worms started 1 day after stress exposure.

4.4. Fluorescence Microscopy Analysis

For the fluorescence observation, the nematodes were placed on a 2% agar pad on a microscope slide and anesthetized with 4 µL NaN$_3$ (1M). The images were taken with the aid of the Axioskop fluorescence microscope (Carl Zeiss, Oberkochen, Germany) and filter set 13 from the Zeiss 4880 series (Carl Zeiss, Oberkochen, Germany). Nematodes with a ruptured vulva phenotype were excluded from analysis. The images were captured and analysed on the 3rd, 7th or 12th day of adulthood, respectively.

Wild type nematodes were analysed to determine and quantify ageing-related pigment accumulation. The images were captured at 100x magnification and a red filterset (TRITC, 545/30 nm ex, 610/70 nm em) was used. It was necessary to acquire additional images in a bright field, because of the poor visibility of the body contour in the dark field. The images in the bright field were used to delineate the perimeter of the worm to which the dark field image was overlayed by the CellProfiler Software (Version 3.1.9; Broad Institute, Cambridge, MA, USA) [122,123]. The quantification of age-related pigment accumulation was expressed by the mean intensity of the red fluorescence per total worm body.

The OW13 transgenic strain features yellow fluorescent protein linked to α-synuclein in the body wall muscle cells. Therefore, the nematodes were monitored using a yellow barrier filter with 100x magnification to quantify the light emission proportional to the amount of accumulation of the pathological protein. The images were processed using the CellProfiler software, and the yellow fluorescence intensity emitted per total worm body was calculated.

The UA44 transgenic strain features GFP linked to the dopamine transporter in the six dopaminergic neurons of the head and two in the tail as well as harmful α-synuclein in dopaminergic neurons. The green fluorescence intensity represents the vitality of the neurons, therefore, the green barrier filter was used for the analysis. The number of detectable anterior neurons was counted at the microscope with 200x magnification. In addition, individual images of the head were captured. The nematodes were assayed for patterns of degeneration at indicated time points, as described previously from Harrington, et al. [68].

4.5. Swim Behaviour Assay

The study of locomotion was realized with a swim assay according to Restif et al. [61] and Ibáñez–Ventoso et al. [124]. Wells with a depth of 0.5 mm and Ø 10 mm were created with two self-adhesive marking films for microscope slides and filled with M9 buffer. Five to 10 nematodes were transferred per well and covered by a cover slip to facilitate visualization and recorded with a connected camera. Nematodes with a ruptured vulva phenotype were excluded from analysis. The analysis was conducted with wild type nematodes at the 3rd, 7th, and 12th day of adulthood. Furthermore, wild type rotenone-treated worms and the strains OW13 and UA44 were analysed at the 3rd and 7th day of adulthood. Each video was converted into single frames and processed with Photoshop (version 19.1.7; Adobe Inc., San José, CA, USA) to meet the required settings. Thereafter, pictures were analysed with the CeleST software (version 3.1; distributed by GitHub Pages, https://github.com/DCS-LCSR/CeleST, accessed on 08 April 2020) as described by Restif et al. [61] and Ibáñez–Ventoso, et al. [124]. The thrashing rate, the body wave number and the activity index were evaluated as representative parameters of motor activity.

4.6. Statistical Analysis

All experiments were independently conducted at least two times. The Online Application for Survival Analysis (OASIS 2; https://sbi.postech.ac.kr/oasis2/, accessed on 8 April 2020) [125] was used for comparing survival differences between two conditions. Fluorescence intensities as well as swim behaviour parameters were calculated as mean ± SEM, and statistical significance was calculated by a two-tailed t-test using GraphPad (https://www.graphpad.com/quickcalcs/, accessed on 8 April 2020). Chi-square test was used to compare the number of worms with intact and degenerated neurons in the UA44 strain.

5. Conclusions

Due to their different actions in terms of lifespan during non-stressful conditions, disparate modes of actions could be the underlying cause of the beneficial characteristics of the olive polyphenols OLE and HT. Accordingly, possible additive or even synergistic effects by combining both polyphenols should be studied in the future. Intense research in the last decade has provided increasing knowledge of the biochemical and cell biology basis of the beneficial effects of plant, notably olive polyphenols. Such growing information indicates that these polyphenols have the potential as promising tools to be used to develop new therapeutic and preventive approaches against ageing and ageing-associated neurodegenerative diseases. However, even though *C. elegans* and mammalian models are frequently used to test possible human treatments, human clinical trials are still needed to verify this assumption.

Author Contributions: Conceptualization, V.C. and N.S.; methodology, N.S. and C.S.-L.; validation, N.S., C.S.-L., V.C.; formal analysis, G.B., G.D.R.; investigation, G.B., G.D.R.; resources, V.C., M.S. (Maria Scuto), M.L., M.S. (Massimo Stefani); writing—original draft preparation, G.B., G.D.R. and N.S.; writing—review and editing, V.C., C.S.-L., M.S. (Massimo Stefani), M.L.; supervision, N.S. and V.C.; funding acquisition, N.S. All authors have read and agreed to the published version of the manuscript.

Funding: This project has received funding from the European Union's Horizon 2020 research and innovation programme under Grant agreement No 633589 (Aging with Elegans). This publication reflects only the authors' views and the Commission is not responsible for any use that may be made of the information it contains. Furthermore, ML is granted by the Italian Airalzh Association (Reg. n° 0043966.30-10- 359 2014-u).

Acknowledgments: We thank the Caenorhabditis Genetics Centre (CGC), which is funded by NIH Office of Research Infrastructure Programs (P40 OD010440) and the Caldwell laboratory, University of Alabama for the supply of the *Caenorhabditis elegans* strains. Furthermore, we thank Thea Böttcher and Shumon Chakrabarti for their support in the lab. Not least, a special thanks goes to Christian E.W. Steinberg who enabled this project and cooperation.

Conflicts of Interest: The authors declare no conflict of interest.

References

1. Deas, E.; Cremades, N.; Angelova, P.R.; Ludtmann, M.H.; Yao, Z.; Chen, S.; Horrocks, M.H.; Banushi, B.; Little, D.; Devine, M.J. Alpha-synuclein oligomers interact with metal ions to induce oxidative stress and neuronal death in Parkinson's disease. *Antioxid. Redox Signal.* **2016**, *24*, 376–391. [CrossRef] [PubMed]

2. Trist, B.G.; Hare, D.J.; Double, K.L. Oxidative stress in the aging substantia nigra and the etiology of Parkinson's disease. *Aging Cell* **2019**, *18*, e13031. [CrossRef] [PubMed]

3. Shankar, G.M.; Li, S.; Mehta, T.H.; Garcia-Munoz, A.; Shepardson, N.E.; Smith, I.; Brett, F.M.; Farrell, M.A.; Rowan, M.J.; Lemere, C.A. Amyloid-β protein dimers isolated directly from Alzheimer's brains impair synaptic plasticity and memory. *Nat. Med.* **2008**, *14*, 837. [CrossRef] [PubMed]

4. Spillantini, M.G.; Schmidt, M.L.; Lee, V.M.; Trojanowski, J.Q.; Jakes, R.; Goedert, M. Alpha-synuclein in Lewy bodies. *Nature* **1997**, *388*, 839–840. [CrossRef] [PubMed]

5. Tysnes, O.-B.; Storstein, A. Epidemiology of Parkinson's disease. *J. Neural Transm.* **2017**, *124*, 901–905. [CrossRef] [PubMed]

6. Jankovic, J. Parkinson's disease: Clinical features and diagnosis. *J. Neurol. Neurosurg. Psychiatry* **2008**, *79*, 368–376. [CrossRef]

7. Deng, H.; Wang, P.; Jankovic, J. The genetics of Parkinson disease. *Ageing Res. Rev.* **2018**, *42*, 72–85. [CrossRef]

8. Hwang, O. Role of oxidative stress in Parkinson's disease. *Exp. Neurobiol.* **2013**, *22*, 11–17. [CrossRef]

9. Dias, V.; Junn, E.; Mouradian, M.M. The role of oxidative stress in Parkinson's disease. *J. Parkinsons Dis.* **2013**, *3*, 461–491. [CrossRef]

10. Schapira, A.; Mann, V.; Cooper, J.; Dexter, D.; Daniel, S.; Jenner, P.; Clark, J.; Marsden, C. Anatomic and disease specificity of NADH CoQ1 reductase (complex I) deficiency in Parkinson's disease. *J. Neurochem.* **1990**, *55*, 2142–2145. [CrossRef]

11. Whitton, P. Inflammation as a causative factor in the aetiology of Parkinson's disease. *Br. J. Pharmacol.* **2007**, *150*, 963–976. [CrossRef] [PubMed]

12. Sampson, T.R.; Debelius, J.W.; Thron, T.; Janssen, S.; Shastri, G.G.; Ilhan, Z.E.; Challis, C.; Schretter, C.E.; Rocha, S.; Gradinaru, V. Gut microbiota regulate motor deficits and neuroinflammation in a model of Parkinson's disease. *Cell* **2016**, *167*, 1469–1480. [CrossRef] [PubMed]

13. Calabrese, V.; Santoro, A.; Monti, D.; Crupi, R.; Di Paola, R.; Latteri, S.; Cuzzocrea, S.; Zappia, M.; Giordano, J.; Calabrese, E.J. Aging and Parkinson's disease: Inflammaging, neuroinflammation and biological remodeling as key factors in pathogenesis. *Free Radic. Biol. Med.* **2018**, *115*, 80–91. [CrossRef] [PubMed]

14. Zhao, B. Natural antioxidants protect neurons in Alzheimer's disease and Parkinson's disease. *Neurochem. Res.* **2009**, *34*, 630–638. [CrossRef] [PubMed]

15. Putteeraj, M.; Lim, W.L.; Teoh, S.L.; Yahaya, M.F. Flavonoids and its neuroprotective effects on brain ischemia and neurodegenerative diseases. *Curr. Drug Targets* **2018**, *19*, 1710–1720. [CrossRef]

16. Morais, L.; Barbosa-Filho, J.; Almeida, R. Plants and bioactive compounds for the treatment of Parkinson's disease. *Arq. Bras. Fitomedicina Científica* **2003**, *1*, 127–132.

17. Caruana, M.; Högen, T.; Levin, J.; Hillmer, A.; Giese, A.; Vassallo, N. Inhibition and disaggregation of α-synuclein oligomers by natural polyphenolic compounds. *FEBS Lett.* **2011**, *585*, 1113–1120. [CrossRef]

18. Féart, C.; Samieri, C.; Alles, B.; Barberger-Gateau, P. Potential benefits of adherence to the Mediterranean diet on cognitive health. *Proc. Nutr. Soc.* **2013**, *72*, 140–152. [CrossRef]

19. Scarmeas, N.; Luchsinger, J.A.; Stern, Y.; Gu, Y.; He, J.; DeCarli, C.; Brown, T.; Brickman, A.M. Mediterranean diet and magnetic resonance imaging–assessed cerebrovascular disease. *Ann. Neurol.* **2011**, *69*, 257–268. [CrossRef]

20. Stevanovic, T.; Diouf, P.N.; Garcia-Perez, M.E. Bioactive polyphenols from healthy diets and forest biomass. *Curr. Nutr. Food Sci.* **2009**, *5*, 264–295. [CrossRef]

21. Parkinson, L.; Cicerale, S. The health benefiting mechanisms of virgin olive oil phenolic compounds. *Molecules* **2016**, *21*, 1734. [CrossRef] [PubMed]

22. Angeloni, C.; Malaguti, M.; Barbalace, M.C.; Hrelia, S. Bioactivity of olive oil phenols in neuroprotection. *Int. J. Mol. Sci.* **2017**, *18*, 2230. [CrossRef] [PubMed]

23. Declerck, K.; Szarc vel Szic, K.; Palagani, A.; Heyninck, K.; Haegeman, G.; Morand, C.; Milenkovic, D.; Vanden Berghe, W. Epigenetic control of cardiovascular health by nutritional polyphenols involves multiple chromatin-modifying writer-reader-eraser proteins. *Curr. Top. Med. Chem.* **2016**, *16*, 788–806. [CrossRef] [PubMed]

24. Ayissi, V.B.O.; Ebrahimi, A.; Schluesenner, H. Epigenetic effects of natural polyphenols: A focus on SIRT1—Mediated mechanisms. *Mol. Nutr. Food Res.* **2014**, *58*, 22–32. [CrossRef]

25. Arias, C.; Zambrano, T.; SP Abdalla, D.; A Salazar, L. Polyphenol-related epigenetic modifications in osteoarthritis: Current therapeutic perspectives. *Curr. Pharm. Des.* **2016**, *22*, 6682–6693. [CrossRef] [PubMed]

26. Rattan, S. Biology of ageing: Principles, challenges and perspectives. *Rom. J. Morphol. Embryol.* **2015**, *56*, 1251–1253.

27. Martucci, M.; Ostan, R.; Biondi, F.; Bellavista, E.; Fabbri, C.; Bertarelli, C.; Salvioli, S.; Capri, M.; Franceschi, C.; Santoro, A. Mediterranean diet and inflammaging within the hormesis paradigm. *Nutr. Rev.* **2017**, *75*, 442–455. [CrossRef]

28. Mattson, M.P.; Son, T.G.; Camandola, S. Mechanisms of action and therapeutic potential of neurohormetic phytochemicals. *Dose-Response* **2007**, *5*. [CrossRef]

29. Mattson, M.P.; Cheng, A. Neurohormetic phytochemicals: Low-dose toxins that induce adaptive neuronal stress responses. *Trends Neurosci.* **2006**, *29*, 632–639. [CrossRef]

30. Joseph, J.A.; Shukitt-Hale, B.; Casadesus, G. Reversing the deleterious effects of aging on neuronal communication and behavior: Beneficial properties of fruit polyphenolic compounds. *Am. J. Clin. Nutr.* **2005**, *81*, 313S–316S. [CrossRef]

31. Visioli, F.; Poli, A.; Gall, C. Antioxidant and other biological activities of phenols from olives and olive oil. *Med. Res. Rev.* **2002**, *22*, 65–75. [CrossRef] [PubMed]

32. Stark, A.H.; Madar, Z. Olive oil as a functional food: Epidemiology and nutritional approaches. *Nutr. Rev.* **2002**, *60*, 170–176. [CrossRef] [PubMed]

33. Rigacci, S.; Stefani, M. Nutraceutical properties of olive oil polyphenols. An itinerary from cultured cells through animal models to humans. *Int. J. Mol. Sci.* **2016**, *17*, 843. [CrossRef] [PubMed]

34. Bendini, A.; Cerretani, L.; Carrasco-Pancorbo, A.; Gómez-Caravaca, A.M.; Segura-Carretero, A.; Fernández-Gutiérrez, A.; Lercker, G. Phenolic molecules in virgin olive oils: A survey of their sensory properties, health effects, antioxidant activity and analytical methods. An overview of the last decade Alessandra. *Molecules* **2007**, *12*, 1679–1719. [CrossRef]

35. Casamenti, F.; Stefani, M. Olive polyphenols: New promising agents to combat aging-associated neurodegeneration. *Expert Rev. Neurother.* **2017**, *17*, 345–358. [CrossRef]

36. Palazzi, L.; Bruzzone, E.; Bisello, G.; Leri, M.; Stefani, M.; Bucciantini, M.; de Laureto, P.P. Oleuropein aglycone stabilizes the monomeric α-synuclein and favours the growth of non-toxic aggregates. *Sci. Rep.* **2018**, *8*, 8337. [CrossRef]

37. Palazzi, L.; Leri, M.; Cesaro, S.; Stefani, M.; Bucciantini, M.; de Laureto, P.P. Insight into the molecular mechanism underlying the inhibition of α-synuclein aggregation by hydroxytyrosol. *Biochem. Pharmacol.* **2020**, *173*, 113722. [CrossRef]

38. Nardiello, P.; Pantano, D.; Lapucci, A.; Stefani, M.; Casamenti, F. Diet supplementation with hydroxytyrosol ameliorates brain pathology and restores cognitive functions in a mouse model of amyloid-β deposition. *J. Alzheimers Dis.* **2018**, *63*, 1161–1172. [CrossRef]

39. Grossi, C.; Rigacci, S.; Ambrosini, S.; Dami, T.E.; Luccarini, I.; Traini, C.; Failli, P.; Berti, A.; Casamenti, F.; Stefani, M. The polyphenol oleuropein aglycone protects TgCRND8 mice against Aß plaque pathology. *PLoS ONE* **2013**, *8*, e71702. [CrossRef]

40. Diomede, L.; Rigacci, S.; Romeo, M.; Stefani, M.; Salmona, M. Oleuropein aglycone protects transgenic C. elegans strains expressing Aβ42 by reducing plaque load and motor deficit. *PLoS ONE* **2013**, *8*, e58893. [CrossRef]

41. De la Torre, R. Bioavailability of olive oil phenolic compounds in humans. *Inflammopharmacology* **2008**, *16*, 245–247. [CrossRef] [PubMed]

42. Yuan, H.; Zhang, Z.-W.; Liang, L.-W.; Shen, Q.; Wang, X.-D.; Ren, S.-M.; Ma, H.-J.; Jiao, S.-J.; Liu, P. Treatment strategies for Parkinson's disease. *Neurosci. Bull.* **2010**, *26*, 66–76. [CrossRef] [PubMed]

43. Jorgensen, E.M.; Mango, S.E. The art and design of genetic screens: Caenorhabditis elegans. *Nat. Rev. Genet.* **2002**, *3*, 356. [CrossRef] [PubMed]

44. Kim, W.; Underwood, R.S.; Greenwald, I.; Shaye, D.D. OrthoList 2: A new comparative genomic analysis of human and caenorhabditis elegans genes. *Genetics* **2018**, *210*, 445–461. [CrossRef] [PubMed]

45. Chen, X.; Barclay, J.W.; Burgoyne, R.D.; Morgan, A. Using C. elegans to discover therapeutic compounds for ageing-associated neurodegenerative diseases. *Chem. Cent. J.* **2015**, *9*, 65. [CrossRef] [PubMed]

46. Wilson, M.A.; Shukitt-Hale, B.; Kalt, W.; Ingram, D.K.; Joseph, J.A.; Wolkow, C.A. Blueberry polyphenols increase lifespan and thermotolerance in Caenorhabditis elegans. *Aging Cell* **2006**, *5*, 59–68. [CrossRef]

47. Papaevgeniou, N.; Chondrogianni, N. Anti-aging and anti-aggregation properties of polyphenolic compounds in C. elegans. *Curr. Pharm. Des.* **2018**, *24*, 2107–2120. [CrossRef]

48. Saul, N.; Pietsch, K.; Stürzenbaum, S.R.; Menzel, R.; Steinberg, C.E. Diversity of polyphenol action in Caenorhabditis elegans: Between toxicity and longevity. *J. Nat. Prod.* **2011**, *74*, 1713–1720. [CrossRef]

49. Pincus, Z.; Mazer, T.C.; Slack, F.J. Autofluorescence as a measure of senescence in C. elegans: Look to red, not blue or green. *Aging (Albany N. Y.)* **2016**, *8*, 889–898. [CrossRef]

50. Keith, S.A.; Amrit, F.R.G.; Ratnappan, R.; Ghazi, A. The C. elegans healthspan and stress-resistance assay toolkit. *Methods* **2014**, *68*, 476–486. [CrossRef]

51. Benedetti, M.G.; Foster, A.L.; Vantipalli, M.C.; White, M.P.; Sampayo, J.N.; Gill, M.S.; Olsen, A.; Lithgow, G.J. Compounds that confer thermal stress resistance and extended lifespan. *Exp. Gerontol.* **2008**, *43*, 882–891. [CrossRef] [PubMed]

52. Sampayo, J.N.; Jenkins, N.L.; Lithgow, G.J. Using stress resistance to isolate novel longevity mutations in Caenorhabditis elegans. *Ann. N. Y. Acad. Sci.* **2000**, *908*, 324–326. [CrossRef] [PubMed]

53. Dues, D.J.; Andrews, E.K.; Schaar, C.E.; Bergsma, A.L.; Senchuk, M.M.; Van Raamsdonk, J.M. Aging causes decreased resistance to multiple stresses and a failure to activate specific stress response pathways. *Aging (Albany N. Y.)* **2016**, *8*, 777. [CrossRef] [PubMed]

54. Johnson, T.; Henderson, S.; Murakami, S.; De Castro, E.; de Castro, S.H.; Cypser, J.; Rikke, B.; Tedesco, P.; Link, C. Longevity genes in the nematode Caenorhabditis elegans also mediate increased resistance to stress and prevent disease. *J. Inherit. Metab. Dis.* **2002**, *25*, 197–206. [CrossRef]

55. Prahlad, V.; Morimoto, R.I. Integrating the stress response: Lessons for neurodegenerative diseases from C. elegans. *Trends Cell Biol.* **2009**, *19*, 52–61. [CrossRef]

56. Rodriguez, M.; Snoek, L.B.; De Bono, M.; Kammenga, J.E. Worms under stress: C. elegans stress response and its relevance to complex human disease and aging. *Trends Genet.* **2013**, *29*, 367–374. [CrossRef]

57. Fernandez del Rio, L.; Gutiérrez-Casado, E.; Varela-López, A.; Villalba, J.M. Olive oil and the hallmarks of aging. *Molecules* **2016**, *21*, 163. [CrossRef]

58. Cañuelo, A.; Gilbert-López, B.; Pacheco-Liñán, P.; Martínez-Lara, E.; Siles, E.; Miranda-Vizuete, A. Tyrosol, a main phenol present in extra virgin olive oil, increases lifespan and stress resistance in Caenorhabditis elegans. *Mech. Ageing Dev.* **2012**, *133*, 563–574. [CrossRef]

59. Herndon, L.A.; Schmeissner, P.J.; Dudaronek, J.M.; Brown, P.A.; Listner, K.M.; Sakano, Y.; Paupard, M.C.; Hall, D.H.; Driscoll, M. Stochastic and genetic factors influence tissue-specific decline in ageing C. elegans. *Nature* **2002**, *419*, 808. [CrossRef]

60. Hahm, J.H.; Kim, S.; DiLoreto, R.; Shi, C.; Lee, S.J.; Murphy, C.T.; Nam, H.G. C. elegans maximum velocity correlates with healthspan and is maintained in worms with an insulin receptor mutation. *Nat. Commun.* **2015**, *6*, 8919. [CrossRef]

61. Restif, C.; Ibáñez-Ventoso, C.; Vora, M.M.; Guo, S.; Metaxas, D.; Driscoll, M. CeleST: Computer vision software for quantitative analysis of C. elegans swim behavior reveals novel features of locomotion. *PLoS Comput. Biol.* **2014**, *10*, e1003702. [CrossRef] [PubMed]

62. Sherer, T.B.; Betarbet, R.; Testa, C.M.; Seo, B.B.; Richardson, J.R.; Kim, J.H.; Miller, G.W.; Yagi, T.; Matsuno-Yagi, A.; Greenamyre, J.T. Mechanism of toxicity in rotenone models of Parkinson's disease. *J. Neurosci.* **2003**, *23*, 10756–10764. [CrossRef] [PubMed]

63. Gaeta, A.L.; Caldwell, K.A.; Caldwell, G.A. Found in translation: The utility of C. elegans alpha-synuclein models of parkinson's disease. *Brain Sci.* **2019**, *9*, 73. [CrossRef] [PubMed]

64. Van Ham, T.J.; Thijssen, K.L.; Breitling, R.; Hofstra, R.M.; Plasterk, R.H.; Nollen, E.A. C. elegans model identifies genetic modifiers of α-synuclein inclusion formation during aging. *PLoS Genet.* **2008**, *4*, e1000027. [CrossRef] [PubMed]

65. Lakso, M.; Vartiainen, S.; Moilanen, A.M.; Sirviö, J.; Thomas, J.H.; Nass, R.; Blakely, R.D.; Wong, G. Dopaminergic neuronal loss and motor deficits in Caenorhabditis elegans overexpressing human α-synuclein. *J. Neurochem.* **2003**, *86*, 165–172. [CrossRef]

66. Adamla, F.; Ignatova, Z. Somatic expression of unc-54 and vha-6 mRNAs declines but not pan-neuronal rgef-1 and unc-119 expression in aging Caenorhabditis elegans. *Sci. Rep.* **2015**, *5*, 10692. [CrossRef]

67. Tucci, M.L.; Harrington, A.J.; Caldwell, G.A.; Caldwell, K.A. Modeling dopamine neuron degeneration in Caenorhabditis elegans. In *Neurodegeneration*; Springer: Berlin, Germany, 2011; pp. 129–148.

68. Harrington, A.J.; Knight, A.L.; Caldwell, G.A.; Caldwell, K.A. Caenorhabditis elegans as a model system for identifying effectors of α-synuclein misfolding and dopaminergic cell death associated with Parkinson's disease. *Methods* **2011**, *53*, 220–225. [CrossRef]

69. Sulston, J.; Dew, M.; Brenner, S. Dopaminergic neurons in the nematode Caenorhabditis elegans. *J. Comp. Neurol.* **1975**, *163*, 215–226. [CrossRef]

70. Collins, J.J.; Evason, K.; Kornfeld, K. Pharmacology of delayed aging and extended lifespan of Caenorhabditis elegans. *Exp. Gerontol.* **2006**, *41*, 1032–1039. [CrossRef]

71. Pallauf, K.; Duckstein, N.; Rimbach, G. A literature review of flavonoids and lifespan in model organisms. *Proc. Nutr. Soc.* **2017**, *76*, 145–162. [CrossRef]

72. Xiong, L.-G.; Chen, Y.-J.; Tong, J.-W.; Gong, Y.-S.; Huang, J.-A.; Liu, Z.-H. Epigallocatechin-3-gallate promotes healthy lifespan through mitohormesis during early-to-mid adulthood in Caenorhabditis elegans. *Redox Biol.* **2018**, *14*, 305–315. [CrossRef] [PubMed]

73. Testa, G.; Biasi, F.; Poli, G.; Chiarpotto, E. Calorie restriction and dietary restriction mimetics: A strategy for improving healthy aging and longevity. *Curr. Pharm. Des.* **2014**, *20*, 2950–2977. [CrossRef] [PubMed]

74. Procházková, D.; Boušová, I.; Wilhelmová, N. Antioxidant and prooxidant properties of flavonoids. *Fitoterapia* **2011**, *82*, 513–523. [CrossRef] [PubMed]

75. Pazoki-Toroudi, H.; Amani, H.; Ajami, M.; Nabavi, S.F.; Braidy, N.; Kasi, P.D.; Nabavi, S.M. Targeting mTOR signaling by polyphenols: A new therapeutic target for ageing. *Ageing Res. Rev.* **2016**, *31*, 55–66. [CrossRef]

76. Martel, J.; Ojcius, D.M.; Ko, Y.-F.; Ke, P.-Y.; Wu, C.-Y.; Peng, H.-H.; Young, J.D. Hormetic effects of phytochemicals on health and longevity. *Trends Endocrinol. Metab.* **2019**, *30*, 335–346. [CrossRef]

77. Hansen, M.; Kennedy, B.K. Does longer lifespan mean longer healthspan? *Trends Cell Biol.* **2016**, *26*, 565–568. [CrossRef]

78. Bansal, A.; Zhu, L.J.; Yen, K.; Tissenbaum, H.A. Uncoupling lifespan and healthspan in Caenorhabditis elegans longevity mutants. *Proc. Natl. Acad. Sci. USA* **2015**, *112*, E277–E286. [CrossRef]

79. Luyten, W.; Antal, P.; Braeckman, B.P.; Bundy, J.; Cirulli, F.; Fang-Yen, C.; Fuellen, G.; Leroi, A.; Liu, Q.; Martorell, P.; et al. Ageing with elegans: A research proposal to map healthspan pathways. *Biogerontology* **2016**, *17*, 771–782. [CrossRef]

80. Fuellen, G.; Jansen, L.; Cohen, A.A.; Luyten, W.; Gogol, M.; Simm, A.; Saul, N.; Cirulli, F.; Berry, A.; Antal, P. Health and aging: Unifying concepts, scores, biomarkers and pathways. *Aging Dis.* **2019**, *10*, 883. [CrossRef]

81. Murakami, H.; Bessinger, K.; Hellmann, J.; Murakami, S. Aging-dependent and -independent modulation of associative learning behavior by insulin/insulin-like growth factor-1 signal in Caenorhabditis elegans. *J. Neurosci.* **2005**, *25*, 10894–10904. [CrossRef]

82. Fischer, K.E.; Hoffman, J.M.; Sloane, L.B.; Gelfond, J.A.; Soto, V.Y.; Richardson, A.G.; Austad, S.N. A cross-sectional study of male and female C57BL/6Nia mice suggests lifespan and healthspan are not necessarily correlated. *Aging (Albany N. Y.)* **2016**, *8*, 2370. [CrossRef] [PubMed]

83. Avanesian, A.; Khodayari, B.; Felgner, J.S.; Jafari, M. Lamotrigine extends lifespan but compromises health span in Drosophila melanogaster. *Biogerontology* **2010**, *11*, 45. [CrossRef] [PubMed]

84. Malik, N.S.; Bradford, J.M. Recovery and stability of oleuropein and other phenolic compounds during extraction and processing of olive (Olea europaea L.) leaves. *J. Food Agric. Environ.* **2008**, *6*, 8.

85. Longo, E.; Morozova, K.; Scampicchio, M. Effect of light irradiation on the antioxidant stability of oleuropein. *Food Chem.* **2017**, *237*, 91–97. [CrossRef] [PubMed]

86. Wang, W.; Sun, C.; Mao, L.; Ma, P.; Liu, F.; Yang, J.; Gao, Y. The biological activities, chemical stability, metabolism and delivery systems of quercetin: A review. *Trends Food Sci. Technol.* **2016**, *56*, 21–38. [CrossRef]

87. Saul, N.; Pietsch, K.; Menzel, R.; Steinberg, C.E. Quercetin-mediated longevity in Caenorhabditis elegans: Is DAF-16 involved? *Mech. Ageing Dev.* **2008**, *129*, 611–613. [CrossRef]

88. Kampkötter, A.; Timpel, C.; Zurawski, R.F.; Ruhl, S.; Chovolou, Y.; Proksch, P.; Wätjen, W. Increase of stress resistance and lifespan of Caenorhabditis elegans by quercetin. *Comp. Biochem. Physiol. Part B Biochem. Mol. Biol.* **2008**, *149*, 314–323. [CrossRef]

89. Brown, M.K.; Evans, J.L.; Luo, Y. Beneficial effects of natural antioxidants EGCG and alpha-lipoic acid on life span and age-dependent behavioral declines in Caenorhabditis elegans. *Pharmacol. Biochem. Behav.* **2006**, *85*, 620–628. [CrossRef]

90. Zhang, L.; Jie, G.; Zhang, J.; Zhao, B. Significant longevity-extending effects of EGCG on Caenorhabditis elegans under stress. *Free Radic. Biol. Med.* **2009**, *46*, 414–421. [CrossRef]

91. Speroni, E.; Guerra, M.; Minghetti, A.; Crespi-Perellino, N.; Pasini, P.; Piazza, F.; Roda, A. Oleuropein evaluated in vitro and in vivo as an antioxidant. *Phytother. Res. Int. J. Devoted Pharmacol. Toxicol. Eval. Nat. Prod. Deriv.* **1998**, *12*, S98–S100. [CrossRef]

92. Pun, P.B.L.; Gruber, J.; Tang, S.Y.; Schaffer, S.; Ong, R.L.S.; Fong, S.; Ng, L.F.; Cheah, I.; Halliwell, B. Ageing in nematodes: Do antioxidants extend lifespan in Caenorhabditiselegans? *Biogerontology* **2010**, *11*, 17–30. [CrossRef] [PubMed]

93. Feldman, N.; Kosolapov, L.; Ben-Zvi, A. Fluorodeoxyuridine improves Caenorhabditis elegans proteostasis independent of reproduction onset. *PLoS ONE* **2014**, *9*, e85964. [CrossRef]

94. Van Raamsdonk, J.M.; Hekimi, S. FUdR causes a twofold increase in the lifespan of the mitochondrial mutant gas-1. *Mech. Ageing Dev.* **2011**, *132*, 519–521. [CrossRef]

95. Wang, H.; Zhao, Y.; Zhang, Z. Age-dependent effects of floxuridine (FUdR) on senescent pathology and mortality in the nematode Caenorhabditis elegans. *Biochem. Biophys. Res. Commun.* **2019**, *509*, 694–699. [CrossRef] [PubMed]

96. Cooper, J.F.; Dues, D.J.; Spielbauer, K.K.; Machiela, E.; Senchuk, M.M.; Van Raamsdonk, J.M. Delaying aging is neuroprotective in Parkinson's disease: A genetic analysis in C. elegans models. *NPJ Parkinsons Dis* **2015**, *1*, 15022. [CrossRef]

97. Sarrafchi, A.; Bahmani, M.; Shirzad, H.; Rafieian-Kopaei, M. Oxidative stress and Parkinson's disease: New hopes in treatment with herbal antioxidants. *Curr. Pharm. Des.* **2016**, *22*, 238–246. [CrossRef]

98. Jin, H.; Kanthasamy, A.; Ghosh, A.; Anantharam, V.; Kalyanaraman, B.; Kanthasamy, A.G. Mitochondria-targeted antioxidants for treatment of Parkinson's disease: Preclinical and clinical outcomes. *Biochim. Biophys. Acta-Mol. Basis Dis.* **2014**, *1842*, 1282–1294. [CrossRef] [PubMed]

99. Snow, B.J.; Rolfe, F.L.; Lockhart, M.M.; Frampton, C.M.; O'Sullivan, J.D.; Fung, V.; Smith, R.A.; Murphy, M.P.; Taylor, K.M.; Group, P.S. A double-blind, placebo-controlled study to assess the mitochondria-targeted antioxidant MitoQ as a disease-modifying therapy in Parkinson's disease. *Mov. Disord.* **2010**, *25*, 1670–1674. [CrossRef]

100. Trinh, K.; Moore, K.; Wes, P.D.; Muchowski, P.J.; Dey, J.; Andrews, L.; Pallanck, L.J. Induction of the phase II detoxification pathway suppresses neuron loss in Drosophila models of Parkinson's disease. *J. Neurosci.* **2008**, *28*, 465–472. [CrossRef]

101. Katsiki, M.; Chondrogianni, N.; Chinou, I.; Rivett, A.J.; Gonos, E.S. The olive constituent oleuropein exhibits proteasome stimulatory properties in vitro and confers life span extension of human embryonic fibroblasts. *Rejuvenation Res.* **2007**, *10*, 157–172. [CrossRef]

102. McKinnon, C.; Tabrizi, S.J. The ubiquitin-proteasome system in neurodegeneration. *Antioxid. Redox Signal.* **2014**, *21*, 2302–2321. [CrossRef] [PubMed]

103. Calabrese, V.C.R. Potential prevention and treatment of neurodegenerative diseases: Olive polyphenols and hydroxytyrosol. *Eur. J. Neurodegener. Dis.* **2016**, *5*, 81–108.

104. Rodríguez-Morató, J.; Xicota, L.; Fito, M.; Farre, M.; Dierssen, M.; de la Torre, R. Potential role of olive oil phenolic compounds in the prevention of neurodegenerative diseases. *Molecules* **2015**, *20*, 4655–4680. [CrossRef] [PubMed]

105. Loru, D.; Incani, A.; Deiana, M.; Corona, G.; Atzeri, A.; Melis, M.; Rosa, A.; Dessi, M. Protective effect of hydroxytyrosol and tyrosol against oxidative stress in kidney cells. *Toxicol. Ind. Health* **2009**, *25*, 301–310. [CrossRef] [PubMed]

106. Morales, J.C.; Lucas, R. Structure–Activity relationship of phenolic antioxidants and olive components. In *Olives and Olive Oil in Health and Disease Prevention*; Elsevier: Amsterdam, the Netherlands, 2010; pp. 905–914.

107. Gems, D.; Doonan, R. Antioxidant defense and aging in C. elegans: Is the oxidative damage theory of aging wrong? *Cell Cycle* **2009**, *8*, 1681–1687. [CrossRef] [PubMed]

108. Ristow, M. Unraveling the truth about antioxidants: Mitohormesis explains ROS-induced health benefits. *Nat. Med.* **2014**, *20*, 709–711. [CrossRef] [PubMed]

109. Ristow, M.; Zarse, K. How increased oxidative stress promotes longevity and metabolic health: The concept of mitochondrial hormesis (mitohormesis). *Exp. Gerontol.* **2010**, *45*, 410–418. [CrossRef]

110. Cypser, J.R.; Johnson, T.E. Multiple stressors in Caenorhabditis elegans induce stress hormesis and extended longevity. *J. Gerontol. Ser. A Biol. Sci. Med Sci.* **2002**, *57*, B109–B114. [CrossRef]

111. Schmeisser, S.; Schmeisser, K.; Weimer, S.; Groth, M.; Priebe, S.; Fazius, E.; Kuhlow, D.; Pick, D.; Einax, J.W.; Guthke, R. Mitochondrial hormesis links low-dose arsenite exposure to lifespan extension. *Aging Cell* **2013**, *12*, 508–517. [CrossRef]

112. Pietsch, K.; Saul, N.; Chakrabarti, S.; Stürzenbaum, S.R.; Menzel, R.; Steinberg, C.E. Hormetins, antioxidants and prooxidants: Defining quercetin-, caffeic acid-and rosmarinic acid-mediated life extension in C. elegans. *Biogerontology* **2011**, *12*, 329–347. [CrossRef]

113. Calabrese, V.; Santoro, A.; Trovato Salinaro, A.; Modafferi, S.; Scuto, M.; Albouchi, F.; Monti, D.; Giordano, J.; Zappia, M.; Franceschi, C. Hormetic approaches to the treatment of Parkinson's disease: Perspectives and possibilities. *J. Neurosci. Res.* **2018**, *96*, 1641–1662. [CrossRef]

114. Leak, R.K.; Calabrese, E.J.; Kozumbo, W.J.; Gidday, J.M.; Johnson, T.E.; Mitchell, J.R.; Ozaki, C.K.; Wetzker, R.; Bast, A.; Belz, R.G. Enhancing and extending biological performance and resilience. *Dose-Response* **2018**, *16*. [CrossRef] [PubMed]

115. Marini, A.M.; Jiang, H.; Pan, H.; Wu, X.; Lipsky, R.H. Hormesis: A promising strategy to sustain endogenous neuronal survival pathways against neurodegenerative disorders. *Ageing Res. Rev.* **2008**, *7*, 21–33. [CrossRef] [PubMed]

116. Matus, S.; Castillo, K.; Hetz, C. Hormesis: Protecting neurons against cellular stress in Parkinson disease. *Autophagy* **2012**, *8*, 997–1001. [CrossRef] [PubMed]

117. Govindan, S.; Amirthalingam, M.; Duraisamy, K.; Govindhan, T.; Sundararaj, N.; Palanisamy, S. Phytochemicals-induced hormesis protects Caenorhabditis elegans against α-synuclein protein aggregation and stress through modulating HSF-1 and SKN-1/Nrf2 signaling pathways. *Biomed. Pharmacother.* **2018**, *102*, 812–822. [CrossRef] [PubMed]

118. Cao, S.; Gelwix, C.C.; Caldwell, K.A.; Caldwell, G.A. Torsin-mediated protection from cellular stress in the dopaminergic neurons of Caenorhabditis elegans. *J. Neurosci.* **2005**, *25*, 3801–3812. [CrossRef] [PubMed]

119. Brenner, S. The genetics of Caenorhabditis elegans. *Genetics* **1974**, *77*, 71–94.

120. Mitchell, D.H.; Stiles, J.W.; Santelli, J.; Sanadi, D.R. Synchronous growth and aging of Caenorhabditis elegans in the presence of fluorodeoxyuridine. *J. Gerontol.* **1979**, *34*, 28–36. [CrossRef]

121. Konno, K.; Hirayama, C.; Yasui, H.; Nakamura, M. Enzymatic activation of oleuropein: A protein crosslinker used as a chemical defense in the privet tree. *Proc. Natl. Acad. Sci. USA* **1999**, *96*, 9159–9164. [CrossRef]

122. McQuin, C.; Goodman, A.; Chernyshev, V.; Kamentsky, L.; Cimini, B.A.; Karhohs, K.W.; Doan, M.; Ding, L.; Rafelski, S.M.; Thirstrup, D. CellProfiler 3.0: Next-generation image processing for biology. *PLoS Biol.* **2018**, *16*, e2005970. [CrossRef]

123. Hamilton, N. Quantification and its applications in fluorescent microscopy imaging. *Traffic* **2009**, *10*, 951–961. [CrossRef] [PubMed]

124. Ibáñez-Ventoso, C.; Herrera, C.; Chen, E.; Motto, D.; Driscoll, M. Automated analysis of C. elegans swim behavior using CeleST software. *J. Vis. Exp.* **2016**, e54359. [CrossRef] [PubMed]

125. Han, S.K.; Lee, D.; Lee, H.; Kim, D.; Son, H.G.; Yang, J.-S.; Lee, S.-J.V.; Kim, S. OASIS 2: Online application for survival analysis 2 with features for the analysis of maximal lifespan and healthspan in aging research. *Oncotarget* **2016**, *7*, 56147. [CrossRef] [PubMed]

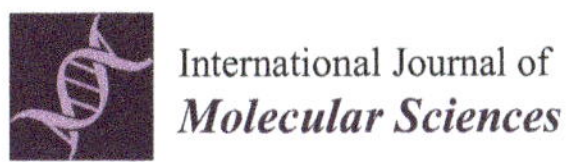

International Journal of
Molecular Sciences

Review

Caenorhabditis Elegans and Probiotics Interactions from a Prolongevity Perspective

Marianna Roselli [1], Emily Schifano [2], Barbara Guantario [1], Paola Zinno [1], Daniela Uccelletti [2,*] and Chiara Devirgiliis [1,*]

[1] Research Centre for Food and Nutrition, CREA (Council for Agricultural Research and Economics), 00178 Rome, Italy; marianna.roselli@crea.gov.it (M.R.); barbara.guantario@crea.gov.it (B.G.); paola.zinno@crea.gov.it (P.Z.)

[2] Department of Biology and Biotechnology "C. Darwin", Sapienza University of Rome, 00185 Rome, Italy; emily.schifano@uniroma1.it

* Correspondence: daniela.uccelletti@uniroma1.it (D.U.); chiara.devirgiliis@crea.gov.it (C.D.); Tel.: +39-064-991-2132 (D.U.); Tel.: +39-065-149-4647 (C.D.)

Received: 16 September 2019; Accepted: 10 October 2019; Published: 10 October 2019

Abstract: Probiotics exert beneficial effects on host health through different mechanisms of action, such as production of antimicrobial substances, competition with pathogens, enhancement of host mucosal barrier integrity and immunomodulation. In the context of ageing, which is characterized by several physiological alterations leading to a low grade inflammatory status called inflammageing, evidences suggest a potential prolongevity role of probiotics. Unraveling the mechanisms underlying anti-ageing effects requires the use of simple model systems. To this respect, the nematode *Caenorhabditis elegans* represents a suitable model organism for the study of both host-microbe interactions and for ageing studies, because of conserved signaling pathways and host defense mechanisms involved in the regulation of its lifespan. Therefore, this review analyses the impact of probiotics on *C. elegans* age-related parameters, with particular emphasis on oxidative stress, immunity, in flammation and protection from pathogen infections. The picture emerging from our analysis highlights that several probiotic strains are able to exert anti-ageing effects in nematodes by acting on common molecular pathways, such as insulin/insulin-like growth factor-1 (IIS) and p38 mitogen-activated protein kinase (p38 MAPK). In this perspective, *C. elegans* appears to be advantageous for shedding light on key mechanisms involved in host prolongevity in response to probiotics supplementation.

Keywords: ageing; nematode; immunosenescence; oxidative stress; lifespan; probiotic bacteria; pathogen protection

---

1. *Caenorhabditis Elegans* as a Model System to Study Prolongevity

Caenorhabditis elegans is a small nematode widely used as a model system for different biological studies because of its many advantages. It is characterized by transparency, a short life cycle, ease of cultivation, and availability of large sets of mutants [1]. Worms can be grown cheaply and in large numbers on agar plates and they are normally fed bacteria, although they can be also fed yeasts. In addition, even if *C. elegans* is considered a simple organism, many of the molecular cascades controlling its development are also found in more complex organisms, like humans [2,3]. Nematode lifespan is a parameter that can be influenced by genetic and environmental factors, in cluding nutritional stimuli. The genes involved in lifetime regulation are associated with different molecular pathways, evolutionarily conserved, that modulate ageing processes [4], such as insulin/insulin-like growth factor-1 (IIS) [5] and p38 mitogen-activated protein kinase (p38 MAPK) pathways [6]. For these reasons *C. elegans* represents a suitable model organism for ageing studies and for evaluating the impact of nutritional stimuli on prolongevity. Indeed, different bacterial feedings can play an important role in the regulation of nematode lifespan by inducing specific host responses [7]. In particular, while some pathogens shorten worm viability, several probiotic strains show beneficial effects, prolonging lifespan and leading to a delay in ageing [8,9]. It has been widely reported that these effects correlate to the host defense responses and stress resistance of *C. elegans* [10]. Indeed, ageing is characterized by progressive damage of the stress response and cellular machine. In nematodes different ageing biomarkers could be studied to evaluate the effects of a diet. Indeed, pharyngeal pumping rate, locomotion ability, body size and intestinal lipofuscin autofluorescent granules are the most examined markers, thanks to the ease of analysis [11,12] (Figure 1). *C. elegans* is therefore considered as one of the best model systems used to study longevity and to screen bacteria showing probiotic properties and anti-ageing effects [13].

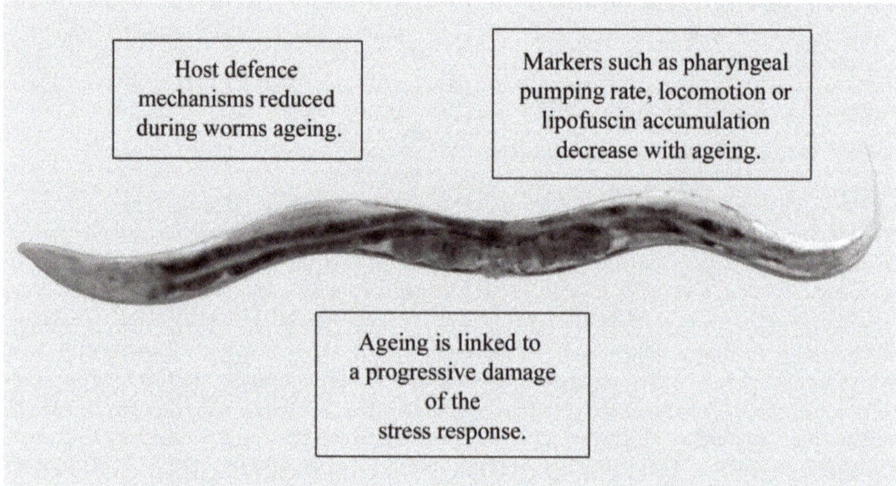

Figure 1. Micrograph representing the model organism *Caenorhabditis elegans* (the head is on the left; the tail is on the right). The principal biomarkers and physiological traits associated with ageing are described in the squares. Magnification: 5×.

2. Probiotics: Characteristics and Relevance to Ageing

Probiotics are commonly defined as "live microorganisms that, when administered in adequate amounts, confer a health benefit to the host" [14]. The majority of species known to have probiotic properties belong to the genera *Lactobacillus* and *Bifidobacterium*, commonly found in the gastrointestinal tract of humans and animals and thus generally regarded as safe. However, also members of other bacterial genera can have documented health benefits, such as *Bacillus*, *Enterococcus* as well as the yeast *Saccharomyces*. It is widely recognized that the health benefits of probiotics are strictly strain-specific, consequently distinct strains belonging to the same species can have different effects. For this reason, accurate characterization of novel potentially probiotic strains is very important. Amongst various possible mechanisms of action, probiotics are believed to exert their effects by production of antimicrobial substances, competition with pathogens for adhesion sites and nutrients, enhancement of host mucosal barrier integrity and immune modulation [15,16]. Thus, the beneficial activities of probiotics are attributable to three main core benefits: supporting a healthy gut microbiota, a healthy digestive tract and a healthy immune system [17].

In the context of ageing, several physiological changes affecting the immune and digestive systems as well as gut microbiota composition lead to a physiological low-grade inflammatory status called, "inflammageing" [18,19], which can be potentially counteracted by probiotic interventions [20]. Indeed, perturbations of gut microbiota composition and immune function associated with ageing can favor the growth of pathogens and increase the susceptibility to gut-related diseases [21], affecting members of the health-promoting bacteria resident in the gut. In particular, a reduction of numbers and species of bifidobacteria has been reported in older persons [22,23]. At this stage of life, probiotics exert several beneficial effects for the host by protecting against pathogenic bacteria and viruses, enhancing immune function, counteracting intestinal inflammatory diseases, and improving metabolic functions and the lipid profile [24,25].

One of the major issues related to the study of probiotics is the need of appropriate, simplified in vivo models representing useful and less expensive screening tools to identify probiotic strains from a large number of microbial candidates. To this respect, the nematode *Caenorhabditis elegans* is becoming an increasingly valuable in vivo model to study host-probiotic interactions to enhance lifespan [26].

3. Review Methodology

We conducted a literature search on PubMed by using the following keywords: (1) *Caenorhabditis elegans* and probiotics; (2) *Caenorhabditis elegans* and lactobacilli; (3) *Caenorhabditis elegans* and bifidobacteria; and (4) *Caenorhabditis elegans* and lactic acid bacteria. Publication dates were restricted to the last ten years. The first search retrieved 46 results. The other searches produced a majority of overlapping results with the first one and some additional results, in particular: the second search retrieved six more results; the third one two more results and the fourth one three more results. Following this initial search, eight articles were excluded, on the basis of their low adherence to the description of probiotic activities in *C. elegans*, which was the main focus of our search. Thus, finally 49 articles were carefully evaluated for the review preparation. Table 1 shows a list of the microorganism strains described in the selected literature to exert a probiotic activity in *C. elegans*. The majority of the tested species belonged to the *Lactobacillus* genus, with 16 different species and 35 strains, followed by the *Bifidobacterium* genus, with 4 species and 6 strains. Then we found 2 *Pediococcus*, 2 *Weissella* and 2 *Enterococcus* species, and finally *Bacillus*, *Butyricicoccus*, *Megasphaera*, *Clostridium*, *Propionibacterium*, *Escherichia* and *Kluyveromyces*, with one species each. Of note, only *Kluyveromyces marxianus* belongs to the Ascomycota phylum within the yeast kingdom.

Among the collected papers, those taking into account the effect of probiotic supplementation on nematode lifespan were further selected and analyzed to evaluate important parameters related to ageing, such as oxidative stress, immune system and susceptibility to pathogen infection, which are known to be involved in immunosenescence. This condition refers to the gradual deterioration of the immune system brought on by age progression. It involves both the host capacity to respond to infections and the development of long-term immunological memory. Immunosenescence can be considered as a crucial contributory factor to the increased occurrence of morbidity and mortality among the elderly. This age-related immune deficiency is ubiquitous and found in both long- and short-living species, and it is characterized by a particular "remodeling" of the immune system, in duced by oxidative stress [27]. Together with inflammageing, immunosenescence is suggested to stand at the origin of the majority of elderly-related alterations, such as infections, cancer, autoimmune disorders, and chronic inflammatory diseases [28]. The present review focuses on the role of probiotics on *C. elegans* age-related parameters, with particular emphasis on oxidative stress, immunity and inflammation, and protection from pathogen infections.

Table 1. List of microbial strains reported in the selected literature to exert a probiotic activity in *C. elegans*.

Genus	Species	Strain(s)	Nematode Signaling Pathway(s) Influenced	References
Lactobacillus	*acidophilus*	NCFM	p38 MAPK beta-catenin	[29]
	brevis	SDL1411	unknown	[30]
	casei	CL11	unknown	[31]
		LAB9	p38 MAPK	[32,33]
	coryniformis	H307.6	unknown	[34]
	delbrueckii	*bulgaricus* ATCC11842; *lactis* LMG6401; *lactis* 23	unknown	[35]
	fermentum	MBC2	unknown	[36]
		JDFM216	p38 MAPK	[37]
		LA12	unknown	[38]
		LF21	IIS	[39]
	gasseri	SBT2055	p38 MAPK	[9]
	helveticus	NBRC15019	unknown	[40]
	murinus	CR147	unknown	[41]
	paracasei	28.4	unknown	[42]
	plantarum	CAU1054; CAU1055; CAU1064; CAU1106	unknown	[43] [44]
		JDFM60; JDFM440; JDFM970; JDFM1000	unknown	[38,39] [45]
		CJLP133	IIS	[40]
		K90	unknown	
		NBRC15891	unknown	
	pentosus	D303.36	unknown	[34]
	reuteri	CL9	unknown	[46]
		S64	unknown	[31]
		DSM 20016	unknown	[47]
	rhamnosus	R4	unknown	[48]
		CNCM I-3690	IIS	[49]
		NBRC14710	unknown	[40]
	salivarius	FDB89	unknown	[50]
		DSM 20555	unknown	[47]
	zeae	LB1	p38 MAPK IIS	[51,52]
Bifidobacterium	*animalis subsp. lactis*	CECT8145	IIS	[53]
	breve	UCC2003	unknown	[54]
	Infantis [1]	ATCC15697	p38 MAPK IIS	[40,55–57]
	longum	ATCC15707	unknown	[29,40]
		BB68	JNK	[58]
		BR-108	IIS	[59]
Bacillus	*licheniformis*	141	unknown	[60]
Butyricicoccus	*pullicaecorum*	KCTC 15070	TGF-beta	[61]
Clostridium	*butyricum*	MIYAIRI 588 (CBM 588)	IIS	[62]
Enterococcus	*faecalis*	MMH594	p38 MAPK beta-catenin	[63]
		Symbioflor®	unknown	[64]
	faecium	L11	TGF-beta p38 MAPK	[65]
		E007	p38 MAPK beta-catenin	[63]
Escherichia	*coli*	Nissle 1917	unknown	[66]

Table 1. *Cont.*

Genus	Species	Strain(s)	Nematode Signaling Pathway(s) Influenced	References
Megasphaera	*elsdenii*	KCTC 5187	TGF-beta	[61]
Pediococcus	*acidilactici*	DSM 20284 DM-9	unknown unknown	[47] [30]
	pentosaceus	SDL1409	unknown	[30]
Propionibacterium	*freudenreichii*	KCTC 1063	p38 MAPK	[67]
Weissella	*cibaria*	KACC11845	JNK AMPK	[68]
	koreensis	KACC11853	JNK AMPK	
Kluyveromyces	*marxianus*	CIDCA 8154	p38 MAPK	[69]

[1] The current adscription is *Bifidobacterium longum* subsp. *infantis* [70].

4. Mechanisms Involved in *C. elegans* Lifespan Extension Induced by Probiotics

4.1. Description of the Main Pathways

The principal pathways involved in lifespan control, oxidative stress, regulation of immune response and defense against pathogen infection in *C. elegans* include the IIS pathway and p38 MAPK pathway [71]. Each pathway is composed of a cascade of signaling molecules that finally activate/regulate the transcription of specific target genes. In particular, the IIS pathway is initiated by the activation of dauer formation (DAF)-2, an insulin/insulin-like growth factor-1 receptor ortholog, subsequently triggering a cascade of phosphorylation events that activate specific kinases and downstream mediators. These include phosphatidylinositol 3-kinase AGE-1, phosphoinositide-dependent kinase (PDK)-1, and various serine/threonine protein kinases (AKT-1, AKT-2, and SGK-1), culminating in phosphorylation of DAF-16, a protein belonging to class O of the forkhead transcription factors (FOXO), resulting in its inactivation [72]. On the contrary, in the presence of heat stress, anoxia, oxidative stress, starvation, and infections, the IIS pathway is down-regulated and DAF-16 migrates to the nucleus, where it switches on the expression of specific target genes, that contribute to several cellular processes, from apoptosis to stress resistance, prolongevity and anti-ageing [58,73]. The nuclear translocation of DAF-16 leads to both up-regulation and down-regulation of large sets of genes, referred to as class I and II, respectively [5]. The IIS signaling pathway transcriptionally regulates many genes involved also in the immune responses, closely linked to longevity in *C. elegans*.

The p38 MAPK pathway is the most ancient signal transduction cascade in nematode immunity and plays a central role in *C. elegans* response against different pathogens, as it does in mammals. The p38 MAPK pathway is required for the activation of a set of immune effectors necessary to maintain a basal level of immune function and it is also involved in lifespan extension [67,74]. A neuronal symmetry (NSY)-1–SAPK/ERK kinase (SEK-1)–p38 mitogen-activated protein kinase ortholog (PMK)-1 p38 MAPK cascade (MAPKKK-MAPKK-MAPK, respectively) was elegantly identified as a key component of the *C. elegans* immune response [6].

Such signaling pathways are evolutionarily conserved in different animal species, from nematodes and flies to higher vertebrates and mammals. Evidence suggests that these pathways are relevant also to mammalian aging [75]. In particular, human studies conducted on centenarians highlighted an important role of the IIS pathway in setting lifespan, since associations have been found between polymorphisms in IIS genes and longevity [76].

Probiotic strains used in *C. elegans* studies have been shown to act through one or more of the above mentioned signaling pathways (Figure 2 and Table 1).

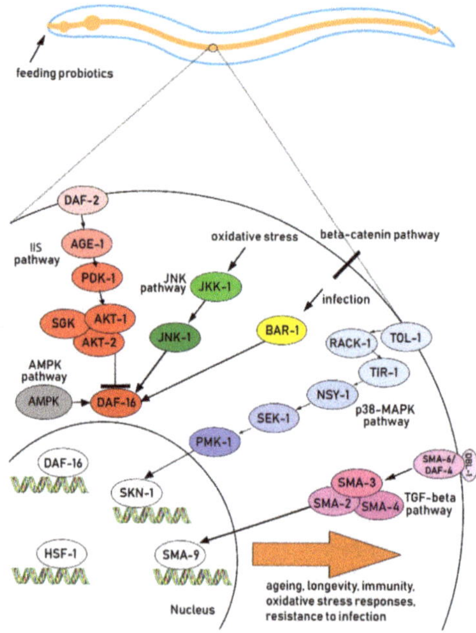

Figure 2. Schematic representation of the most common signaling pathways influenced by probiotic strains employed in *C. elegans* studies. Each pathway is represented by a distinct color gradient. The list of single bacterial strains influencing the different pathways is reported in Table 1. Abbreviations used: AMPK: 5′ AMP-activated protein kinase; AKT-1/2: serine/threonine protein kinase orthologs; BAR-1: beta-catenin/armadillo Related-1; DAF: dauer formation; DBL-1: DPP/BMP-Like-1; HSF-1: heat-shock transcription factor-1; IIS: insulin/insulin-like growth factor-1; JKK-1: c-Jun N-terminal kinase kinase; JNK-1: c-Jun N-terminal kinase; MAPK: mitogen-activated protein kinase; NSY-1: neuronal symmetry-1; PDK-1: phosphoinositide-dependent kinase 1; PMK-1: p38 mitogen-activated protein kinase-1 ortholog; RACK-1: receptor activated protein C kinase; SEK-1: SAPK/ERK kinase-1; SGK-1: serine/threonine protein kinase ortholog; SKN-1: skinhead family member-1; SMA: small; TIR-1: toll interleukin-1 receptor-1.

4.2. Oxidative Stress Response

Oxidative stress plays a detrimental role in different organisms. Normally, antioxidant defenses protect cells by removing reactive oxygen species (ROS). During ageing, on the other hand, ROS and other products of oxygen metabolism accumulate damaging proteins, lipids and DNA, and weaken antioxidant defenses [77]. In *C. elegans* animal model, several studies have been carried out to understand the mechanisms through which probiotics can influence ageing, by activating different longevity signaling pathways related to oxidative stress resistance. Prolongevity and oxidative stress responses in *C. elegans* fed probiotics are induced via mechanism(s) that can be DAF-2/DAF-16-dependent or a result of a cross talk among different pathways.

Among the tested strains, *Bifidobacterium longum* strain BR-108 has been shown to increase worm lifespan following H_2O_2-induced oxidative stress, through activation of IIS pathway. After the cascade activation, DAF-16 seems to co-localize with the heat-shock transcription factor (HSF)-1 in the nucleus, inducing the transcription of *hsp-16.2* and *hsp-70* that are involved in stress responses and longevity [59].

Similarly, *Lactobacillus rhamnosus* CNCM I-3690 and *Bifidobacterium animalis* subsp. *lactis* CECT8145 strains stimulated a strong resistance to oxidative stress in *C. elegans*, which was in part dependent on the IIS pathway [49,53]. On the other hand, *Lactobacillus gasseri* SBT2055 (LG2055) has been reported to promote a prolongevity effect in *daf-2* and *daf-16* mutant worms [9], thus demonstrating that prolongevity and enhancement of stress resistance were DAF-16-independent. These phenotypes occur rather by triggering the p38 MAPK pathway, which culminates with the nuclear translocation of the transcriptional factor skinhead family member (SKN)-1 [9]. SKN-1, an ortholog of the mammalian Nrf2, in duces the expression of target genes involved in oxidative stress resistance and it is responsible for the beneficial effect exerted by several other probiotic microbes [37,62,69].

Bifidobacterium longum subsp. infantis (formerly *B. infantis*) strain ATCC15697 resulted to extend wild-type nematode lifespan, but it failed to prolong the lifespan of *pmk-1*, *skn-1* and also *daf-2* mutants, demonstrating the involvement of p38 MAPK and IIS signaling pathways, both modulating SKN-1 activation [56,78].

It has been reported that the c-Jun N-terminal kinase (JNK) family, a subgroup of the MAPK superfamily, phosphorylates DAF-16 at a different site with respect to DAF-2-mediated phosphorylation, resulting in its nuclear translocation [79].

Similarly, AAK-2, which is one of the two alpha-catalytic subunits of 5′-AMP-activated protein kinase (AMPK), can directly phosphorylate DAF-16, triggering prolongevity and oxidative stress responses. Analysis of lifespan and gene expression of worms fed *Weissella koreensis* or *W. cibaria*, demonstrates that some *Weissella* species promote longevity in *C. elegans* by inducing oxidative stress responses through activation of DAF-16 via the JNK and AMPK pathways [68].

Several detoxifying enzymes are induced by different transcription factors in response to oxidative stress. Two of these are the superoxide dismutase (SOD) and the glutathione S-transferase (GST), which detoxify ROS [80]. Moreover, oxidative stress causes the activation of the transcriptional factor HSF-1, which also regulates lifespan, and activation of JNK pathway. Zhao and coworkers demonstrated that *Bifidobacterium longum* BB68, isolated from a centenarian subject, was able to increase lifespan and oxidative responses in *C. elegans*, through increased expression of *sod-3* gene, mediated by the toll interleukin-1 receptor (TIR)–JNK signal transduction pathway resulting in DAF-16 nuclear translocation [58]. Specifically, this highly conserved pathway consists of a TIR-domain protein, TIR-1, activating JNK-1 through phosphorylation. In turn, JNK-1 phosphorylates DAF-16, which migrates to the nucleus.

As stated above, in tracellular ROS represent an important marker to analyze the extent of oxidative stress, and some probiotic strains have been shown to reduce their level in *C. elegans*. *L. fermentum* MBC2, in addition to lifespan extension and anti-ageing effects, in duced a reduction of ROS levels and an increased expression of detoxifying enzymes, such as GST-4, paralleled by an amelioration of the other ageing biomarkers, such as locomotion activity, pumping rate and lipofuscin granules [36].

4.3. Immune Response and Pathogen Protection

Several candidate probiotic bacteria analyzed in this review have also been demonstrated to affect immunity and inflammation pathways in *C. elegans*. Immune response and lifespan are tightly linked in *C. elegans*. Among different molecular pathways shared with higher organisms, in nate immunity of *C. elegans* shows many aspects similar to humans. Although the nematode does not have a cell-mediated immune system, it possesses innate immune defense mechanisms that are evolutionarily conserved [74]. In particular, *C. elegans* possesses different pathways associated with immunity, in cluding the above mentioned p38 MAPK and IIS pathways, but also the trasforming growth factor-beta (TGF-beta) [52,71,74,81] and the beta-catenin signaling pathways [82], which can be induced by probiotics (Figure 2).

Immune responses to bacteria are mediated by interaction of specific microbial cell wall structures, (microbial associated molecular patterns, MAMPs), such as peptidoglycan, teichoic acids and lipopolysaccharides, with host receptors, in particular toll-like receptors (TLRs). In mammals different TLRs have selective specificity for the different MAMPs, while in *C. elegans* a unique TLRs homolog, TOL-1, has been identified so far [83]. The interaction of a TLR with its microbial ligand activates several signaling pathways, in cluding p38 MAPK, resulting in the transcription of genes necessary to mount the defense mechanism in the host. The main cell wall MAMPs share a common basic structure among different bacterial species, both pathogens and probiotics, but various subtle chemical modifications present in the different species or strains can contribute to the strain-specific properties of probiotics. This also implies that the final outcome of the TLR activation depends on the type of interacting microorganism, meaning that a MAMP from one bacterial species can activate a certain TLR, while a similar MAMP from another species, or strain, can down-regulate the same TLR signaling [84]. The involvement of nematode TOL-1 in the regulation of prolongevity effect exerted by *Bifidobacterium longum* subsp. *infantis* (formerly *B. infantis*) strain ATCC15697 was recently demonstrated [57].

The majority of probiotic strains were employed in *C. elegans* to verify their prolongevity effects in the context of protection from pathogen infection, through killing assays. Some others were tested on lifespan extension in normal conditions. The use of nematode functional mutants or the analysis of gene expression profile by RT-qPCR/microarray allowed the elucidation of the molecular players acting as targets of probiotic action. Many human pathogens, such as *Pseudomonas aeruginosa*, *Salmonella enterica*, *Staphylococcus aureus*, *Klebsiella pneumoniae*, enterotoxigenic *Escherichia coli*, *Yersinia enterocolitica* and *Listeria monocytogenes*, can cause nematode death. It is known that pathogen infection induces worm innate immune responses, consisting in the production of several antimicrobial proteins, whose expressions are regulated by signaling pathways involved in the defense against harmful bacteria [74]. Such antimicrobial proteins include lysozyme (LYS) family, and C-type lectins (CLEC). As mentioned above, *C. elegans* lacks a cell-mediated immune system and the production of antimicrobial peptides is, therefore, the outcome of its innate immunity to counteract infections [85]. As an example, *Bacillus subtilis* NCIB3610, which forms a biofilm contributing to nematode prolongevity, specifically stimulated *lys-2* expression, in creasing *C. elegans* resistance to *P. aeruginosa* infection [86].

Zhou and coworkers reported that *L. reuteri* CL9 induced the expression of antimicrobial peptide genes *clec-60* and *clec-85*, in volved in the protection of nematodes against enterotoxigenic *Escherichia coli* (ETEC) infection [51]. Similarly, *L. zeae* LB1 induced the production of antimicrobial peptides and defensive molecules, such as LYS-7 and CLEC-85, through the p38 MAPK and IIS pathways, enhancing resistance of *C. elegans* to ETEC infection [52]. The p38 MAPK pathway was also activated by *L. acidophilus* NCFM, employed for protecting nematodes against the Gram-positive pathogens *Staphylococcus aureus* and *Enterococcus faecalis* [29], as well as by *L. casei* LAB9, which displayed protection against *Klebsiella pneumoniae* infection. In particular *L. casei* LAB9 activates TLR and triggers the PMK-1/p38 MAPK pathway through the up-regulation of receptor activated protein C kinase (RACK)-1, an adaptor molecule that plays a critical role in the host defence and survival [32].

The p38 MAPK pathway is also involved in protection against *Legionella pneumophila* infection promoted by *Bifidobacterium longum* subsp. *infantis* (formerly *B. infantis*) ATCC 15697 via PMK-1 [55], as well as in stimulation of *C. elegans* host defense by six foodborne strains of *Bacillus licheniformis* [87]. Moreover, Kwon et al. (2016) described that *Propionibacterium freudenreichii* KCTC 1063, isolated from a dairy product, in creased resistance against *Salmonella* typhimurium, through the activation of SKN-1, upon phosphorylation by PMK-1 [67].

L. acidophilus NCFM immune stimulation involved also the beta-catenin pathway through the beta-catenin/armadillo related (BAR)-1 mediator, in dicating that different signaling pathways can act in parallel to promote immunity [29].

On the other hand, the IIS signaling pathway was influenced by *Clostridium butyricum* MIYAIRI 588 (CBM 588), which was able to confer resistance to *S. aureus* and *S. enteric* infection through DAF-16-dependent class II genes [62]. Two other genes implicated in the defense response and the innate immune response through the IIS pathway, *acdh-1* and *cnc-2*, were up-regulated by heat-killed *L. plantarum* LP133 and *L. fermentum* LF21, protecting worms against Gram-negative pathogens *Salmonella* typhimurium and *Yersinia enterocolitica* [39].

The evidence that the protective activity of different *Lactobacillus* species can be directed either to Gram-positive or Gram-negative bacteria indicates that probiotic effects are species- and strain-specific, as explained above, concerning TLR-MAMP interactions. To this respect, the above-mentioned

Lactobacillus acidophilus strain NCFM, while active against Gram-positive bacteria, displayed a minimal inhibitory effect on Gram-negative infection with *P. aeruginosa* or *S. enterica* [29].

In the absence of pathogen infection, lifespan extension exerted by *L. salivarius* DSM 20555 resulted to be dependent on the up-regulation of *lys-7* and *thn-2* genes, encoding LYS and an immune effector member of the thaumatin family, respectively, in a DAF-16-independent manner, suggesting the involvement of pathways other than IIS signaling [47]. In line with this evidence, *Butyricicoccus pullicaecorum* KCTC 15070 and *Megasphaera elsdenii* KCTC 5187 prolonged *C. elegans* lifespan by activating nuclear receptor signaling and the innate immune system in a TGF-beta pathway-dependent, but IIS pathway-independent manner. The signaling involves DPP/BMP-Like (DBL)-1, a mediator of the TGF-beta pathway, which binds to TGF-beta receptors, such as SMA-6/DAF-4, on the cell membrane and activates transcription of specific target genes related to antibacterial defense, in ducing the production of antimicrobial peptides, such as CLEC, LYS and lipase [61].

5. Conclusions

C. elegans represents a valuable in vivo model for studying how probiotics interact with the host and what mechanisms are involved in prolongevity. Its numerous tools and the possibility of genetic approaches has allowed advances in understanding these interactions. *C. elegans* shows highly conserved pathways through which the host responds to microbes, revealing cross-talk regulating longevity, ageing, stress resistance, and innate immune responses. The unique opportunity to manipulate its diet renders *C. elegans* a powerful model organism for understanding the effect of bacteria on these interconnected processes. Moreover, it also represents a low expensive screening tool to identify novel probiotic strains from a large number of microbial candidates. The picture emerging from this review evidences that several probiotic strains are able to exert anti-ageing effects in nematodes, by acting on common, conserved molecular pathways, such as IIS and p38 MAPK. The cell wall components of probiotic bacteria are thought to be primarily responsible for immunostimulation, as clearly demonstrated for TLRs-MAMPs interactions. Moreover, key ageing- and stress-associated regulatory elements, such as DAF-16/FOXO and HSF-1 transcription factors, also emerged as common targets of probiotic activities. However, other important mediators still need to be identified and characterized. In this perspective, *C. elegans* therefore appears to be advantageous to unravel key mechanisms involved in host prolongevity in response to probiotics supplementation. Since these mechanisms appear to be conserved across several species, the possibility of promoting the longevity of humans through the consumption of probiotics is gaining increasing attention. Indeed, probiotic supplementation has been suggested to slow or reverse the age-related changes in intestinal microbiota composition, as well as the gradual decline of immune function in elderly, thus lowering the risk of associated age-related morbidities. Nevertheless, future studies are needed to deepen insight on the effect of probiotics on longevity in mammals.

Author Contributions: Review structuring: D.U. and C.D.; Literature searching: C.D. and M.R.; Manuscript drafting: D.U., C.D., M.R., B.G., P.Z. and E.S.; Manuscript critical revision: D.U., C.D. and M.R.; Figures preparation: E.S.

Funding: This research received no external funding.

Conflicts of Interest: The authors declare no conflicts of interest.

Abbreviations

AMPK	5′-AMP-activated protein kinase
CLEC	C-type lectins
DAF	Dauer Formation
ETEC	Enterotoxigenic *E. coli*
FOXO	Forkhead box O
GST	Glutathione S-transferase
HSF-1	Heat-Shock transcription Factor-1
IIS	Insulin/insulin-like growth factor-1

JNK	c-Jun N-terminal kinase
LYS	Lysozyme
MAMP	Microbial associated molecular pattern
p38 MAPK	p38 mitogen-activated protein kinase
PMK-1	Mitogen-activated protein kinase-1
ROS	Reactive oxygen species;
SKN-1	Skinhead family member-1
SOD	Superoxide dismutase
TGF-beta	Transforming growth factor-beta
TIR-1	Toll interleukin-1 receptor-1
TLR	Toll-like receptor

References

1. Brenner, S. The genetics of *Caenorhabditis elegans*. *Genetics* **1974**, *77*, 71–94. [PubMed]
2. Kenyon, C. A conserved regulatory system for aging. *Cell* **2001**, *105*, 165–168. [CrossRef]
3. O'Kane, C.J. Modelling human diseases in *Drosophila* and *Caenorhabditis*. *Semin. Cell Dev. Biol.* **2003**, *14*, 3–10. [CrossRef]
4. Fontana, L.; Partridge, L.; Longo, V.D. Extending healthy life span—From yeast to humans. *Science* **2010**, *328*, 321–326. [CrossRef] [PubMed]
5. Murphy, C.T.; McCarroll, S.A.; Bargmann, C.I.; Fraser, A.; Kamath, R.S.; Ahringer, J.; Li, H.; Kenyon, C. Genes that act downstream of DAF-16 to influence the lifespan of *Caenorhabditis elegans*. *Nature* **2003**, *424*, 277–283. [CrossRef] [PubMed]
6. Kim, D.H.; Feinbaum, R.; Alloing, G.; Emerson, F.E.; Garsin, D.A.; Inoue, H.; Tanaka-Hino, M.; Hisamoto, N.; Matsumoto, K.; Tan, M.W.; et al. A conserved p38 MAP kinase pathway in *Caenorhabditis elegans* innate immunity. *Science* **2002**, *297*, 623–626. [CrossRef] [PubMed]
7. So, S.; Tokumaru, T.; Miyahara, K.; Ohshima, Y. Control of lifespan by food bacteria, nutrient limitation and pathogenicity of food in *C. elegans*. *Mech. Ageing Dev.* **2011**, *132*, 210–212. [CrossRef] [PubMed]
8. Troemel, E.R.; Chu, S.W.; Reinke, V.; Lee, S.S.; Ausubel, F.M.; Kim, D.H. p38 MAPK regulates expression of immune response genes and contributes to longevity in *C. elegans*. *PLoS Genet.* **2006**, *2*, e183. [CrossRef]
9. Nakagawa, H.; Shiozaki, T.; Kobatake, E.; Hosoya, T.; Moriya, T.; Sakai, F.; Taru, H.; Miyazaki, T. Effects and mechanisms of prolongevity induced by *Lactobacillus gasseri* SBT2055 in *Caenorhabditis elegans*. *Aging Cell* **2016**, *15*, 227–236. [CrossRef]
10. Johnson, T.E.; Henderson, S.; Murakami, S.; de Castro, E.; de Castro, S.H.; Cypser, J.; Rikke, B.; Tedesco, P.; Link, C. Longevity genes in the nematode *Caenorhabditis elegans* also mediate increased resistance to stress and prevent disease. *J. Inherit. Metab. Dis.* **2002**, *25*, 197–206. [CrossRef]
11. Pincus, Z.; Mazer, T.C.; Slack, F.J. Autofluorescence as a measure of senescence in *C. elegans*: Look to red, not blue or green. *Aging* **2016**, *8*, 889–898. [CrossRef] [PubMed]
12. Son, H.G.; Altintas, O.; Kim, E.J.E.; Kwon, S.; Lee, S.V. Age-dependent changes and biomarkers of aging in *Caenorhabditis elegans*. *Aging Cell* **2019**, *18*, e12853. [CrossRef]
13. Clark, L.C.; Hodgkin, J. Commensals, probiotics and pathogens in the *Caenorhabditis elegans* model. *Cell Microbiol.* **2014**, *16*, 27–38. [CrossRef] [PubMed]
14. FAO/WHO. Probiotics in Food. Health and nutritional properties and guidelines for evaluation. Available online: http://www.fao.org/3/a-a0512e.pdf (accessed on 7 February 2019).
15. O'Hara, A.M.; Shanahan, F. Mechanisms of action of probiotics in intestinal diseases. *Sci. World J.* **2007**, *7*, 31–46. [CrossRef] [PubMed]
16. Bermudez-Brito, M.; Plaza-Diaz, J.; Munoz-Quezada, S.; Gomez-Llorente, C.; Gil, A. Probiotic mechanisms of action. *Ann. Nutr. Metab.* **2012**, *61*, 160–174. [CrossRef]
17. Hill, C.; Guarner, F.; Reid, G.; Gibson, G.R.; Merenstein, D.J.; Pot, B.; Morelli, L.; Canani, R.B.; Flint, H.J.; Salminen, S.; et al. Expert consensus document. The International scientific association for probiotics and prebiotics consensus statement on the scope and appropriate use of the term probiotic. *Nat. Rev. Gastroenterol. Hepatol.* **2014**, *11*, 506–514. [CrossRef]

18. Calder, P.C.; Bosco, N.; Bourdet-Sicard, R.; Capuron, L.; Delzenne, N.; Dore, J.; Franceschi, C.; Lehtinen, M.J.; Recker, T.; Salvioli, S.; et al. Health relevance of the modification of low grade inflammation in ageing (inflammageing) and the role of nutrition. *Ageing Res. Rev.* **2017**, *40*, 95–119. [CrossRef]

19. Ferrucci, L.; Fabbri, E. Inflammageing: Chronic inflammation in ageing, cardiovascular disease, and frailty. *Nat. Rev. Cardiol.* **2018**, *15*, 505–522. [CrossRef]

20. Biagi, E.; Candela, M.; Turroni, S.; Garagnani, P.; Franceschi, C.; Brigidi, P. Ageing and gut microbes: Perspectives for health maintenance and longevity. *Pharm. Res.* **2013**, *69*, 11–20. [CrossRef]

21. Neish, A.S. Microbes in gastrointestinal health and disease. *Gastroenterology* **2009**, *136*, 65–80. [CrossRef]

22. Hopkins, M.J.; Macfarlane, G.T. Changes in predominant bacterial populations in human faeces with age and with *Clostridium difficile* infection. *J. Med. Microbiol.* **2002**, *51*, 448–454. [CrossRef]

23. Claesson, M.J.; Jeffery, I.B.; Conde, S.; Power, S.E.; O'Connor, E.M.; Cusack, S.; Harris, H.M.; Coakley, M.; Lakshminarayanan, B.; O'Sullivan, O.; et al. Gut microbiota composition correlates with diet and health in the elderly. *Nature* **2012**, *488*, 178–184. [CrossRef] [PubMed]

24. Aureli, P.; Capurso, L.; Castellazzi, A.M.; Clerici, M.; Giovannini, M.; Morelli, L.; Poli, A.; Pregliasco, F.; Salvini, F.; Zuccotti, G.V. Probiotics and health: An evidence-based review. *Pharm. Res.* **2011**, *63*, 366–376. [CrossRef] [PubMed]

25. Finamore, A.; Roselli, M.; Donini, L.; Brasili, D.E.; Rami, R.; Carnevali, P.; Mistura, L.; Pinto, A.; Giusti, A.; Mengheri, E. Supplementation with *Bifidobacterium longum* Bar33 and *Lactobacillus helveticus* Bar13 mixture improves immunity in elderly humans (over 75 years) and aged mice. *Nutrition* **2019**, *63*, 184–192. [CrossRef] [PubMed]

26. Zhang, R.; Hou, A. Host-Microbe Interactions in *Caenorhabditis elegans*. *ISRN Microbiol.* **2013**, *2013*. [CrossRef] [PubMed]

27. Ventura, M.T.; Casciaro, M.; Gangemi, S.; Buquicchio, R. Immunosenescence in aging: Between immune cells depletion and cytokines up-regulation. *Clin. Mol. Allergy* **2017**, *15*, 21. [CrossRef]

28. Fulop, T.; Larbi, A.; Dupuis, G.; Le Page, A.; Frost, E.H.; Cohen, A.A.; Witkowski, J.M.; Franceschi, C. Immunosenescence and inflamm-aging as two sides of the same coin: Friends or foes? *Front. Immunol.* **2017**, *8*, 1960. [CrossRef]

29. Kim, Y.; Mylonakis, E. *Caenorhabditis elegans* immune conditioning with the probiotic bacterium *Lactobacillus acidophilus* strain NCFM enhances Gram-positive immune responses. *Infect. Immun.* **2012**, *80*, 2500–2508. [CrossRef]

30. Chelliah, R.; Choi, J.G.; Hwang, S.B.; Park, B.J.; Daliri, E.B.; Kim, S.H.; Wei, S.; Ramakrishnan, S.R.; Oh, D.H. In vitro and in vivo defensive effect of probiotic LAB against *Pseudomonas aeruginosa* using *Caenorhabditis elegans* model. *Virulence* **2018**, *9*, 1489–1507. [CrossRef]

31. Wang, C.; Wang, J.; Gong, J.; Yu, H.; Pacan, J.C.; Niu, Z.; Si, W.; Sabour, P.M. Use of *Caenorhabditis elegans* for preselecting *Lactobacillus* isolates to control *Salmonella* Typhimurium. *J. Food Prot.* **2011**, *74*, 86–93. [CrossRef]

32. Kamaladevi, A.; Balamurugan, K. *Lactobacillus casei* triggers a TLR mediated RACK-1 dependent p38 MAPK pathway in *Caenorhabditis elegans* to resist *Klebsiella pneumoniae* infection. *Food Funct.* **2016**, *7*, 3211–3223. [CrossRef] [PubMed]

33. Kamaladevi, A.; Ganguli, A.; Balamurugan, K. *Lactobacillus casei* stimulates phase-II detoxification system and rescues malathion-induced physiological impairments in *Caenorhabditis elegans*. *Comp. Biochem. Physiol. C Toxicol. Pharm.* **2016**, *179*, 19–28. [CrossRef] [PubMed]

34. Guantario, B.; Zinno, P.; Schifano, E.; Roselli, M.; Perozzi, G.; Palleschi, C.; Uccelletti, D.; Devirgiliis, C. In vitro and in vivo selection of potentially probiotic lactobacilli from nocellara del belice table olives. *Front. Microbiol.* **2018**, *9*, 595. [CrossRef] [PubMed]

35. Zanni, E.; Schifano, E.; Motta, S.; Sciubba, F.; Palleschi, C.; Mauri, P.; Perozzi, G.; Uccelletti, D.; Devirgiliis, C.; Miccheli, A. Combination of metabolomic and proteomic analysis revealed different features among *Lactobacillus delbrueckii* subspecies *bulgaricus* and *lactis* strains while in vivo testing in the model organism *Caenorhabditis elegans* highlighted probiotic properties. *Front. Microbiol.* **2017**, *8*, 1206. [CrossRef] [PubMed]

36. Schifano, E.; Zinno, P.; Guantario, B.; Roselli, M.; Marcoccia, S.; Devirgiliis, C.; Uccelletti, D. The foodborne strain *Lactobacillus fermentum* MBC2 triggers pept-1-dependent pro-longevity effects in *Caenorhabditis elegans*. *Microorganisms* **2019**, *7*, 45. [CrossRef] [PubMed]

37. Park, M.R.; Ryu, S.; Maburutse, B.E.; Oh, N.S.; Kim, S.H.; Oh, S.; Jeong, S.Y.; Jeong, D.Y.; Kim, Y. Probiotic *Lactobacillus fermentum* strain JDFM216 stimulates the longevity and immune response of *Caenorhabditis elegans* through a nuclear hormone receptor. *Sci. Rep.* **2018**, *8*, 7441. [CrossRef] [PubMed]

38. Lee, J.; Yun, H.S.; Cho, K.W.; Oh, S.; Kim, S.H.; Chun, T.; Kim, B.; Whang, K.Y. Evaluation of probiotic characteristics of newly isolated *Lactobacillus* spp.: Immune modulation and longevity. *Int. J. Food Microbiol.* **2011**, *148*, 80–86. [CrossRef]

39. Lee, J.; Choe, J.; Kim, J.; Oh, S.; Park, S.; Kim, S.; Kim, Y. Heat-killed *Lactobacillus* spp. cells enhance survivals of *Caenorhabditis elegans* against *Salmonella* and *Yersinia* infections. *Lett. Appl. Microbiol.* **2015**, *61*, 523–530. [CrossRef]

40. Ikeda, T.; Yasui, C.; Hoshino, K.; Arikawa, K.; Nishikawa, Y. Influence of lactic acid bacteria on longevity of *Caenorhabditis elegans* and host defense against *Salmonella enterica* serovar enteritidis. *Appl. Environ. Microbiol.* **2007**, *73*, 6404–6409. [CrossRef]

41. Pan, F.; Zhang, L.; Li, M.; Hu, Y.; Zeng, B.; Yuan, H.; Zhao, L.; Zhang, C. Predominant gut *Lactobacillus murinus* strain mediates anti-inflammaging effects in calorie-restricted mice. *Microbiome* **2018**, *6*, 54. [CrossRef]

42. De Barros, P.P.; Scorzoni, L.; Ribeiro, F.C.; Fugisaki, L.R.O.; Fuchs, B.B.; Mylonakis, E.; Jorge, A.O.C.; Junqueira, J.C.; Rossoni, R.D. *Lactobacillus paracasei* 28.4 reduces in vitro hyphae formation of *Candida albicans* and prevents the filamentation in an experimental model of *Caenorhabditis elegans*. *Microb. Pathog.* **2018**, *117*, 80–87. [CrossRef] [PubMed]

43. Lee, H.K.; Choi, S.H.; Lee, C.R.; Lee, S.H.; Park, M.R.; Kim, Y.; Lee, M.K.; Kim, G.B. Screening and characterization of lactic acid bacteria strains with anti-inflammatory activities through in vitro and *Caenorhabditis elegans* model testing. *Korean J. Food Sci. Anim. Resour.* **2015**, *35*, 91–100. [CrossRef] [PubMed]

44. Park, M.R.; Yun, H.S.; Son, S.J.; Oh, S.; Kim, Y. Short communication: Development of a direct in vivo screening model to identify potential probiotic bacteria using *Caenorhabditis elegans*. *J. Dairy Sci.* **2014**, *97*, 6828–6834. [CrossRef] [PubMed]

45. Sharma, K.; Pooranachithra, M.; Balamurugan, K.; Goel, G. Probiotic mediated colonization resistance against *E. coli* infection in experimentally challenged *Caenorhabditis elegans*. *Microb. Pathog.* **2019**, *127*, 39–47. [CrossRef] [PubMed]

46. Zhou, M.; Zhu, J.; Yu, H.; Yin, X.; Sabour, P.M.; Zhao, L.; Chen, W.; Gong, J. Investigation into *in vitro* and *in vivo* models using intestinal epithelial IPEC-J2 cells and *Caenorhabditis elegans* for selecting probiotic candidates to control porcine enterotoxigenic *Escherichia coli*. *J. Appl. Microbiol.* **2014**, *117*, 217–226. [CrossRef] [PubMed]

47. Fasseas, M.K.; Fasseas, C.; Mountzouris, K.C.; Syntichaki, P. Effects of *Lactobacillus salivarius*, *Lactobacillus reuteri*, and *Pediococcus acidilactici* on the nematode *Caenorhabditis elegans* include possible antitumor activity. *Appl. Microbiol. Biotechnol.* **2013**, *97*, 2109–2118. [CrossRef]

48. Azat, R.; Liu, Y.; Li, W.; Kayir, A.; Lin, D.B.; Zhou, W.W.; Zheng, X.D. Probiotic properties of lactic acid bacteria isolated from traditionally fermented Xinjiang cheese. *J. Zhejiang Univ. Sci. B* **2016**, *17*, 597–609. [CrossRef]

49. Grompone, G.; Martorell, P.; Llopis, S.; Gonzalez, N.; Genoves, S.; Mulet, A.P.; Fernandez-Calero, T.; Tiscornia, I.; Bollati-Fogolin, M.; Chambaud, I.; et al. Anti-inflammatory *Lactobacillus rhamnosus* CNCM I-3690 strain protects against oxidative stress and increases lifespan in *Caenorhabditis elegans*. *PLoS ONE* **2012**, *7*, e52493. [CrossRef]

50. Zhao, Y.; Zhao, L.; Zheng, X.; Fu, T.; Guo, H.; Ren, F. *Lactobacillus salivarius* strain FDB89 induced longevity in *Caenorhabditis elegans* by dietary restriction. *J. Microbiol.* **2013**, *51*, 183–188. [CrossRef]

51. Zhou, M.; Yu, H.; Yin, X.; Sabour, P.M.; Chen, W.; Gong, J. *Lactobacillus zeae* protects *Caenorhabditis elegans* from enterotoxigenic *Escherichia coli*-caused death by inhibiting enterotoxin gene expression of the pathogen. *PLoS ONE* **2014**, *9*, e89004. [CrossRef]

52. Zhou, M.; Liu, X.; Yu, H.; Yin, X.; Nie, S.P.; Xie, M.Y.; Chen, W.; Gong, J. Cell signaling of *Caenorhabditis elegans* in response to enterotoxigenic *Escherichia coli* infection and *Lactobacillus zeae* protection. *Front. Immunol.* **2018**, *9*, 1745. [CrossRef] [PubMed]

53. Martorell, P.; Llopis, S.; Gonzalez, N.; Chenoll, E.; Lopez-Carreras, N.; Aleixandre, A.; Chen, Y.; Karoly, E.D.; Ramon, D.; Genoves, S. Probiotic strain *Bifidobacterium animalis* subsp. *lactis* CECT 8145 reduces fat content and modulates lipid metabolism and antioxidant response in *Caenorhabditis elegans*. *J. Agric. Food Chem.* **2016**, *64*, 3462–3472. [CrossRef] [PubMed]

54. Christiaen, S.E.; O'Connell Motherway, M.; Bottacini, F.; Lanigan, N.; Casey, P.G.; Huys, G.; Nelis, H.J.; van Sinderen, D.; Coenye, T. Autoinducer-2 plays a crucial role in gut colonization and probiotic functionality of *Bifidobacterium breve* UCC2003. *PLoS ONE* **2014**, *9*, e98111. [CrossRef] [PubMed]

55. Komura, T.; Yasui, C.; Miyamoto, H.; Nishikawa, Y. *Caenorhabditis elegans* as an alternative model host for *Legionella pneumophila*, and protective effects of *Bifidobacterium infantis*. *Appl. Environ. Microbiol.* **2010**, *76*, 4105–4108. [CrossRef] [PubMed]

56. Komura, T.; Ikeda, T.; Yasui, C.; Saeki, S.; Nishikawa, Y. Mechanism underlying prolongevity induced by bifidobacteria in *Caenorhabditis elegans*. *Biogerontology* **2013**, *14*, 73–87. [CrossRef] [PubMed]

57. Sun, S.; Mizuno, Y.; Komura, T.; Nishikawa, Y.; Kage-Nakadai, E. Toll-like receptor homolog TOL-1 regulates *Bifidobacterium infantis*-elicited longevity and behavior in *Caenorhabditis elegans*. *Biosci. Microbiota. Food Health* **2019**, *38*, 105–110. [CrossRef] [PubMed]

58. Zhao, L.; Zhao, Y.; Liu, R.; Zheng, X.; Zhang, M.; Guo, H.; Zhang, H.; Ren, F. The transcription factor DAF-16 is essential for increased longevity in *C. elegans* exposed to *Bifidobacterium longum* BB68. *Sci. Rep.* **2017**, *7*, 7408. [CrossRef]

59. Sugawara, T.; Sakamoto, K. Killed *Bifidobacterium longum* enhanced stress tolerance and prolonged life span of *Caenorhabditis elegans* via DAF-16. *Br. J. Nutr.* **2018**, *120*, 872–880. [CrossRef]

60. Park, M.R.; Oh, S.; Son, S.J.; Park, D.J.; Kim, S.H.; Jeong, D.Y.; Oh, N.S.; Lee, Y.; Song, M.; Kim, Y. *Bacillus icheniformis* isolated from traditional Korean food resources enhances the longevity of *Caenorhabditis elegans* through serotonin signaling. *J. Agric. Food Chem.* **2015**, *63*, 10227–10233. [CrossRef]

61. Kwon, G.; Lee, J.; Koh, J.H.; Lim, Y.H. Lifespan Extension of *Caenorhabditis elegans* by *Butyricicoccus pullicaecorum* and *Megasphaera elsdenii* with probiotic potential. *Curr. Microbiol.* **2018**, *75*, 557–564. [CrossRef]

62. Kato, M.; Hamazaki, Y.; Sun, S.; Nishikawa, Y.; Kage-Nakadai, E. *Clostridium butyricum* MIYAIRI 588 increases the lifespan and multiple-stress resistance of *Caenorhabditis elegans*. *Nutrients* **2018**, *10*, 1921. [CrossRef] [PubMed]

63. Yuen, G.J.; Ausubel, F.M. Both live and dead enterococci activate *Caenorhabditis elegans* host defense via immune and stress pathways. *Virulence* **2018**, *9*, 683–699. [CrossRef] [PubMed]

64. Neuhaus, K.; Lamparter, M.C.; Zolch, B.; Landstorfer, R.; Simon, S.; Spanier, B.; Ehrmann, M.A.; Vogel, R.F. Probiotic *Enterococcus faecalis* Symbioflor((R)) down regulates virulence genes of EHEC in vitro and decrease pathogenicity in a *Caenorhabditis elegans* model. *Arch. Microbiol.* **2017**, *199*, 203–213. [CrossRef] [PubMed]

65. Sim, I.; Park, K.T.; Kwon, G.; Koh, J.H.; Lim, Y.H. Probiotic potential of *Enterococcus faecium* isolated from chicken cecum with immunomodulating activity and promoting longevity in *Caenorhabditis elegans*. *J. Microbiol. Biotechnol.* **2018**, *28*, 883–892. [CrossRef] [PubMed]

66. Hwang, I.Y.; Koh, E.; Wong, A.; March, J.C.; Bentley, W.E.; Lee, Y.S.; Chang, M.W. Engineered probiotic *Escherichia coli* can eliminate and prevent *Pseudomonas aeruginosa* gut infection in animal models. *Nat. Commun.* **2017**, *8*, 15028. [CrossRef] [PubMed]

67. Kwon, G.; Lee, J.; Lim, Y.H. Dairy Propionibacterium extends the mean lifespan of *Caenorhabditis elegans* via activation of the innate immune system. *Sci. Rep.* **2016**, *6*, 31713. [CrossRef] [PubMed]

68. Lee, J.; Kwon, G.; Lim, Y.H. Elucidating the Mechanism of *Weissella*-dependent lifespan extension in *Caenorhabditis elegans*. *Sci. Rep.* **2015**, *5*, 17128. [CrossRef]

69. Romanin, D.E.; Llopis, S.; Genoves, S.; Martorell, P.; Ramon, V.D.; Garrote, G.L.; Rumbo, M. Probiotic yeast *Kluyveromyces marxianus* CIDCA 8154 shows anti-inflammatory and anti-oxidative stress properties in in vivo models. *Benef. Microbes.* **2016**, *7*, 83–93. [CrossRef]

70. Mattarelli, P.; Bonaparte, C.; Pot, B.; Biavati, B. Proposal to reclassify the three biotypes of *Bifidobacterium longum* as three subspecies: *Bifidobacterium longum* subsp. longum subsp. nov., *Bifidobacterium longum* subsp. infantis comb. nov. and *Bifidobacterium longum* subsp. suis comb. *Nov. Int. J. Syst. Evol. Microbiol.* **2008**, *58*, 767–772. [CrossRef]

71. Kurz, C.L.; Ewbank, J.J. *Caenorhabditis elegans*: An emerging genetic model for the study of innate immunity. *Nat. Rev. Genet.* **2003**, *4*, 380–390. [CrossRef]

72. Sun, X.; Chen, W.D.; Wang, Y.D. DAF-16/FOXO transcription factor in aging and longevity. *Front. Pharm.* **2017**, *8*, 548. [CrossRef] [PubMed]

73. Dues, D.J.; Andrews, E.K.; Schaar, C.E.; Bergsma, A.L.; Senchuk, M.M.; Van Raamsdonk, J.M. Aging causes decreased resistance to multiple stresses and a failure to activate specific stress response pathways. *Aging* **2016**, *8*, 777–795. [CrossRef] [PubMed]

74. Irazoqui, J.E.; Urbach, J.M.; Ausubel, F.M. Evolution of host innate defence: Insights from *Caenorhabditis elegans* and primitive invertebrates. *Nat. Rev. Immunol.* **2010**, *10*, 47–58. [CrossRef] [PubMed]

75. Pan, H.; Finkel, T. Key proteins and pathways that regulate lifespan. *J. Biol. Chem.* **2017**, *292*, 6452–6460. [CrossRef] [PubMed]

76. Jesko, H.; Stepien, A.; Lukiw, W.J.; Strosznajder, R.P. The Cross-talk between sphingolipids and insulin-like growth factor signaling: significance for aging and neurodegeneration. *Mol. Neurobiol.* **2019**, *56*, 3501–3521. [CrossRef] [PubMed]

77. Ristow, M.; Schmeisser, S. Extending life span by increasing oxidative stress. *Free Radic. Biol. Med.* **2011**, *51*, 327–336. [CrossRef]

78. Tullet, J.M.; Hertweck, M.; An, J.H.; Baker, J.; Hwang, J.Y.; Liu, S.; Oliveira, R.P.; Baumeister, R.; Blackwell, T.K. Direct inhibition of the longevity-promoting factor SKN-1 by insulin-like signaling in *C. elegans*. *Cell* **2008**, *132*, 1025–1038. [CrossRef]

79. Oh, S.W.; Mukhopadhyay, A.; Svrzikapa, N.; Jiang, F.; Davis, R.J.; Tissenbaum, H.A. JNK regulates lifespan in *Caenorhabditis elegans* by modulating nuclear translocation of forkhead transcription factor/DAF-16. *Proc. Natl. Acad. Sci. USA* **2005**, *102*, 4494–4499. [CrossRef]

80. Hu, Q.; D'Amora, D.R.; MacNeil, L.T.; Walhout, A.J.M.; Kubiseski, T.J. The *Caenorhabditis elegans* oxidative stress response requires the NHR-49 transcription factor. *G3* **2018**, *8*, 3857–3863. [CrossRef]

81. Zugasti, O.; Ewbank, J.J. Neuroimmune regulation of antimicrobial peptide expression by a noncanonical TGF-beta signaling pathway in *Caenorhabditis elegans* epidermis. *Nat. Immunol.* **2009**, *10*, 249–256. [CrossRef]

82. Irazoqui, J.E.; Ng, A.; Xavier, R.J.; Ausubel, F.M. Role for beta-catenin and HOX transcription factors in *Caenorhabditis elegans* and mammalian host epithelial-pathogen interactions. *Proc. Natl. Acad. Sci. USA* **2008**, *105*, 17469–17474. [CrossRef] [PubMed]

83. Tenor, J.L.; Aballay, A. A conserved Toll-like receptor is required for *Caenorhabditis elegans* innate immunity. *EMBO Rep.* **2008**, *9*, 103–109. [CrossRef] [PubMed]

84. Lebeer, S.; Vanderleyden, J.; De Keersmaecker, S.C. Host interactions of probiotic bacterial surface molecules: Comparison with commensals and pathogens. *Nat. Rev. Microbiol.* **2010**, *8*, 171–184. [CrossRef] [PubMed]

85. Alper, S.; McBride, S.J.; Lackford, B.; Freedman, J.H.; Schwartz, D.A. Specificity and complexity of the *Caenorhabditis elegans* innate immune response. *Mol. Cell Biol.* **2007**, *27*, 5544–5553. [CrossRef] [PubMed]

86. Smolentseva, O.; Gusarov, I.; Gautier, L.; Shamovsky, I.; DeFrancesco, A.S.; Losick, R.; Nudler, E. Mechanism of biofilm-mediated stress resistance and lifespan extension in *C. elegans*. *Sci. Rep.* **2017**, *7*, 7137. [CrossRef] [PubMed]

87. Yun, H.S.; Heo, J.H.; Son, S.J.; Park, M.R.; Oh, S.; Song, M.H.; Kim, J.N.; Go, G.W.; Cho, H.S.; Choi, N.J.; et al. *Bacillus licheniformis* isolated from Korean traditional food sources enhances the resistance of *Caenorhabditis elegans* to infection by *Staphylococcus aureus*. *J. Microbiol. Biotechnol.* **2014**, *24*, 1105–1108. [CrossRef] [PubMed]

International Journal of
Molecular Sciences

Review

Royal Jelly and Its Components Promote Healthy Aging and Longevity: From Animal Models to Humans

Hiroshi Kunugi [1] and Amira Mohammed Ali [1,2,*]

1 Department of Mental Disorder Research, National Institute of Neuroscience, National Center of Neurology and Psychiatry, Tokyo 187-8551, Japan; hkunugi@ncnp.go.jp
2 Department of Psychiatric Nursing and Mental Health, Faculty of Nursing, Alexandria University, Alexandria 21527, Egypt
* Correspondence: mercy.ofheaven2000@gmail.com; Tel.: +81-042-346-1714

Received: 3 September 2019; Accepted: 18 September 2019; Published: 20 September 2019

Abstract: Aging is a natural phenomenon that occurs in all living organisms. In humans, aging is associated with lowered overall functioning and increased mortality out of the risk for various age-related diseases. Hence, researchers are pushed to find effective natural interventions that can promote healthy aging and extend lifespan. Royal jelly (RJ) is a natural product that is fed to bee queens throughout their entire life. Thanks to RJ, bee queens enjoy an excellent reproductive function and lengthened lifespan compared with bee workers, despite the fact that they have the same genome. This review aimed to investigate the effect of RJ and/or its components on lifespan/healthspan in various species by evaluating the most relevant studies. Moreover, we briefly discussed the positive effects of RJ on health maintenance and age-related disorders in humans. Whenever possible, we explored the metabolic, molecular, and cellular mechanisms through which RJ can modulate age-related mechanisms to extend lifespan. RJ and its ingredients—proteins and their derivatives e.g., royalactin; lipids e.g., 10-hydroxydecenoic acid; and vitamins e.g., pantothenic acid—improved healthspan and extended lifespan in worker honeybees *Apis mellifera*, *Drosophila Melanogaster* flies, *Gryllus bimaculatus* crickets, silkworms, *Caenorhabditis elegans* nematodes, and mice. The longevity effect was attained via various mechanisms: downregulation of insulin-like growth factors and targeting of rapamycin, upregulation of the epidermal growth factor signaling, dietary restriction, and enhancement of antioxidative capacity. RJ and its protein and lipid ingredients have the potential to extend lifespan in various creatures and prevent senescence of human tissues in cell cultures. These findings pave the way to inventing specific RJ anti-aging drugs. However, much work is needed to understand the effect of RJ interactions with microbiome, diet, activity level, gender, and other genetic variation factors that affect healthspan and longevity.

Keywords: aging; alternative therapy; composition of royal jelly; dietary interventions; healthspan; lifespan; longevity; royal jelly; *IGF-1*; oxidative stress

---

1. Introduction

Reduced mortality from both communicable and noncommunicable diseases accelerates the growth of aging populations worldwide. However, long living entails reduced healthspan out of increased burden of noncommunicable diseases [1]. Aging, healthspan, and lifespan are tightly related concepts. Though aging is a natural phenomenon, numerous factors such as stress, poor nutrition, and pollution are associated with increased internal production of free radicals, which enhance chronic subclinical inflammation and lead to faster aging [2,3]. Aging-related oxidative damage and inflammation contribute to a trail of cellular and molecular alterations: telomere attrition, epigenetic alterations, genome instability, reduced proteostasis, disturbed nutrient sensing, mitochondrial dysfunction, cellular senescence, stem cell exhaustion, and altered intercellular communication. In this respect, aging represents a process of random cellular degradation [4–6], which is associated with increased likelihood of progressive loss of function, decreased fertility, and increased risk of various diseases (both physical and cognitive)—contributing to premature death and increased morbidity in aging populations (Figure 1) [2,7].

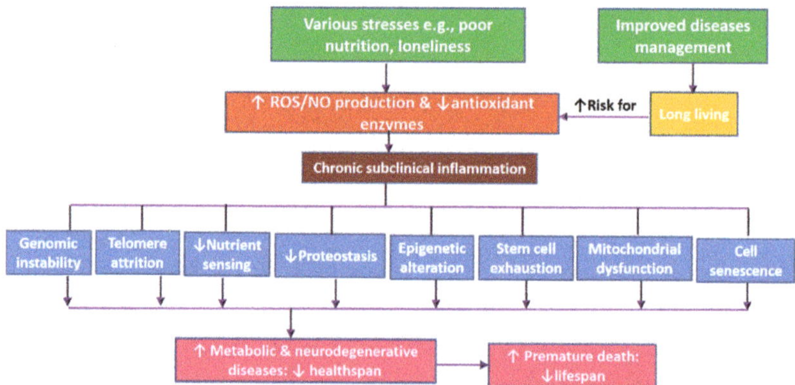

Figure 1. The dynamics of aging-related decrease of healthspan and lifespan. The production of free radicals increases with aging (which is on the rise out of improved disease management) and also with exposure to various stresses (e.g., pollution, poor nutrition, and psychological stress) leading to oxidative stress. This is associated with reduced production of antioxidant enzymes and activation of inflammatory pathways. Both oxidative stress and chronic inflammation lead to a trail of cellular and molecular alterations: telomere attrition, epigenetic alterations, genome instability, reduced proteostasis, disturbed nutrient sensing, mitochondrial dysfunction, cellular senescence, stem cell exhaustion, and altered intercellular communication. Such age-related physiological alterations result in poor health and loss of function out of increased occurrence of various metabolic and neurodegenerative disorders. Age-related disorders are associated with increased mortality and premature death i.e., decreased lifespan.

A recent review indicates that family profiles of centenarians emphasize the contribution of "protective or longevity genes" to extreme longevity as well as to low incidence of age-related diseases in this group [5]. Apart from genetic factors, favorable environmental conditions and healthy lifestyles contribute to low morbidity in people who survive to extreme ages [8,9]. Therefore, the worldwide substantial increase of aging populations, who encounter a trail of debilitating illnesses, has driven researchers to find strategies to slow aging and support a healthy lifespan by delaying the onset of age-related diseases [5]. Research denotes that the natural decline of function that occurs with aging, as well as age-related disorders, can be prevented and/or reversed through environmental modifications such as dietary interventions [10,11]. Accumulating evidence indicates that functional foods can promote healthy aging, reduce morbidity, and lengthen healthy lifespans [12]. Royal jelly (RJ) is considered a functional food (with a documented safety profile) because it has a range of pharmacological activities: antioxidant, anti-inflammatory, antitumor, antimicrobial, anti-hypercholesterolemic, vasodilative, and hypotensive. Indeed, RJ has been widely used to treat several health conditions such as diabetes mellitus, cardiovascular diseases, and cancer, to name a few [13,14].

RJ is a creamy substance secreted from the mandibular and hypopharyngeal glands of the worker honeybee *Apis mellifera*. It is the food of all bee larvae during their first 3 days of life; workers then shift to worker jelly (WJ), composed mainly of honey and pollen, while bee queens continue to consume RJ [15–17]. It is suggested that RJ is a potent promoter of healthy aging and longevity because it enhances overall health and fertility of queen bees, who may lay up to 3000 eggs a day and survive as long as five years compared with infertile workers that live up to 45 days only [18–20]. However, studies of the effect RJ on longevity in humans are scarce, presumably due to technical and ethical reasons. In this respect, this paper reviews, synthesizes, and discusses the most relevant studies that used RJ to enhance healthspan and extend lifespan in different species. We also examined the metabolic, molecular, and cellular mechanisms associated with the longevity-promoting properties of RJ.

2. Chemical Composition of RJ

Water constitutes 60–70% (*w/w*) of fresh RJ; pH of fresh RJ usually ranges between 3.6 and 4.2. The various pharmacological properties of RJ are attributed to its unique and rich composition of proteins, carbohydrates, vitamins, lipids, minerals, flavonoids, polyphenols, as well as several biologically active substances [21]. Herein, we provide a detailed description of most ingredients of RJ.

2.1. Sugars

Sugars comprise 7.5–15% of RJ content. Fructose and glucose constitute 90% of the total sugar fraction of RJ, whereas sucrose accounts for 0.8–3.6%. RJ contains very small amounts of other sugars such as maltose, trehalose, melibiose, ribose, and erlose [22]. RJ sugar content varies remarkably from one sample to another based on season, geographical location, botanical origin, bee species, and method of production. For example, French RJ contents of sucrose and erlose are less than 1.8% and 0.4%, respectively, whereas their concentrations in RJ produced by sugarcane feeding are comparatively higher (7.7% and 1.7%) [23]. Sugars of RJ are thought to contribute to its epigenetic effects, given that RJ sugar content is extremely high compared with WJ (the main food of bee workers). Meanwhile, supplementing WJ with fructose and glucose (4%) fosters the development of larvae to adult workers. In the same way, a gradual increase of WJ sugar concentrations (up to 20%) increases larval consumption of WJ and eventually results in the development of intercastes (midway between queens and workers) and queens at a rate similar to that obtained by pure RJ for in vitro rearing [24,25]. Thus, sugars of RJ represent a phagostimulant that functions through the *insulin/insulin-like* signaling cascades and the nutrient sensing *mTOR* pathway to derive larval development by increasing quantities of ingested food and increasing intake of nutrients necessary for queen development [25].

2.2. Lipids

Among the principal bioactive constituents of RJ, lipids constitute 7–18% of RJ content; 90% of these lipids are rare and unique short hydroxy fatty acids with 8–12 carbon atoms in the chain and dicarboxylic acids. The most prominent RJ fatty acids in order are 10-hydroxydecanoic acid (10-HDA), 10-hydroxy-2-decenoic acid (10H2DA), and sebacic acid (SA) [18]. The 10-HDA exerts epigenetic control over caste differentiation of *Apis mellifera* by inhibiting histone deacetylases, which catalyze the hydrolysis of ε-acetyl-lysine residues of histones [25,26]. Due to the low pH of RJ, 10-HDA acts as a strong bactericidal. It therefore, protects bee larvae against virulent bacterial infections that affect bee hives such as those caused by some strains of *Paenibacillus* larvae [27]. In mammals, it protects mice against pulmonary damage induced by lipoteichoic acid, a toxin from *Staphylococcus aureus* [28]. It also exerts an antibacterial effect against various pathogenic bacterial species in human cancer colon cells [29]. The 10-HDA may be used to treat age-related neurodegenerative disorders given its documented neurogenic activity—it stimulates neuronal differentiation from progenitor cells (PC12) cells through mimicking the effect of brain-derived neurotrophic factor [30]. In addition, it possesses neuroprotective effects against glutamate- and hypoxia-induced neurotoxicity [31]. The 10-HDA may also be used for manufacturing cosmetics and anticancer drugs, given that it increases skin-whitening and exerts antiproliferative effects on B16F10 melanoma cells by inhibiting the expression of microphthalmia-associated transcription factor and tyrosinase-related protein 1 (*TRP-1*) and *TRP-2* [32]. Moreover, 10-HDA has been identified as an inhibiting factor of matrix metalloproteinases (*MMPs*)—proenzymes activated by proteolytic cleavage under inflammatory conditions, which degrade matrix and non-matrix proteins and contribute to tissue aging (e.g., skin) and cause serious disabling diseases such as rheumatoid arthritis [33,34].

SA, 10-HDA, and 10H2DA, demonstrate anti-inflammatory effects through regulation of several proteins involved in the mitogen-activated protein kinase (*MAPK*) and nuclear factor kappa-B signaling [29,35]. Moreover, these acids mediate estrogen signaling by enhancing the activity of estrogen receptors (*ERs*) *ERα*, *ERβ* [36], which can benefit bone, muscle, and adipose tissue in a sex-dependent manner [31]. A derivative of 4-hydroperoxy-2-decenoic acid known as 4-hydroperoxy-2-decenoic acid ethyl ester (HPO-DAEE) prevents 6-hydroxydopamine-induced cell death in human neuroblastoma SH-SY5Y cells through triggering slight emission of reactive oxygen species (ROS), which stimulates the production of antioxidants via activation of antioxidant pathways: nuclear factor erythroid 2 (*NRF2*)-antioxidant response element (*ARE*) and eukaryotic initiation factor 2 (*eIF2α*), an upstream effector of the activating transcription factor-4 (*ATF4*) [37]. HPO-DAEE also demonstrates anticancer effects through accumulation of intracellular ROS and activation of proapoptotic CCAAT-enhancer-binding protein homologous protein expression [38]

2.3. Proteins

Proteins are the dominant ingredient of RJ (50% of its dry matter) and more than 80% of total RJ proteins are composed of nine major RJ proteins (MRJPs, 49–87 kDa)—the first five MRJPs constitute up to 82–90% of MRJPs. Glycosylation and phosphorylation of MRJPs is essential for biological processes that involve glycoproteins, such as cell adhesion, cell differentiation, cell growth, and immunity. MRJPs modulate the development of female larvae, not only through their high nutritional value but mainly through physiological activity of their highly homologous 400–578 amino acids that contribute to RJ's role in cell proliferation, cytokine suppression, and antimicrobial activity [12].

Research documents anti-senescence activity of MRJPs for human cells in vitro [39]. MRJP1 is the most dominant among all MRJPs; essential amino acids constitute 48% of its content. Circular dichroism measurements indicate that the secondary structure of MRJP1 consists of 9.6% α-helices, 38.3% β-sheets, and 20% β-turns [22]. MRJP1 exists in two distinct forms: oligomer and monomer. Oligomer MRJP1 is highly heat-resistant and it is considered a predominant proliferation factor compared with MRJP2 and MRJP3 [40]. High-performance liquid chromatography and SDS-PAGE analyses of MRJP1 revealed the presence of a 57 kDa monomeric glycoprotein, which can be degraded at 40 °C, known as royalactin. Royalactin mimics the effect of epidermal growth factor (*EGF*) in rat hepatocytes and modulates the development of bee larvae [41]. Royalactin is reported to bind with the most sensitive regions in mouse embryonic stem cell culture, resulting in activation of a pluripotency gene network that enables self-renewal of stem cells [42]. MRJP1 exerts nematicidal activity against *C. elegans* via constant downregulation of a rate-controlling enzyme of the citric acid cycle known as isocitrate dehydrogenase encoding the idhg-1 gene [43]. On the other side, MRJP2 and its isoform X1 exhibit potent anticancer effects and protect hepatocytes against CCl4 toxicity by inducing caspase-dependent apoptosis, scavenging intracellular free radicals, inhibiting tumor necrosis factor (*TNF*)-α, and mixed lineage kinase domain-like protein [44].

RJ contains proteins other than MRJPs, albeit in small amounts, such as royalisin, jelleines, and aspimin. Royalisin and jelleines are common RJ antimicrobial peptides that enhance efficiency of the immune response of bee larvae to various infections. The structure of royalisin is highly compact due to its high cysteine content, which boosts its stability at low pH and high temperature. On the other hand jelleines are thought to stem from trypsin digestion of MRJP1 by the action of exo-proteinase of the hypopharyngeal glands on C-terminal to N-terminal tryptic fragment. Peptides of royalisin and jelleines possess hydrophobic residues, which contribute to their antimicrobial properties by affecting functions of bacterial membranes. RJ also contains apolipophorin III-like protein, a lipid binding protein that exerts antimicrobial effect by carrying lipids into aqueous environments through the formation of protein–lipid complexes. In addition, the antibacterial effect of RJ is partially attributed to its glucose oxidase enzyme, which catalyzes the oxidation of glucose to hydrogen peroxide [45].

Recently, examination of whole RJ and single protein bands by off-line LC-MALDI-TOF MS glycomic analyses, complemented by permethylation, Western blotting, and arraying data, revealed the presence of glucuronic acid termini, sulfation of mannose residues, core β-mannosylation of the N-glycans, and a fairly scarce zwitterionic modification with phosphoethanolamine, which may contribute to the development of honey bees and their innate immunity [46].

2.4. Phenols, Flavonoids, and Free Amino Acids

The antioxidant potency of RJ, at least in part, is attributed to its polyphenolic compounds and flavonoids, which are measured based on gallic acid and rutin equivalent (GAE and RE), respectively [47]. RJ contains 23.3 ± 0.92 GAE µg/mg and 1.28 ± 0.09 RE µg/mg of total phenolics and flavonoids, respectively. The vast phenolic content of RJ consists of pinobanksin, organic acids (e.g., octanoic acids, dodecanoic acid, 1,2-benzenedicarboxylic acid), and their esters. Meanwhile, flavonoids of RJ can be differentiated into four groups: (1) flavanones e.g., hesperetin, isosakuranetin, and naringenin; (2) flavones e.g., acacetin, apigenin and its glucoside, chrysin, and luteolin glucoside; (3) flavonols e.g., isorhamnetin and kaempferol glucosides; and (4) isoflavonoids e.g., coumestrol, formononetin, and genistein. The antioxidant activity of these components contributes to the antiapoptotic and anti-inflammatory properties of RJ [18]. Age of larvae that produce RJ (1–14 days) and harvesting time affect RJ content of phenols and amino acids; RJ harvested from the youngest larvae (one-day-old) within 24 h contains higher proteins and polyphenolic compounds and exhibits stronger free radical scavenging effect compared with RJ harvested from older larvae or later than 24 h [47]. Small peptides such as di-peptides (Lys-Tyr, Arg-Tyr, and Tyr-Tyr) obtained from RJ proteins hydrolyzed by protease N possess high antioxidative activity owing to their phenolic hydroxyl groups, which scavenge free radicals by releasing a hydrogen atom [48].

RJ is rich in amino acids, including essential ones [22]. Concentrations of free amino acids in RJ increase with harvesting time from 4.30 mg/g at 24 h to 9.48 mg/g at 72 h, whereas levels of total amino acids decrease by time from 197.96 mg/g at 24 h to 121.32 mg/g at 72 h [49]. LC/MS method analysis and hydrophilic interaction liquid chromatography-tandem mass spectrometry indicate that lysine is the most prominent free amino acid in RJ (62.43 mg/100 g), followed by proline (58.76 mg/100 g), cystine (21.76 mg/100 g), and aspartic acid (17.33 mg/100 g). RJ contains less than 5 mg/100 g of other amino acids such as valine, glutamic acid, serine, glycine, cysteine, threonine, alanine, tyrosine, phenylalanine, hydroxyproline, leucine-isoleucine, and glutamine [18,50].

The amino acid content in protease-treated RJ (pRJ)—obtained by removal of two allergen proteins to convert RJ hydrolysates into shorter chain monomers that are easy to absorb—is greater than in crude RJ [51]. Moreover, eluted water pRJ with 30% MeOH contains higher levels of dipeptides and tripeptides than pRJ. These components are likely to possess a lifespan-prolonging activity since pRJ increases lifespan of *C. elegans* compared with RJ [52]. Evidence signifies that amino acids of RJ can prolong lifespan in mammals. Long-term dietary supplementation of branched-chain amino acid-enriched mixture (BCAAem), which contains 3 branched-chain amino acids that can be found in RJ (leucine, isoleucine, and valine), enhances mitochondrial biogenesis and sirtuin 1 expression, and reduces ROS production in cardiac and skeletal muscle, which is associated with alleviation of age-related muscle dysfunction, resulting in an increase of the average lifespan of male mice. The effect of these amino acids on mitochondrial biogenesis involved activation of signaling pathways of endothelial nitric oxide synthase (*eNOs*) and *mTOR* and its substrates: *S6K* and eukaryotic translation initiation factor (*eIF4E*)-binding protein (*4E-BP1*) [53].

2.5. Vitamins, Minerals, and Bioactive Substances

Pantothenic acid (vitamin B5) is the most abundant vitamin in RJ (52.8 mg/100 g), followed by niacin (42.42 mg/100 g). RJ contains small amounts of various B group vitamins (B1, B2, B6, B8, B9, and B12), ascorbic acid (vitamin C), vitamin E, and vitamin A [18,22]. Gardner (1948) noted that pantothenic acid of RJ is a lifespan-extending agent [54].

Mineral salts constitute 1.5% of RJ content [18]. Inductively coupled plasma optical emission spectroscopy and double focusing magnetic sector field inductively coupled plasma mass spectrometry indicate that RJ contains small amounts of various minerals and trace elements such as K, Na, Mg, Ca, P, S, Cu, Fe, Zn, Al, Ba, Sr, Bi, Cd, Hg, Pb, Sn, Te, Tl, W, Sb, Cr, Mn, Ni, Ti, V, Co, and Mo. Whereas concentrations of trace and mineral elements in honey vary according to botanical origin, RJ content of trace elements and minerals is highly constant. In this respect, RJ can be considered a form of larval lactation that possesses homeostatic adjustment, same as mammalian and human breast milk [55].

RJ contains high amounts of acetylcholine (Ach, 4–8 mM), and RJ concentration of Ach is highly conserved because of its acidic pH [56]. It is well-known that Ach acts as a neurotransmitter that plays a major role in memory formation and cognitive functioning. Glucose metabolism and insulin contribute to Ach synthesis by controlling the activity of choline acetyltransferase [57]. In this respect, consumption of RJ may prevent the development of cognitive dysfunction thanks to its Ach content. In addition, Ach content of RJ has a survival promoting effect [56].

RJ is rich in nucleotides such as free bases (e.g., adenosine, uridine, guanosine, iridin, and cytidine) and phosphates (e.g., adenosine diphosphate (ADP), adenosine triphosphate (ATP), and adenosine monophosphate (AMP)). Nucleotides constitute 2682.9 mg/kg and 3152.8 mg/kg in fresh and commercial RJ, respectively. Levels of ADP, ATP, and AMP are higher in fresh RJ, and therefore, they can signify RJ freshness. These compounds are necessary for organisms' physiological activities e.g., metabolic degradation of intracellular ATP is essential to provide cells with energy necessary for transport systems and enzymatic activities of proteins [22,58]. Among all nucleotides, AMP N1-oxide is considered a unique active component that exists nowhere in nature except in RJ. AMP N1-oxide demonstrates neurogenic and neurotrophic activities: it stimulates neurite outgrowth and induces differentiation of PC12 cells into neurons similar to sympathetic neurons. This action is similar to that of nerve growth factor, which functions through activation of two cascades of cellular signaling *MAPK*/extracellular signal-regulated kinase 1 or 2 (*ERK1/2*) and phosphatidylinositol 3-kinase/*Akt* pathways. The neurite outgrowth-promoting activity of AMP N1-oxide is mediated by adenylate cyclase-coupled adenosine A2A receptors, which are highly expressed in the brain (striatum in particular)—adenosine A2A receptors prevent radical formation and apoptosis, and they contribute to early neuronal development and regulation of synaptic plasticity [18,59].

3. Healthspan and Longevity Effects of RJ in Various Species

3.1. RJ Enhances Fertility and Longevity in Population of the Beehive

Numerous bioactive elements are abundant in RJ, making it an optimal food and a well-balanced nutrient-rich diet [25] that, when consumed by bee larvae, induces their development into queens. Meanwhile, larvae that consume honey or pollen grow into workers, which live shorter and are unable to reproduce. On the other hand, emerging bee workers fed RJ-rich diet acquire queen-like morphogenic characteristics such as increased body size and ovary development [41,60,61]. Real-time RT-PCR and HPLC-ECD analysis denoted that augmentation of fertility (ovarian development) in bee workers by RJ is the result of enhancement of brain levels of tyrosine, dopamine, and tyramine by RJ [61], as well as activation of epidermal growth factor receptor (*EGFR*), which increases the production of the juvenile hormone, titre, known to regulate growth [41]. RJ also improves the memory of bee workers [62] and increases their survival [15,41,56,62,63], owing to its high Ach concentration [56], MRJP1 content of royalactin [41], MRJPs 2, 3, and 5 [25], and a water soluble RJ protein extracted by precipitation with 60% ammonium sulfate (RJP$_{60}$) [15].

3.2. RJ Enhances Healthspan and Longevity in Other Species

Based on the assumption that RJ is the main factor contributing to long survival of bee queens, several studies examined the lifespan-expanding effects of RJ in diverse species (Figure 2), and results seem to be consistent with the naturally occurring model of bee queens.

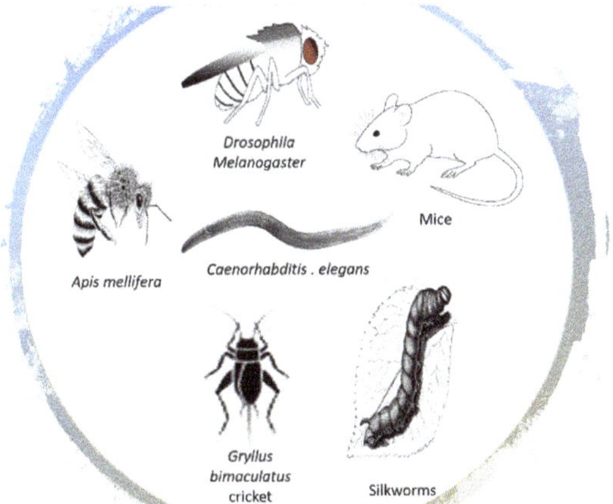

Figure 2. Royal jelly and its components improve healthspan and extend lifespan in different species.

3.2.1. Drosophila Melanogaster (Fruit Fly)

The fruit fly *Drosophila melanogaster* (M.) has been widely used in the literature as an invertebrate model to understand the pathology of numerous diseases and to examine the effect of various agents, including RJ. The earliest study of the longevity effect of RJ dates back to 1948. That study reported that large doses of dehydrated RJ, its pantothenic acid, as well as its water soluble and insoluble organic acids significantly extended lifespan of *Drosophila M.* compared with controls. The author suggested that pantothenic acid of RJ may have an anti-aging effect by itself or by synergizing the action of other vitamins [54]. In a relatively recent study, RJ (0.1, 0.3, and 0.5 g) significantly increased the average lifespan of both male and female flies—the effect was attributable to enhancement of the antioxidation capacity noticed in flies treated by high doses of RJ: increased superoxide dismutase (SOD) and catalase levels [64].

As a replication of his experiment of RJ and royalactin in bees, Kamakura (2011) treated *Drosophila Canton-S.* larvae with royalactin and 20% fresh RJ. RJ treatment increased body size, cell size, and fertility; prolonged lifespan; and shortened developmental time from larva to adult compared with control. Same as in bees, RT-PCR and enzyme immunoassay showed that both RJ and royalactin induced *EGFR* signaling (not insulin signaling), which activated *S6K* in the fat body, which further increased the body size. *EGFR* also activated the *MAPK* pathway, which increased the synthesis of a biologically active ecdysteroid known as 20-hydroxyecdysone (*20E*) and juvenile hormone—an effect that was demonstrated by growth regulation—reducing the developmental time [41]. A subsequent study stated that supplementing *Drosophila M.* with 1% freeze-dried RJ (FDRJ) powder shortened developmental time, prolonged the lifespan of adult males, and increased females' egg production without any morphological changes. In female flies, RT-PCR indicated that FDRJ was significantly associated with heightened gene expression of an insulin-like peptide known as dilp5, its insulin receptor (*InR*), and the nutrient sensing molecule *mTOR*, the mechanistic target of rapamycin—all these molecules are known to affect growth and reproduction. However, stimulation of the *insulin/TOR* signaling pathways was not associated with extension of lifespan of FDRJ-fed female flies [17]. Another study noted that supplementation of MRJPs, especially MRJP 1 and MRJP3, at an optimal dose of 2.5% (*w/w*) of diet significantly lengthened the mean lifespan of both male and female *Drosophila*. The longevity effect of MRJPs was positively associated with increased feeding and fertility. Microarray data and gene ontology enrichment analyses revealed that the molecular mechanism underlying increased lifespan and fertility was similar to that discovered by Kamakura (2011): MRJP supplementation upregulated the gene expression of *S6K*, *MAPK*, and *EGFR* in *EGFR*-mediated signaling. In addition, MRJPs improved the anti-oxidation capacity of flies by increasing the expression of *CuZn-SOD* gene i.e., SOD levels were higher, while malonaldehyde (MDA) levels were lower than control flies [12]. Another study replicated the size/growth enhancing effect of RJ at low concentrations (10–30%) in *Drosophila M.* [16]. However, males' lifespan decreased by 20% RJ treatment whereas none of these concentrations affected the lifespan of females. Findings provided no evidence of RJ activation of insulin and *EGFR* signaling pathways, yet RJ regulated the gene expression related to oxidative stress and catabolism. The authors attributed the discrepancy noted between their results and findings of other studies to employing a slightly different strain of *Drosophila* (*Canton-S*), using a commercially available source of RJ, and difference in the nutrient contents of the control culture medium. On the other hand, higher concentrations (40–70%) of RJ had adverse effects: prolonged development time, shortened lifespan, increased mortality, and reduced productivity in both sexes. Data on global gene expression indicated that excess nutrients in high doses of RJ altered cellular processes as a result of altering genes involved in amino acid metabolism and encoding glutathione S transferases, which detoxify xenobiotic compounds [16].

3.2.2. Gryllus Bimaculatus Cricket and Silkworms

The two-spotted cricket *Gryllus (G.) bimaculatus*, a member of primitive group *Polyneoptera*, is another kind of species that were used to examine the longevity effect of RJ. RJ dietary supplementation to *G. bimaculatus* during early nymph stage significantly decreased developmental time, extended lifespan, and increased the body size of both males and females in a dose-dependent manner (8–15% *w/w*) compared with the control high protein, sugar, and lipid diet. The prolonged lifespan of RJ was not due to an extended nymph stage since the adult stage emerged earlier in crickets fed RJ than those fed the control diet. Similarly, RJ administration increased body size and egg size in female silkworms. The authors concluded that the effects of RJ were not attributed to the nutritional supplement itself; however, they did not examine the molecular mechanism behind them [65].

3.2.3. Caenorhabditis (C.) Elegans Nematodes

Caenorhabditis (C.) elegans nematodes have been used in several instances to test the longevity-promoting activity of RJ. Japanese researchers found that RJ, protease-treated RJ (pRJ), pRJ-Fraction 5 (pRJ-Fr.5), and a derivative of pRJ-Fr.5—10-HDA (the main lipid of RJ)—extended the lifespan of *C. elegans*. RJ 10 µg/mL was an optimal dose for enhancing longevity by 7–9%, while RJ 1 or 100 µg/mL had no effect. Meanwhile, all pRJ concentrations significantly prolonged the mean lifespan—though the greatest effect was noticed at 10 µg/mL, which increased mean lifespan by 7–18%. The longevity effect of combined pRJ-Fr.5 and 10-HDA was greater than that induced by each treatment on its own. DNA microarray and RT-PCR showed that the longevity-promoting effect was attributed to reduction of the *insulin/IGF-1* signaling—pRJ-Fr.5 upregulated the expression of dod-3 gene and downregulated the expression of *ins-9*, an insulin-like peptide gene, along with *dod-19*, *dao-4*, and *fkb-4* genes (further details are shown below in the mechanism section) [52]. In two subsequent studies RJ, pRJ, and 10-HDA enhanced longevity and increased stress resistance of *C. elegans* against thermal, irradiation, and oxidative stress [66,67]. Intact, deglycosylated, and mildly heat-treated royalactin extended mean lifespan of *C. elegans* by 18–34%—higher concentrations produced the vastest lifespan-extending effect. Royalactin also enhanced locomotion, which indicates promotion of healthy aging [68].

3.2.4. Mice

Few studies investigated the lifespan extending activity of RJ in mice. In an early study, intermediate and high doses (50 and 500 ppm) of RJ significantly prolonged the mean lifespan of C3H/HeJ mice by 25%, whereas RJ at a low dose (5 ppm) yielded no significant effect. Meanwhile, all doses of powdered RJ, contrary to bee and *Drosophila* studies, had no effect on mice growth, food intake, or appearance compared with control mice. RJ treatment significantly lowered kidney DNA and serum levels of 8-hydroxy-2-deoxyguanosine, a marker of oxidative stress that increases with aging [2]. Similarly, long-term intragastric administration of RJ and pRJ to a d-galactose-induced aging mice model resulted in numerous anti-aging and healthspan effects: preventing aging-related weight loss, improving memory and motor performance, and delaying aging-related atrophy of thymus, thus preventing diminution of the immune function compared with control animals. The effects were attributed to inhibition of lipid peroxidation and improvement of levels of antioxidant enzymes [69]. Likewise, dietary supplementation of RJ and pRJ (0.05% or 0.5%, *v/v*) to genetically heterogeneous head tilt mice—which exhibit vestibular dysfunction, imbalanced position, and inability to swim—could not prolong lifespan but significantly delayed age-related impairment of motor functions, positively improved physical performance of treated mice on four tests (grip strength, wire hang, horizontal bar, and rotarod), lowered age-related muscular atrophy, increased markers of satellite cells (muscle stem cells), and suppressed catabolic genes [70]. Another study examined the survival-expanding time (not lifespan extension) of oral RJ treatment (75, 150, and 300 mg/kg body wt/day for 13 consecutive days) following NaNO2 intraperitoneal injection or decapitation as models of brain hypoxia and complete brain ischemia [71]. Findings indicated that the intermediate dose of RJ (150 mg) significantly expanded survival time, whereas RJ 75 mg had no significant effect, meanwhile RJ 300 mg significantly decreased survival time. The author suggested that low pH of RJ (3.6 to 4.2) in mice treated with high doses of RJ induced activation of acid-sensing channels, increased acidosis of extracellular fluid, and aggravated brain ischemia [71].

4. RJ Might Enhance Longevity in Humans by Promoting General Health

It is an important question whether observed effects of RJ on healthspan and longevity in model organisms (e.g., bees, fruit flies, *C. elegans*, silkworms, crickets, and mice) can be generalized to humans. A small number of in vitro studies examined the antiaging activity of RJ, 10-HDA, and MRJPs on human cell lines, and results support findings reported from studies of model organisms. In two experiments, normal human skin fibroblasts were ultraviolet-irradiated (as a model of skin photoaging) and treated with RJ and 10-HDA. Results revealed that RJ and 10-HDA protected cells against ultraviolet A- and B-induced ROS-related oxidative damage, decreased cellular senescence, stimulated the production of procollagen type I and transforming growth factor-β1 [33,72], and suppressed the expression of *MMP-1* and *MMP-3* at transcriptional and protein levels [33]. In another study, human embryonic lung fibroblast cells were treated with different concentrations of MRJPs (0.1–0.3 mg/mL) versus bovine serum albumin. MRJP-treated cells showed the highest proliferation activity, the lowest senescence, and the longest telomeres. Such effects were associated with upregulation of SOD1 and downregulation of *mTOR*, catenin beta like-1, and tumor protein p53 [39]. Apart from these limited in vitro studies, we could not locate any study that investigated the effect of RJ on lifespan in humans in vivo. However, several studies assessed the effect of RJ on promotion of wellbeing and prevention of severe diseases associated with increased early death—these studies may mirror the healthspan effects of RJ. Herein, we explore how RJ may support healthy aging in the general population.

It is becoming clear that certain metabolic pathways reduce longevity in humans by increasing the risk of serious illnesses that contribute to mortality e.g., diabetes mellitus, metabolic syndrome, cardiovascular diseases, and cancer [73]. The life-expanding effect of RJ possibly originates from its antioxidant and anti-inflammatory properties, which can promote healthy aging by improving glycemic status, lipid profiles, and oxidative stress—and hence can prevent the occurrence of various debilitating metabolic diseases [13,14]. In accordance, administration of RJ in healthy volunteers was associated with improved indicators of physical wellbeing (erythropoiesis and glucose tolerance) [74].

Rheumatoid arthritis (RA) is one of the most common disabling disorders that seriously endanger healthspan. It is a chronic systemic inflammatory arthritis that occurs at an age of onset of 55 years and increases with age i.e., it predominantly affects the older population. Meanwhile, treatment is complicated by age-related decline in organ function (e.g., renal and metabolic), comorbidities, and changes in body composition (e.g., decreased lean mass), which make outcomes of RA treatment (steroids and anti-*TNF* agents) in the elderly highly disappointing [75]. On the other hand, a relatively large number of in vitro studies indicate that 10-HDA can be a safe treatment of RA. The 10-HDA is likely to prevent joint destruction by inhibiting *MMP* production from rheumatoid arthritis synovial fibroblasts through blockage of p38 kinase and c-Jun N-terminal kinase-AP-1 signaling pathways [34,76]. The 10-HDA also prevents cell proliferation of fibroblast-like synoviocytes by inhibiting target genes of *PI3K–AKT* pathway and genes of cytokine-cytokine receptor interaction [77].

It is believed that hormones play a major role in healthspan and longevity by regulating cellular responses, metabolism, and growth [78]. Insulin is one of such hormones that affect every single cell in the body; its signaling pathway interacts with a variety of other pathways and affects their functioning [79]. For instance, overexpression of insulin-like growth factors (*IGFs*) and receptors for insulin is associated with increased cell proliferation and risk of cancer [80]. Meanwhile, downregulation of insulin signaling contributes to proper mitochondrial function, suppression of inflammatory mediators, regulation of cellular metabolism, cellular resistance to stress, activation of DNA repair genes, and reduction of oxidative damage of macromolecules and cellular senescence, which all further enhance health and longevity, both in humans and other species [79,81,82]. In this respect, most anti-aging dietary interventions target growth-promoting pathways i.e., they function by downregulating *IGF-1* and *mTOR-S6K* pathway and activating nutrient sensors (*MAPK* and sirtuins), which signal nutrient scarcity and stimulate catabolism [83]. A majority of the above-reviewed studies indicate that RJ enhanced longevity by targeting insulin signaling; this mechanism is likely to work in humans. In fact, RJ has an insulin-like activity—an insulin-like peptide had been purified from RJ and it is similar to insulin of vertebrates in solubility, chromatographic, immunological, and biological characteristics [84]. Administration of RJ to healthy athletes significantly decreased their insulin and increased thyroxine (T4) hormone levels in plasma [85]. In line with this, healthy adults who consumed a single oral dose of RJ (20 g) immediately before oral glucose tolerance test showed significantly reduced glucose level [86].

Aging is usually associated with reduction of sex hormones [87]. The genetic switch model of aging postulates that end of reproduction entails a genetically programmed inactivation of survival and maintenance pathways, which causes a progressive age-related decline of function [88]. In fact, sex hormones are considered markers of longevity since they have a neuroprotective effect, which originates from their ability to improve insulin resistance and enhance the DNA repair capacity of neurons [74,89]. It is becoming clear that RJ modulates sex hormones. The endocrine stimulation exerted by RJ is necessary for ovary development in bee queens and it increased egg-laying in *Drosophila M.* and silkworms [41,63,65]. RJ and royalactin increased juvenile hormone titre (a fertility hormone) downstream of *EGFR* signaling in *Drosophila M.* [12,17,41]. Several studies indicated that RJ and its lipids exert estrogenic activity both by binding with estrogen receptors and activating the expression of endogenous genes [87,90]. Therefore, sex hormones represent another facet through which RJ may enhance longevity. In humans, RJ supplementation to healthy volunteers increased serum testosterone and the ratio of testosterone/dehydroepiandrosterone sulfate, indicating that RJ accelerates conversion of dehydroepiandrosterone sulfate into testosterone. The authors suggested that RJ-induced improvement of erythropoiesis in their sample could be ascribed to the anabolic effect of testosterone [74].

Menopause is a common inevitable age-related phenomenon that results from reduction of estrogen production in women at an age range of 45 to 55 years [91]. It is often associated with various uncomfortable physical and psychological symptoms. Therefore, menopausal women may receive hormone therapy to alleviate discomfort and lower their risk for various serious diseases such as osteoporosis and cardiovascular disorders. However, hormone therapy is associated with increased risk of cancer [92,93]. Several natural alternatives (including RJ) have been under investigation as safer alternatives. Animal studies document that consumption of RJ by ovariectomized rats improved bone strength [94] and prevented bone loss in a fashion similar to the effect of 17β-estradiol [95]. Cell culture models revealed that the role of RJ in bone formation is due to upregulation of procollagen I α1 gene expression [96,97], enhancement of intestinal calcium absorption, and increased bone calcium content [95]. Human studies lend further support to this evidence; pRJ at a dose of 800 mg/day significantly decreased lower back pain in healthy menopausal women [93]. Likewise, daily consumption of 150 mg of RJ for three months exerted cardioprotective effects by improving lipid profile of postmenopausal women [92].

The healthspan-inducing effects of RJ in humans are not only limited to enhancement of physical health but also include improvement of general mental health [74], reduction of anxiety symptoms [93], improvement of mood, and mild cognitive impairment in the elderly (>60 years old) [98,99]. The psychological and neurological effects of RJ reflect improvement of biomarkers of physical health such as cholesterol [98]. Hypercholesterolemia, in particular, is documented to foster aggregation of β-amyloid around neurons, which causes neuronal loss—a characteristic feature of Alzheimer's disease. Meanwhile, the reduction of plasma lipids induced by RJ has been associated with enhancement of antioxidative capacities, reduction of β-amyloid deposition, and prevention of neuronal damage [100].

5. Healthspan and Longevity Enhancing Mechanisms

It is now evident that RJ contains compounds (MRJPs, royalactin, amino acids, 10-HDA, pantothenic acid, Ach) that can modulate major mechanisms of aging. However, the exact mechanism through which RJ may extend lifespan is not well-understood. The mechanisms through which RJ extend lifespan can differ even within single species (Table 1). Several facets through which RJ can function have been proposed (Figure 3).

Table 1. Effect of Royal jelly and its components on lifespan and healthspan in different species.

RJ/RJ Components	Species	Effect on Lifespan	Effect on Healthspan	Main Mechanism of Action	References
RJ	*Apis mellifera* (worker)	No effect on lifespan	↑ Ovarian activation	NI—Inability to defecate in 100% RJ-fed bees	[60]
RJ	*Apis mellifera* (worker)	↑ Mean lifespan	↑ Ovarian activation	NI	[63]
RJ	*Apis mellifera* (worker)	–	↑ Ovarian activation	↑ Brain levels of tyrosine, dopamine, and tyramine	[61]
RJ + Ach	*Apis mellifera* (larvae)	↑ Mean lifespan	–	NI—possibly the trophic effects of Ach mediated via muscarinic or nicotinic receptors	[56]
RJ RJP$_{60}$	*Apis mellifera* (worker)	↑ Mean and maximum lifespan	–	NI Lower DNA methylation levels	[15]
RJ	*Apis mellifera* (worker)	↑ Mean lifespan	↑ Expression of memory genes	NI	[62]
RJ, heat-treated RJ, pKRJ, RJ plus MRJP1 vs RJ plus MRJPs2,3,5	*Apis mellifera* (larvae)	↑ Mean lifespan (except MRJP1)	↑ Ovarian activation	NI	[25]
Dehydrated RJ RJ pantothenic acid RJ organic acids	*Drosophila M.*	↑ Mean lifespan	–	NI Synergizing action of other vitamins	[54]
RJ	*Drosophila M.*	↑ Mean lifespan (both sexes)	–	↑ Anti-oxidation capacity–SOD and CAT levels	[64]
RJ, royalactin	*Drosophila Canton-S* *Apis mellifera* (larvae)	↑ Mean lifespan	↑ Body size ↑ Cell size ↑ Eggs laying ↓ Developmental time	↑ *EGFR*-mediated signaling pathway, *S6K, MAPK*, juvenile hormone titre, *20E* titre	[41]
Freeze-dried RJ	*Drosophila M.*	↑ Mean lifespan (males only)	↓ Developmental time ↑ Eggs laying	↓ *Insulin/IGF-1* (*dilp5*) signaling ↓ Target of Rapamycin signaling	[17]

Table 1. *Cont.*

RJ/RJ Components	Species	Effect on Lifespan	Effect on Healthspan	Main Mechanism of Action	References
RJ	*Drosophila Canton S*	No effect on lifespan	↑ Body size	↑ Gene expression related to oxidative stress and catabolism	[16]
MRJPs	*Drosophila M.*	↑ Mean and maximum lifespan (both sexes)	↑ Feeding ↑ Eggs laying	↑ Anti-oxidation capacity—*CuZn-SOD* signaling ↑ *EGFR*-mediated signaling	[12]
RJ	*Gryllus bimaculatus* crickets silkworms	↑ Mean lifespan in crickets (both sexes)	↑ Eggs size (silkworms) ↑ Body size ↓ Developmental time,	NI	[65]
RJ, pRJ, pRJ-Fr.5, 10-HDA	*C. elegans*	↑ Mean lifespan	–	↓ *Insulin/IGF-1* and *ins-9* signaling	[52]
Royalactin	*C. elegans*	↑ Mean lifespan	↑ Locomotion in early and mid-adulthood	↑ *EGFR*-mediated signaling pathway	[68]
10-HDA	*C. elegans*	↑ Mean lifespan	↑ Stress resistance	↓ *Insulin/IGF-1* signaling ↓ Target of Rapamycin signaling ↑ Dietary Restriction	[66]
RJ, pRJ	*C. elegans*	↑ Mean lifespan	↑ Stress resistance	↓ *Insulin/IGF-1* signaling	[67]
Powdered RJ	C3H/HeJ mice	↑ Mean lifespan	–	↓ Oxidative stress and DNA damage	[2]
RJ	Male Swiss albino mice	↑ Survival time after NaNO2 IP injection	–	NI	[71]
RJ, pRJ	D-galactose induced aging mice model	–	↓ Atrophy of thymus ↓ Weight loss ↓ Locomotor decline ↑ Learning and memory	↓ Oxidative stress	[69]
RJ, pRJ	HET mice	No effect on lifespan	↓ Muscle atrophy ↓ Age-related motor impairment	↑ Muscle satellite cell (muscle stem cell) markers Suppression of catabolic genes	[70]

Abbreviations. RJ: royal jelly; ↑: increase; ↓: decrease; NI: not investigated; Ach: acetylcholine; RJP$_{60}$: RJ protein attained by precipitation with 60% ammonium sulfate; EGRF: epidermal growth factor receptor; MRJPs: major royal jelly proteins; pKRJ: protease K-treated RJ; SOD: superoxide dismutase; CAT: catalase; *20E*: 20-hydroxyecdysone; pRJ: protease-treated RJ; pRJ-Fr.5: protease-treated-fraction 5; 10-HDA: 10-hydroxy-2-decenoic acid; IP: intraperitoneal; HET: genetically heterogeneous head tilt.

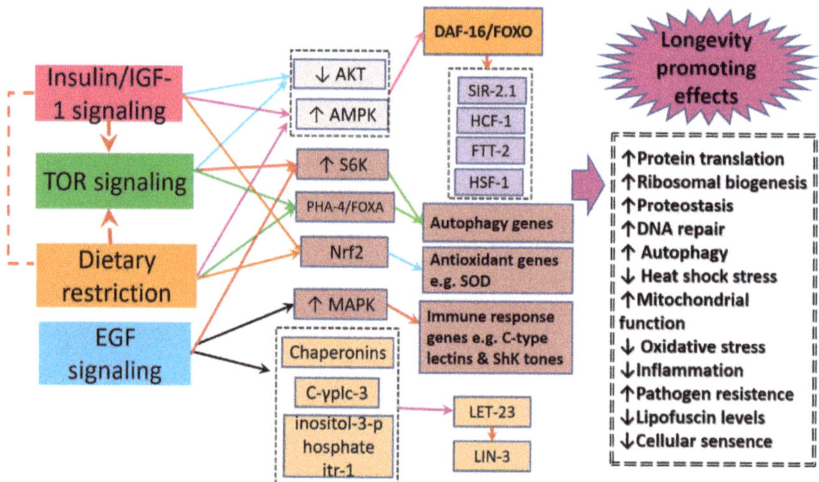

Figure 3. Probable mechanisms through which royal jelly (RJ) and its components extend lifespan. RJ reduces insulin/insulin-like growth factor-1 signaling (*IIS*), which heightens the activity of *DAF-16*—the *C. elegans* counterpart to the mammalian Forkhead Box O transcription factor (*FOXO*). Activation of *DAF-16/FOXO* promotes its translocation from the cytoplasm to the nucleus and fosters interactions with its associated transcriptional co-regulator proteins: host cell factor (*HCF-1*), sirtuin homologue (*SIR-2.1*), and *FTT-2* (a 14-3-3 protein). This results in formation of protein complex inside cells and enhancement of the expression of multiple longevity promoting genes as well as regulation of downstream process of dietary restriction signaling (which lowers food-intake) and the mechanistic target of rapamycin (*mTOR*) pathway by extending the lifespan of the control *unc-24/+* mutants. The interplay among several genes involved in *IIS*, *mTOR*, and dietary restriction signaling boosts key longevity-related cellular processes: DNA repair, autophagy, antioxidant activity, anti-inflammatory activity, stress resistance, and cell proliferation. On the other hand, RJ activates epidermal growth factor receptor (*EGFR*) signaling mainly by activating its receptor (*LET-23*). *EGFR* functions via downstream phospholipase C-γplc-3 and inositol-3-phosphate receptor itr-1 to upregulate elongation factors and chaperonins, which increase protein translation and proteasome activity—a mechanism that entails rebuilding cellular components; enhancement of cellular detoxification, ribosomal function, and muscle maintenance; and reduction of lipofuscin levels (age-pigments that accumulate during senescence).

5.1. Insulin-Signaling/insulin like Growth Factor-1 Signaling

The role of lowered insulin signaling/insulin-like growth factor-1 pathway (*IIS*) in longevity is well documented [101,102]. Findings from studies of *C. elegans* show that *IIS* does not only contribute to prolonged of lifespan, but it also promotes healthspan by resisting different types of stress [102]. The underlying mechanisms include improvement of insulin sensitivity and decrease of insulin levels, suppression of inflammation, reduction of adipose tissue, and increase of adiponectin levels [103]. The lifespan-expanding effect of RJ is associated with an interplay of several genes—expressed in some studies mainly by extending the lifespan of the insulin-like receptor *daf-2* mutants. *daf-2* demonstrates a life-expanding effect by regulating the downstream process of dietary restriction signaling (which lowers food-intake) as well as *TOR* signaling by extending the lifespan of the control *unc-24/+* mutants [66]. Given that *daf-2* is a key upstream component of *IIS*, downregulation of *daf-2* stimulates a signaling cascade that leads to phosphorylation and fine tuning of the main downstream transcription factors of *IIS* known to promote lifespan: *DAF-16*—the *C. elegans* counterpart to the mammalian Forkhead Box O transcription factor (*FOXO*), heat shock transcription factor 1 (*HSF-1*), and *SKN-1/NRF2* [66,67,104]. Among all these pathways, RJ exerted its effect by targeting the activity of *DAF-16/FOXO*, which is a key longevity factor in various species ranging from worms to humans [52,67].

The interactions between *DAF-16* and its associated transcriptional coregulator proteins—host cell factor (*HCF-1*), sirtuin homologue (*SIR-2.1*), and *FTT-2* (a 14-3-3 protein)—lead to formation of a protein complex inside cells [67]. Moreover, activation of *DAF-16* promotes its translocation from the cytoplasm to the nucleus, where it stimulates the expression of multiple genes, which regulate processes that promote longevity: DNA repair, autophagy, antioxidant activity, anti-inflammatory activity, stress resistance, and cell proliferation [10,104]. Aging is associated with increased aggregation and insolubility of various RNA granule proteins (e.g., stress granule proteins). Reduction of *daf-2* receptor signaling heightens *DAF-16* activity, which can efficiently extend lifespan in *C. elegans* by preventing the buildup of misfolded proteins that seed the aggregation of insoluble stress granule proteins [105].

5.2. The Mechanistic Target of Rapamycin Signaling

The mechanistic target of rapamycin (*mTOR*) pathway is another signaling cascade that contributes to RJ enhancement of lifespan extension [17,39,66]. The underlying mechanism involves suppression of *mTOR* gene expression by MRJPs and freeze-dried RJ [17,39]. The nutrient sensor *mTOR* is a serine/threonine protein kinase that comes in two structurally and functionally distinct forms: *TOR* Complex 1 (*TORC1*) and *TORC2*. Mammalian *TORC1* (*mTORC1*)—not *TORC2*—plays a major role in aging. *mTORC1* comprises three components, the catalytic subunit mammalian *TOR* (*mTOR*), regulatory-associated protein of target of rapamycin (*RAPTOR*), and mammalian lethal SEC13 protein 8 (*mLST8*) [106,107]. Evidence denotes that inhibition of *mTORC1* with rapamycin extends lifespan up to the double in model organisms [4]. *mTORC1* signaling is regulated by endocrine signaling, especially growth factors and *IGF-1*, as well as nutrients e.g., amino acids, lipids, and glucose, in addition to cellular energy and oxygen levels [107,108]. In this respect, 10-HDA supplementation to *C. elegans* extended the lifespan of the control *unc-24/+* mutants but did not extend the lifespan of the long-lived heterozygous mutants in *daf-15*, which encode *RAPTOR* [66]. The anti-aging activity of the protein kinase *mTOR* originates from its ability to function both in a cell autonomous manner and a non-cell autonomous manner to regulate growth, protein translation, ribosomal biogenesis, autophagy, and cellular metabolism in response to both environmental and hormonal signals [106,108]. It is thought that *mTORC1* inhibition restores cellular physiological integrity, and hence delays age-related pathologies. In details, *mTORC1* promotes translation initiation of mRNAs of metabolism-related genes and ribosomal-related proteins via phosphorylation of two main ribosomal proteins: *S6K1* and *S6K2*. Thus, *mTORC1* inhibition is associated with enhanced endogenous protein degradation as well as less aggregation of proteotoxic and oxidative stress wastes, which in turn preserve homeostasis in the face of oxidative damage. Autophagy genes play a major role in these processes [4,109]. For a detailed description of mechanisms through which *mTOR* functions, we refer readers to these reviews [4,107,108].

5.3. Dietary Restriction Signaling

Dietary restriction is another mechanism that promotes longevity in various species. It involves prolonged reduction of intake of most dietary elements except vitamins and minerals (without getting into malnutrition)—it is equivalent to voluntary intermittent fasting in humans, which is reported to prevent numerous debilitating disorders such as abdominal obesity, diabetes, hypertension, and cardiovascular diseases [10]. Recent reviews point out that dietary restriction in humans exhibits similar effects to model organisms in terms of body composition, circulating lipoprotein, inflammatory and metabolite profiles, energy expenditure, and oxidative stress [110,111]. *C. elegans* *eat-2* mutant is considered a model of dietary restriction; its acetylcholine receptor mutation hinders pharyngeal pumping and limits intake of nutrients [109]. The main lipid of RJ, 10-HDA, which exhibited a lifespan-extending effect did not extend the lifespan of the *eat-2* mutants in *C. elegans*, which denotes that feeding impairment-related dietary restriction signaling was involved in the lifespan extending mechanism. However, progeny production was not delayed in 10-HDA-treated worms—unlike dietary-restricted worms—which indicated that the lifespan-extending effect of 10-HDA was related to the downstream process of the dietary restriction signaling [66]. Indeed, the longevity effect of *eat-2* mutant is mediated by compensatory changes of several energy sensing effectors: dietary restriction downregulates *IIS*, which activates *FOXO*, whereas the energy sensor for cellular AMP/ATP ratio known as *AMPK* gets activated to stimulate catabolic reactions for energy gain by phosphorylating *DAF-16/FOXO*. It also downregulates *AKT/mTOR* through *PHA-4/FOXA* transcription factor and *S6K* (in a *DAF-16/FOXO*-independent manner), and stimulates the expression of autophagy genes—*unc-51/ULK1*, *bec-1/Beclin1*, *vps-34*, *atg-18*, and *atg-7*—which inhibit general protein translation and simultaneously stimulate the translation of specific mRNAs involved in cellular homeostasis [5,10,109]. In addition, dietary restriction causes activation of *NRF2* transcription factor, which suppresses inflammation and counteracts oxidative damage [112].

5.4. Epidermal Growth Factor Signaling

RJ, royalactin in particular, extended lifespan of various species by activating the epidermal growth factor receptor (*EGFR*) signaling. As shown in Table 1, RJ activation of *EGFR* signaling involves upregulation of *S6K* and *MAPK*, which results in enhanced locomotor activity and antioxidant capacity; increased *20E* titre, which stimulates growth (increased body size); and increased juvenile titre, which increased fertility—all are effects that indicate healthspan. Royalactin might interact with *LIN-3* to promote the binding of the ligand to the extracellular domain of the *EGFR*, which stimulate *EGF* signaling [68]. Still, the exact mechanism through which RJ affects *EGFR* to promote longevity is not clear. However, it has been recently reported that, royalactin-related *EGFR* signaling induces longevity in *C. elegans* via upregulation of elongation factors and chaperonins, which increase protein translation and proteasome activity—a mechanism that entails rebuilding cellular components and enhancement of cellular detoxification, ribosomal function, and muscle maintenance—rather than stabilizing the existing proteome [113]. Furthermore, activation of *MAPK*—which is stimulated by *EGF*—increases the lifespan of *daf-2* mutants and thus controls the expression of pathogen response genes (*C-type lectins*, *ShK toxins*, and *CUB-like* genes); such increased resistance to pathogens increases lifespan in *C. elegans* [114]. *EGF* pathway functions in a manner that is independent of the *insulin/IGF-like* pathway i.e., activation of its receptor (*LET-23*) is necessary for its action [68]. In addition, it is not affected by deficiency of the *DAF-16/FOXO* transcription factor and it exerts more effects when *daf-2/InR* activity is inhibited [115]. Moreover, royalactin is thought to affect regulators of *EGF* signaling—such as high performance in advanced age genes (*HPA-1* and *HPA-2*)—to stimulate the release of *LIN-3* [68]. In fact, *HPA-1* and *HPA-2* genes negatively regulate *EGF* signaling by binding and sequestering *EGF*. While *HPA-1* is thought to contribute to longevity, *HPA-2* is reported to induce healthspan benefits in *C. elegans* by encoding secreted proteins similar in sequence to extracellular domains of insulin receptor. *EGF* signaling functions via downstream phospholipase C-γplc-3 and inositol-3-phosphate receptor itr-1 to promote healthy aging associated with low lipofuscin levels (age pigments that accumulate during senescence), enhance physical performance, and extend lifespan [115,116]. *EGFR*-mediated antiaging effects seem to be evolutionarily conserved from worms to humans. In humans, a recent study tested single-nucleotide polymorphisms (*SNPs*) in *EGFR* for association with longevity. Comparison of genotype frequencies of 41 *EGFR SNPs* between 440 American males of Japanese ancestry aged ≥95 years and 374 men of average lifespan (whites and Koreans) revealed a significant association with longevity for seven *SNPs* in *EGFR*—evidence that genetic variation in *EGFR* contributes to lifespan extension in Japanese people [101].

5.5. Oxidative Stress

According to the free radical theory of aging, aging is the result of accumulation of molecular damages that are induced by reactions of free radicals and reactive species that inevitably form during the course of metabolism, which cause errors in cellular processes that are conducive to various age-related disorders [117]. This notion is supported by findings from bee queen studies.

Despite the fact that bee queens live longer compared with worker bees, queens enjoy youthful and vigorous cellular functioning. This is attributed to the peroxidation-resistant cell membrane and the lower expression of oxidative stress genes [118,119]. The latter has been attributed, at least in part, to the effect of queen food consumption on microbiota composition and microbiota-derived metabolites, whereas gut microbiota of the short-lived workers—that feast on honey or pollens—is aging i.e., deficient in bacteria that produce metabolites that prevent the expression of oxidative stress genes such as lactobacillus and bifidobacterium [119]. Hence, another possible contributor to the longevity effect of RJ is its enhancement of antioxidation capacity and resistance to oxidative stress, which foster scavenging of free radicals as well as phosphorylation and retention of *DAF-16/FOXO* in the cytoplasm, through the involvement of 3-phosphoinositide-dependent kinase-1. As noted above, RJ supplementation to *C. elegans* increased their resistance to various types of stress [66,67]. Similarly, RJ supplementation to *Drosophila* at low concentrations regulated the gene expression related to oxidative stress and catabolism [16]. In particular, MRJPs fostered the gene expression of *CuZn-SOD*, which was reflected by increased activity of the antioxidant SOD and decreased levels of MDA [12]. Similarly, RJ supplementation to mice decreased kidney DNA and serum levels of 8-hydroxy-2-deoxyguanosine, an age-related marker of oxidative stress [2].

6. Discussion

Aging is characterized by progressive functional decline and increased vulnerability to various pathologies, which result from long-term alterations of numerous physiological processes [4]. Therefore, apart from prolonging life, it is necessary to find intelligent anti-aging agents that target age-related genetic pathways and biochemical processes in order to prevent or delay the detrimental effect of age-related physiological changes. Despite the slight discrepancy across studies examined in this review—possibly out of differences in model organisms and experimental designs—the findings indicate that consumption of an appropriate dose of RJ and its ingredients exerts proliferative effects that promote health, increase stress resistance, and prolong lifespan in numerous diverse species. RJ seems to have a potent effect on the functioning of various healthspan and longevity pathways: *IIS, mTOR, EGFR*, oxidative stress, and dietary restriction eat-2. It might be necessary to further examine the effect of RJ on mechanisms related to microbiota since microbiota composition can interact with pathways of oxidative stress to exert longevity effects [119]. It is also necessary to examine the possibility that such mechanisms may be applicable to humans.

Though several animal models demonstrate extended lifespan as a result of RJ treatment, it is not clear which constituents of RJ are responsible for the longevity effect. MRJPs and royalactin appear to be probable candidates [41,68]. Yet, longevity effects were obtained in many studies that did not include MRJPs, which signifies that other ingredients also enhance health and prolong lifespan. It seems that pRJ [51,67], as well as certain RJ constituents such as RJP_{60} [15]; lipids e.g., 10-HDA [66]; and vitamins e.g., pantothenic acid [54] exhibit healthspan and lifespan-extending activities greater than crude RJ.

For RJ to exert an anti-aging effect, it should be regularly used for long periods of time. Therefore, safety of prolonged use of RJ represents another issue of concern. Bee products such as RJ, honey, pollen, and propolis have long been used as multifunctional substances with various biological activities. Despite the limited possibility of occurrence of allergic reactions, most bee products are relatively nontoxic e.g., high doses of propolis (1400 mg/kg/day) in mice had no effect level [120], whereas oral consumption of RJ in high doses (20 g) in humans produced no adverse effects [86]. Nonetheless, most of the observed positive anti-aging effects of RJ in model organisms were dose-dependent—extremely high doses were associated with unfavorable outcomes in some studies. Yet, with the exception of one study that supplemented bee larvae with RJ 100% [60], the definition of "a high dose" varied between studies and outcomes also varied: no effect [52], increased fertility [60] and survival [64], increased developmental time, and decreased survival [16]. Apart from variation of the studied organisms, as well as nutrient contents of culture media used for in vitro breeding [16,25,41] (which may, in part, explain the discrepancy noticed between these studies), very few attempts were made to explore mechanisms underlying the occurrence of adverse effects when high doses of RJ are used. In this respect, unphysiological exposure of *Drosophila* to high levels of RJ (up to 70%) prolonged development time and shortened lifespan because excess nutrients in high concentrations of RJ altered the expression of genes responsible for the metabolism of amino acids and encoding of glutathione S transferases, which detoxify xenobiotic compounds resulting in distortion of cellular processes [16]. Furthermore, evidence documents that uncontrollably high intake of antioxidants (e.g., vitamin C, vitamin E, N-acetyl cysteine) disturbs the redox balance between processes of oxidation and reduction and induces reductive stress—a shift of body redox levels into an extra reduced state, which may cause severe alterations of cellular functions and lead to pathologies in the same way as oxidative stress [121]. Therefore, adverse effects associated with high doses of RJ may be the result of reductive stress induced by antioxidants in RJ (e.g., phenols, amino acids, peptides, fatty acids, and vitamins). Accordingly, from a cost-effectiveness-oriented point of view, determining an optimal RJ dose would be an important issue in future studies. On the other hand, accumulation of neonicotinoid insecticides and their metabolites in the body and products of honey bees as a result of environmental pollution is associated with occurrence of adverse effects in bees such as oral toxicity, especially during winter—when bee consumption of honey and pollen increases [122]. Nonetheless, several reports indicate that compared with pollen and honey [123,124], RJ contamination with neonicotinoids is very limited (1 to 9.5 µg/kg) [125,126], equivalent to 0.016% of the original concentrations of pesticide fed to bee workers during in vitro breeding [125]. Therefore, RJ maybe considered a safe an anti-aging agent compared with other bee products such as bee pollen.

Preserving the biological activity of RJ is a prerequisite for its use to efficiently promote health. RJ seems to be sensitive to temperature and other methods of handling, which may affect its ingredients and potency e.g., RJ lost most of its bioactive components after 30 days of storage at 40 °C [41] and pRJ had a better effect than crude RJ [52]. Further, the effect of commercial RJ was suboptimal compared with RJ from reliable sources [16]. Another concern is the effect of digestive enzymes on bioactive ingredients of RJ when it is orally ingested. For instance, MRJPs prevent senescence of human cells in vitro [39]; however, MRJPs are rapidly digested in the stomach and small intestine except for MRJP2, which can remain in the intestine as a full-length protein for 40 min and it should be resorbed quickly were it to produce any biological effect [127].

Genetic variation is another challenge if we are to identify candidates for pathology prevention in humans. RJ as a dietary supplement may prevent some of the main age-related diseases. However, various types of genetic variation may affect response to RJ treatment. Taking gender as an example, RJ increased the body size of females only in some species such as silkworms [65]—an effect of increased levels of fertility hormones [17,41]. In humans, women live longer than men, yet they have higher genetic risk for some age-related diseases (e.g., Alzheimer's disease) than men, which indicates the possibility for gender-specific treatment targets [128]. Other types of genetic variation are also important; for instance, RJ affects the *EGFR* pathway, which has been shown to be associated with prolonged lifespan in Japanese—but not in whites or Koreans—which highlights a role of ethnic difference in genotype and epigenotype [101]. Other factors, such as other genetic variation factors, diet, and activity level (which affects healthspan and longevity) should be considered in future clinical trials.

7. Conclusions

Accumulating evidence from studies of honey bees, fruit flies, crickets, silkworms, mice, and humans indicates that RJ has an obvious role in modulating the mechanisms of aging, which can promote healthspan and longevity. Although most studies relate the anti-aging properties of RJ to MRJPs and 10-HDA, it is not exactly clear which fractions or doses are most beneficial. The discovery of predictive biomarkers that take into account individual variation in genotype and epigenotype is necessary in order to conduct sound clinical trials that can test the efficacy of RJ on the rate of biological aging as well as the risk of age-related diseases.

Funding: This study was supported by the Strategic Research Program for Brain Sciences from Japan Agency for Medical Research and Development, AMED, Japan (Grant No. 18dm0107100h0003).

Conflicts of Interest: The authors declare no conflict of interest.

Abbreviations

4E-BP1	Binding protein
10-HDA	10-hydroxydecanoic acid
10H2DA	10-hydroxy-2-decenoic acid
Ach	Acetylcholine
ATF4	Activating transcription factor-4
ADP	Adenosine diphosphate
AMP	Adenosine monophosphate
ATP	Adenosine triphosphate
EGFR	Epidermal growth factor receptor
eIF4E	Eukaryotic translation initiation factor
eNOs	Endothelial nitric oxide synthase
ERK	Extracellular signal-regulated kinase
ERs	Estrogen receptors
FDRJ	Freeze-dried RJ
FOXO	Forkhead box O
HCF-1	Host cell factor-1
HPO-DAEE	4-Hydroperoxy-2-decenoic acid ethyl ester
IGFs	Insulin-like growth factors
IIS	*Insulin/IGF-1* signaling
InR	Insulin receptor
MAPK	Mitogen-activated protein kinase
MDA	Malonaldehayde
MMPs	Matrix metalloproteinases
MRJPs	Major royal jelly proteins
mTOR	Mechanistic target of rapamycin
NRF2	Nuclear factor erythroid 2
pRJ	Protease-treated RJ
pRJ-Fr.5	pRJ-Fraction 5
RA	Rheumatoid arthritis
RAPTOR	Regulatory-associated protein of target of rapamycin
RJ	Royal jelly
ROS	Reactive oxygen species
S6K	Ribosomal proteins S6 kinase
SA	Sebacic acid
SIR-2.1	Sirtuin homologue-2.1
SNPs	Single-nucleotide polymorphisms
SOD	Superoxide dismutase
TNF	Tumor necrosis factor
TRP	Tyrosinase-related protein
w/w	Weight/weight
WJ	Worker jelly

References

1. Cao, B.; Bray, F.; Ilbawi, A.; Soerjomataram, I. Effect on longevity of one-third reduction in premature mortality from non-communicable diseases by 2030: A global analysis of the Sustainable Development Goal health target. *Lancet Glob. Health* **2018**, *6*, E1288–E1296. [CrossRef]
2. Inoue, S.-i.; Koya-Miyata, S.; Ushio, S.; Iwaki, K.; Ikeda, M.; Kurimoto, M. Royal Jelly prolongs the life span of C3H/HeJ mice: Correlation with reduced DNA damage. *Exp. Gerontol.* **2003**, *38*, 965–969. [CrossRef]

3. Hassanzadeh, K.; Rahimmi, A. Oxidative stress and neuroinflammation in the story of Parkinson's disease: Could targeting these pathways write a good ending? *J. Cell Physiol.* **2019**, *234*, 23–32. [CrossRef]

4. Weichhart, T. *mTOR* as Regulator of Lifespan, Aging, and Cellular Senescence: A Mini-Review. *Gerontology* **2018**, *64*, 127–134. [CrossRef] [PubMed]

5. Martins, R.; Lithgow, G.J.; Link, W. Long live *FOXO*: Unraveling the role of *FOXO* proteins in aging and longevity. *Aging Cell* **2016**, *15*, 196–207. [CrossRef]

6. Ali, A.M.; Kunugi, H. Bee honey protects astrocytes against oxidative stress: A preliminary in vitro investigation. *Neuropsychopharmacol. Rep.* **2019**, 1–3. [CrossRef] [PubMed]

7. Niraula, A.; Sheridan, J.F.; Godbout, J.P. Microglia Priming with Aging and Stress. *Neuropsychopharmacology* **2017**, *42*, 318–333. [CrossRef] [PubMed]

8. Sebastiani, P.; Bae, H.; Sun, F.X.; Andersen, S.L.; Daw, E.W.; Malovini, A.; Kojima, T.; Hirose, N.; Schupf, N.; Puca, A.; et al. Meta-analysis of genetic variants associated with human exceptional longevity. *Aging (Albany Ny)* **2013**, *5*, 653–661. [CrossRef] [PubMed]

9. Sebastiani, P.; Solovieff, N.; Dewan, A.T.; Walsh, K.M.; Puca, A.; Hartley, S.W.; Melista, E.; Andersen, S.; Dworkis, D.A.; Wilk, J.B.; et al. Genetic signatures of exceptional longevity in humans. *PLoS ONE* **2012**, *7*, e29848. [CrossRef] [PubMed]

10. Fontana, L.; Partridge, L. Promoting Health and Longevity through Diet: From Model Organisms to Humans. *Cell* **2015**, *161*, 106–118. [CrossRef]

11. Hook, M.; Roy, S.; Williams, E.G.; Bou Sleiman, M.; Mozhui, K.; Nelson, J.F.; Lu, L.; Auwerx, J.; Williams, R.W. Genetic cartography of longevity in humans and mice: Current landscape and horizons. *Biochim. Biophys. Acta Mol. Basis Dis.* **2018**, *1864*, 2718–2732. [CrossRef] [PubMed]

12. Xin, X.-x.; Chen, Y.; Chen, D.; Xiao, F.; Parnell, L.D.; Zhao, J.; Liu, L.; Ordovas, J.M.; Lai, C.-Q.; Shen, L.-r. Supplementation with Major Royal-Jelly Proteins Increases Lifespan, Feeding, and Fecundity in Drosophila. *J. Agric. Food Chem.* **2016**, *64*, 5803–5812. [CrossRef] [PubMed]

13. Maleki, V.; Jafari-Vayghan, H.; Saleh-Ghadimi, S.; Adibian, M.; Kheirouri, S.; Alizadeh, M. Effects of Royal jelly on metabolic variables in diabetes mellitus: A systematic review. *Complement. Ther. Med.* **2019**, *43*, 20–27. [CrossRef] [PubMed]

14. Pasupuleti, V.R.; Sammugam, L.; Ramesh, N.; Gan, S.H. Honey, Propolis, and Royal Jelly: A Comprehensive Review of Their Biological Actions and Health Benefits. *Oxid. Med. Cell Longev.* **2017**, *2017*, 1259510. [CrossRef] [PubMed]

15. Yang, W.; Tian, Y.; Han, M.; Miao, X. Longevity extension of worker honey bees (*Apis mellifera*) by royal jelly: Optimal dose and active ingredient. *PeerJ* **2017**, *5*, e3118. [CrossRef] [PubMed]

16. Shorter, J.R.; Geisz, M.; Özsoy, E.; Magwire, M.M.; Carbone, M.A.; Mackay, T.F.C. The Effects of Royal Jelly on Fitness Traits and Gene Expression in *Drosophila melanogaster*. *PLoS ONE* **2015**, *10*, e0134612. [CrossRef] [PubMed]

17. Kayashima, Y.; Yamanashi, K.; Sato, A.; Kumazawa, S.; Yamakawa-Kobayashi, K. Freeze-dried royal jelly maintains its developmental and physiological bioactivity in *Drosophila melanogaster*. *Biosci. Biotechnol. Biochem.* **2012**, *76*, 2107–2111. [CrossRef] [PubMed]

18. Kocot, J.; Kielczykowska, M.; Luchowska-Kocot, D.; Kurzepa, J.; Musik, I. Antioxidant Potential of Propolis, Bee Pollen, and Royal Jelly: Possible Medical Application. *Oxid. Med. Cell Longev.* **2018**, *2018*, 7074209. [CrossRef] [PubMed]

19. Wang, X.; Cao, M.; Dong, Y. Royal jelly promotes *DAF-16*-mediated proteostasis to tolerate β-amyloid toxicity in C. elegans model of Alzheimer's disease. *Oncotarget* **2016**, *7*, 54183–54193. [CrossRef] [PubMed]

20. You, M.; Pan, Y.; Liu, Y.; Chen, Y.; Wu, Y.; Si, J.; Wang, K.; Hu, F. Royal Jelly Alleviates Cognitive Deficits and β-Amyloid Accumulation in APP/PS1 Mouse Model Via Activation of the *cAMP/PKA/CREB/BDNF* Pathway and Inhibition of Neuronal Apoptosis. *Front. Aging Neurosci.* **2019**, *10*, 428. [CrossRef] [PubMed]

21. Ramanathan, A.N.K.G.; Nair, A.J.; Sugunan, V.S. A review on Royal Jelly proteins and peptides. *J. Funct. Foods* **2018**, *44*, 255–264. [CrossRef]

22. Xue, X.; Wu, L.; Wang, K. Chemical Composition of Royal Jelly. In *Bee Products—Chemical and Biological Properties*; Alvarez-Suarez, J.M., Ed.; Springer International Publishing: Cham, Switzerland, 2017; pp. 181–190.

23. Wytrychowski, M.; Chenavas, S.; Daniele, G.; Casabianca, H.; Batteau, M.; Guibert, S.; Brion, B. Physicochemical characterisation of French royal jelly: Comparison with commercial royal jellies and royal jellies produced through artificial bee-feeding. *J. Food Compos. Anal.* **2013**, *29*, 126–133. [CrossRef]

24. Asencot, M.; Lensky, Y. The effect of sugars and Juvenile Hormone on the differentiation of the female honeybee larvae (*Apis mellifera L.*) to queens. *Life Sci.* **1976**, *18*, 693–699. [CrossRef]

25. Buttstedt, A.; Ihling, C.H.; Pietzsch, M.; Moritz, R.F.A. Royalactin is not a royal making of a queen. *Nature* **2016**, *537*, E10. [CrossRef] [PubMed]

26. Polsinelli, G.A.; Yu, H.D. Regulation of histone deacetylase 3 by metal cations and 10-hydroxy-2E-decenoic acid: Possible epigenetic mechanisms of queen-worker bee differentiation. *PLoS ONE* **2018**, *13*, e0204538. [CrossRef] [PubMed]

27. Sediva, M.; Laho, M.; Kohutova, L.; Mojzisova, A.; Majtan, J.; Klaudiny, J. 10-HDA, A Major Fatty Acid of Royal Jelly, Exhibits pH Dependent Growth-Inhibitory Activity Against Different Strains of *Paenibacillus larvae*. *Molecules* **2018**, *23*, 3236. [CrossRef] [PubMed]

28. Chen, Y.-F.; You, M.-M.; Liu, Y.-C.; Shi, Y.-Z.; Wang, K.; Lu, Y.-Y.; Hu, F.-L. Potential protective effect of Trans-10-hydroxy-2-decenoic acid on the inflammation induced by Lipoteichoic acid. *J. Funct. Foods* **2018**, *45*, 491–498. [CrossRef]

29. Yang, Y.C.; Chou, W.M.; Widowati, D.A.; Lin, I.P.; Peng, C.C. 10-hydroxy-2-decenoic acid of royal jelly exhibits bactericide and anti-inflammatory activity in human colon cancer cells. *BMC Complement. Altern. Med.* **2018**, *18*, 202. [CrossRef]

30. Hattori, N.; Nomoto, H.; Fukumitsu, H.; Mishima, S.; Furukawa, S. Royal jelly-induced neurite outgrowth from rat pheochromocytoma PC12 cells requires integrin signal independent of activation of extracellular signal regulated kinases. *Biomed. Res.* **2007**, *28*, 139–146. [CrossRef]

31. Weiser, M.J.; Grimshaw, V.; Wynalda, K.M.; Mohajeri, M.H.; Butt, C.M. Long-Term Administration of Queen Bee Acid (QBA) to Rodents Reduces Anxiety-Like Behavior, Promotes Neuronal Health and Improves Body Composition. *Nutrients* **2017**, *10*, 13. [CrossRef]

32. Peng, C.C.; Sun, H.T.; Lin, I.P.; Kuo, P.C.; Li, J.C. The functional property of royal jelly 10-hydroxy-2-decenoic acid as a melanogenesis inhibitor. *BMC Complement. Altern. Med.* **2017**, *17*, 392. [CrossRef] [PubMed]

33. Zheng, J.; Lai, W.; Zhu, G.; Wan, M.; Chen, J.; Tai, Y.; Lu, C. 10-Hydroxy-2-decenoic acid prevents ultraviolet A-induced damage and matrix metalloproteinases expression in human dermal fibroblasts. *J. Eur. Acad. Derm. Venereol.* **2013**, *27*, 1269–1277. [CrossRef] [PubMed]

34. Wang, J.G.; Ruan, J.; Li, C.Y.; Wang, J.M.; Li, Y.; Zhai, W.T.; Zhang, W.; Ye, H.; Shen, N.H.; Lei, K.F.; et al. Connective tissue growth factor, a regulator related with 10-hydroxy-2-decenoic acid down-regulate *MMPs* in rheumatoid arthritis. *Rheumatol. Int.* **2012**, *32*, 2791–2799. [CrossRef] [PubMed]

35. Chen, Y.F.; Wang, K.; Zhang, Y.Z.; Zheng, Y.F.; Hu, F.L. *In Vitro* Anti-Inflammatory Effects of Three Fatty Acids from Royal Jelly. *Mediat. Inflamm.* **2016**, *2016*, 3583684. [CrossRef] [PubMed]

36. Moutsatsou, P.; Papoutsi, Z.; Kassi, E.; Heldring, N.; Zhao, C.; Tsiapara, A.; Melliou, E.; Chrousos, G.P.; Chinou, I.; Karshikoff, A.; et al. Fatty acids derived from royal jelly are modulators of estrogen receptor functions. *PLoS ONE* **2010**, *5*, e15594. [CrossRef] [PubMed]

37. Inoue, Y.; Hara, H.; Mitsugi, Y.; Yamaguchi, E.; Kamiya, T.; Itoh, A.; Adachi, T. 4-Hydroperoxy-2-decenoic acid ethyl ester protects against 6-hydroxydopamine-induced cell death via activation of *Nrf2-ARE* and *eIF2 alpha-ATF4* pathways. *Neurochem. Int.* **2018**, *112*, 288–296. [CrossRef] [PubMed]

38. Kamiya, T.; Watanabe, M.; Hara, H.; Mitsugi, Y.; Yamaguchi, E.; Itoh, A.; Adachi, T. Induction of Human-Lung-Cancer-A549-Cell Apoptosis by 4-Hydroperoxy-2-decenoic Acid Ethyl Ester through Intracellular ROS Accumulation and the Induction of Proapoptotic *CHOP* Expression. *J. Agric. Food Chem.* **2018**, *66*, 10741–10747. [CrossRef]

39. Jiang, C.M.; Liu, X.; Li, C.X.; Qian, H.C.; Chen, D.; Lai, C.Q.; Shen, L.R. Anti-senescence effect and molecular mechanism of the major royal jelly proteins on human embryonic lung fibroblast (HFL-I) cell line. *J. Zhejiang Univ. Sci. B* **2018**, *19*, 960–972. [CrossRef]

40. Moriyama, T.; Ito, A.; Omote, S.; Miura, Y.; Tsumoto, H. Heat Resistant Characteristics of Major Royal Jelly Protein 1 (MRJP1) Oligomer. *PLoS ONE* **2015**, *10*, e0119169. [CrossRef]

41. Kamakura, M. Royalactin induces queen differentiation in honeybees. *Nature* **2011**, *473*, 478–483. [CrossRef]

42. Wan, D.C.; Morgan, S.L.; Spencley, A.L.; Mariano, N.; Chang, E.Y.; Shankar, G.; Luo, Y.; Li, T.H.; Huh, D.; Huynh, S.K.; et al. Honey bee Royalactin unlocks conserved pluripotency pathway in mammals. *Nat. Commun.* **2018**, *9*, 5078. [CrossRef] [PubMed]

43. Bilal, B.; Azim, M.K. Nematicidal activity of 'major royal jelly protein'-containing glycoproteins from Acacia honey. *Exp. Parasitol.* **2018**, *192*, 52–59. [CrossRef] [PubMed]

44. Abu-Serie, M.M.; Habashy, N.H. Two purified proteins from royal jelly with *in vitro* dual anti-hepatic damage potency: Major royal jelly protein 2 and its novel isoform X1. *Int. J. Biol. Macromol.* **2019**, *128*, 782–795. [CrossRef] [PubMed]

45. Fratini, F.; Cilia, G.; Mancini, S.; Felicioli, A. Royal Jelly: An ancient remedy with remarkable antibacterial properties. *Microbiol. Res.* **2016**, *192*, 130–141. [CrossRef] [PubMed]

46. Hykollari, A.; Malzl, D.; Eckmair, B.; Vanbeselaere, J.; Scheidl, P.; Jin, C.; Karlsson, N.G.; Wilson, I.B.H.; Paschinger, K. Isomeric Separation and Recognition of Anionic and Zwitterionic N-glycans from Royal Jelly Glycoproteins. *Mol. Cell Proteom.* **2018**, *17*, 2177–2196. [CrossRef] [PubMed]

47. Liu, J.-R.; Yang, Y.-C.; Shi, L.-S.; Peng, C.-C. Antioxidant Properties of Royal Jelly Associated with Larval Age and Time of Harvest. *J. Agric. Food Chem.* **2008**, *56*, 11447–11452. [CrossRef]

48. Guo, H.; Kouzuma, Y.; Yonekura, M. Structures and properties of antioxidative peptides derived from royal jelly protein. *Food Chem* **2009**, *113*, 238–245. [CrossRef]

49. Jie, H.; Li, P.M.; Zhao, G.J.; Feng, X.L.; Zeng, D.J.; Zhang, C.L.; Lei, M.Y.; Yu, M.; Chen, Q. Amino acid composition of royal jelly harvested at different times after larval transfer. *Genet. Mol. Res.* **2016**, *15*. [CrossRef]

50. Pina, A.; Begou, O.; Kanelis, D.; Gika, H.; Kalogiannis, S.; Tananaki, C.; Theodoridis, G.; Zotou, A. Targeted profiling of hydrophilic constituents of royal jelly by hydrophilic interaction liquid chromatography-tandem mass spectrometry. *J. Chromatogr. A* **2018**, *1531*, 53–63. [CrossRef]

51. Gu, H.; Song, I.-B.; Han, H.-J.; Lee, N.-Y.; Cha, J.-Y.; Son, Y.-K.; Kwon, J. Antioxidant Activity of Royal Jelly Hydrolysates Obtained by Enzymatic Treatment. *Korean J. Food Sci. Anim. Resour.* **2018**, *38*, 135–142. [CrossRef]

52. Honda, Y.; Fujita, Y.; Maruyama, H.; Araki, Y.; Ichihara, K.; Sato, A.; Kojima, T.; Tanaka, M.; Nozawa, Y.; Ito, M.; et al. Lifespan-extending effects of royal jelly and its related substances on the nematode Caenorhabditis elegans. *PLoS ONE* **2011**, *6*, e23527. [CrossRef] [PubMed]

53. D'Antona, G.; Ragni, M.; Cardile, A.; Tedesco, L.; Dossena, M.; Bruttini, F.; Caliaro, F.; Corsetti, G.; Bottinelli, R.; Carruba, M.O.; et al. Branched-chain amino acid supplementation promotes survival and supports cardiac and skeletal muscle mitochondrial biogenesis in middle-aged mice. *Cell Metab.* **2010**, *12*, 362–372. [CrossRef] [PubMed]

54. Gardner, T.S. The use of Drosophila Melanogaster as a Screening Agent for Longevity FactorsII. The Effects of Biotin, Pyridoxine, Sodium Yeast Nucleate, and Pantothenic Acid on the Life Span of the Fruit Fly. *J. Gerontol.* **1948**, *3*, 9–13. [CrossRef] [PubMed]

55. Stocker, A.; Schramel, P.; Kettrup, A.; Bengsch, E. Trace and mineral elements in royal jelly and homeostatic effects. *J. Trace Elem. Med. Biol.* **2005**, *19*, 183–189. [CrossRef] [PubMed]

56. Wessler, I.; Gartner, H.A.; Michel-Schmidt, R.; Brochhausen, C.; Schmitz, L.; Anspach, L.; Grunewald, B.; Kirkpatrick, C.J. Honeybees Produce Millimolar Concentrations of Non-Neuronal Acetylcholine for Breeding: Possible Adverse Effects of Neonicotinoids. *PLoS ONE* **2016**, *11*, e0156886. [CrossRef] [PubMed]

57. Zamani, Z.; Reisi, P.; Alaei, H.; Asghar Pilehvarian, A. Effect of Royal Jelly on spatial learning and memory in rat model of streptozotocin-induced sporadic Alzheimer's disease. *Adv. Biomed. Res.* **2012**, *1*, 1–10. [CrossRef]

58. Wu, L.; Chen, L.; Selvaraj, J.N.; Wei, Y.; Wang, Y.; Li, Y.; Zhao, J.; Xue, X. Identification of the distribution of adenosine phosphates, nucleosides and nucleobases in royal jelly. *Food Chem.* **2015**, *173*, 1111–1118. [CrossRef] [PubMed]

59. Hattori, N.; Nomoto, H.; Fukumitsu, H.; Mishima, S.; Furukawa, S. AMP N1-Oxide, a Unique Compound of Royal Jelly, Induces Neurite Outgrowth from PC12 Vells via Signaling by Protein Kinase A Independent of that by Mitogen-Activated Protein Kinase. *Evid. Based Complement. Altern. Med.* **2010**, *7*, 63–68. [CrossRef] [PubMed]

60. Lin, H.; Winston, M.L. The role of nutrition and temperature in the ovarian development of the worker honey bee (*Apis mellifera*). *Can. Entomol.* **1998**, *130*, 883–891. [CrossRef]

61. Matsuyama, S.; Nagao, T.; Sasaki, K. Consumption of tyrosine in royal jelly increases brain levels of dopamine and tyramine and promotes transition from normal to reproductive workers in queenless honey bee colonies. *Gen. Comp. Endocrinol.* **2015**, *211*, 1–8. [CrossRef] [PubMed]

62. Shi, J.-l.; Liao, C.-h.; Wang, Z.-l.; Wu, X.-b. Effect of royal jelly on longevity and memory-related traits of *Apis mellifera* workers. *J. Asia Pac. Entomol.* **2018**, *21*, 1430–1433. [CrossRef]

63. Altaye, S.Z.; Pirk, C.W.; Crewe, R.M.; Nicolson, S.W. Convergence of carbohydrate-biased intake targets in caged worker honeybees fed different protein sources. *J. Exp. Biol.* **2010**, *213*, 3311–3318. [CrossRef] [PubMed]

64. Chen, C.-y.; Chen, X.-y.; Lin, C.-y.; Sun, L.-x. The Effects of Royal jelly on the Life Span and Antioxidative Enzyme Activity of Fruit fly. *J. Bee* **2009**, *29*, 3–5.

65. Miyashita, A.; Kizaki, H.; Sekimizu, K.; Kaito, C. Body-enlarging effect of royal jelly in a non-holometabolous insect species, *Gryllus bimaculatus*. *Biol. Open* **2016**, *5*, 770–776. [CrossRef] [PubMed]

66. Honda, Y.; Araki, Y.; Hata, T.; Ichihara, K.; Ito, M.; Tanaka, M.; Honda, S. 10-Hydroxy-2-decenoic Acid, the Major Lipid Component of Royal Jelly, Extends the Lifespan of *Caenorhabditis elegans* through Dietary Restriction and Target of Rapamycin Signaling. *J. Aging Res.* **2015**, *2015*, 425261. [CrossRef] [PubMed]

67. Wang, X.; Cook, L.F.; Grasso, L.M.; Cao, M.; Dong, Y. Royal Jelly-Mediated Prolongevity and Stress Resistance in *Caenorhabditis elegans* Is Possibly Modulated by the Interplays of *DAF-16, SIR-2.1, HCF-1*, and *14-3-3* Proteins. *J. Gerontol A Biol. Sci. Med. Sci.* **2015**, *70*, 827–838. [CrossRef] [PubMed]

68. Detienne, G.; De Haes, W.; Ernst, U.R.; Schoofs, L.; Temmerman, L. Royalactin extends lifespan of *Caenorhabditis elegans* through epidermal growth factor signaling. *Exp. Gerontol.* **2014**, *60*, 129–135. [CrossRef] [PubMed]

69. Ji, W.Z.; Zhang, C.P.; Wei, W.T.; Hu, F.L. The in vivo antiaging effect of enzymatic hydrolysate from royal jelly in d-galactose induced aging mouse. *J. Chin. Inst. Food Sci. Technol.* **2016**, *16*, 18–25.

70. Okumura, N.; Toda, T.; Ozawa, Y.; Watanabe, K.; Ikuta, T.; Tatefuji, T.; Hashimoto, K.; Shimizu, T. Royal Jelly Delays Motor Functional Impairment During Aging in Genetically Heterogeneous Male Mice. *Nutrients* **2018**, *10*, 1191. [CrossRef]

71. Alaraj, M.d. Royal Jelly Pretreatment Can Either Protect or Aggravate Brain Damage Induced by Hypoxia-Ischemia in Mice, Depending on its Dose. *Int. J. Sci. Basic Appl. Res.* **2015**, *19*, 338–346.

72. Park, H.M.; Hwang, E.; Lee, K.G.; Han, S.M.; Cho, Y.; Kim, S.Y. Royal jelly protects against ultraviolet B-induced photoaging in human skin fibroblasts via enhancing collagen production. *J. Med. Food* **2011**, *14*, 899–906. [CrossRef] [PubMed]

73. Cheng, S.; Larson, M.G.; McCabe, E.L.; Murabito, J.M.; Rhee, E.P.; Ho, J.E.; Jacques, P.F.; Ghorbani, A.; Magnusson, M.; Souza, A.L.; et al. Distinct metabolomic signatures are associated with longevity in humans. *Nat. Commun.* **2015**, *6*, 6791. [CrossRef] [PubMed]

74. Morita, H.; Ikeda, T.; Kajita, K.; Fujioka, K.; Mori, I.; Okada, H.; Uno, Y.; Ishizuka, T. Effect of royal jelly ingestion for six months on healthy volunteers. *Nutr. J.* **2012**, *11*, 77. [CrossRef] [PubMed]

75. Dalal, D.S.; Duran, J.; Brar, T.; Alqadi, R.; Halladay, C.; Lakhani, A.; Rudolph, J.L. Efficacy and safety of biological agents in the older rheumatoid arthritis patients compared to Young: A systematic review and meta-analysis. *Semin. Arthritis Rheum.* **2019**, *48*, 799–807. [CrossRef] [PubMed]

76. Yang, X.Y.; Yang, D.S.; Wei, Z.; Wang, J.M.; Li, C.Y.; Hui, Y.; Lei, K.F.; Chen, X.F.; Shen, N.H.; Jin, L.Q.; et al. 10-Hydroxy-2-decenoic acid from Royal jelly: A potential medicine for RA. *J. Ethnopharmacol.* **2010**, *128*, 314–321. [CrossRef]

77. Wang, J.; Zhang, W.; Zou, H.; Lin, Y.; Lin, K.; Zhou, Z.; Qiang, J.; Lin, J.; Chuka, C.M.; Ge, R.; et al. 10-Hydroxy-2-decenoic acid inhibiting the proliferation of fibroblast-like synoviocytes by *PI3K-AKT* pathway. *Int. Immunopharmacol.* **2015**, *28*, 97–104. [CrossRef]

78. Vitale, G.; Pellegrino, G.; Vollery, M.; Hofland, L.J. Role of *IGF-1* system in the modulation of longevity: Controversies and new insights from a centenarians' perspective. *Front. Endocrinol.* **2019**, *10*, 27. [CrossRef]

79. Barbieri, M.; Bonafè, M.; Franceschi, C.; Paolisso, G. *Insulin/IGF-I*-signaling pathway: An evolutionarily conserved mechanism of longevity from yeast to humans. *Am. J. Physiol. Endocrinol. Metab.* **2003**, *285*, E1064–E1071. [CrossRef]

80. Vella, V.; Malaguarnera, R. The Emerging Role of Insulin Receptor Isoforms in Thyroid Cancer: Clinical Implications and New Perspectives. *Int. J. Mol. Sci.* **2018**, *19*, 3814. [CrossRef]

81. Wardelmann, K.; Blümel, S.; Rath, M.; Alfine, E.; Chudoba, C.; Schell, M.; Cai, W.; Hauffe, R.; Warnke, K.; Flore, T.; et al. Insulin action in the brain regulates mitochondrial stress responses and reduces diet-induced weight gain. *Mol. Metab.* **2019**, *21*, 68–81. [CrossRef]

82. Bartke, A. Growth Hormone and Aging: Updated Review. *World J. Mens. Health* **2019**, *37*, 19–30. [CrossRef] [PubMed]

83. De Medina, P. Deciphering the metabolic secret of longevity through the analysis of metabolic response to stress on long-lived species. *Med. Hypotheses* **2019**, *122*, 62–67. [CrossRef] [PubMed]

84. Kramer, K.J.; Childs, C.a.N.; Spiers, R.D.; Jacobs, R.M. Purification of insulin-like peptides from insect haemolymph and royal jelly. *Insect Btochem.* **1982**, *12*, 91–98. [CrossRef]

85. Büyükipekçi, S.; Sarıtaş, N.; Soylu, M.; Mıstık, S.; Silici, S. Effects of royal jelly and honey mixture on some hormones in young males performing maximal strength workout. *Phys. Educ. Stud.* **2018**, *22*, 308–315. [CrossRef]

86. Münstedt, K.; Bargello, M.; Hauenschild, A. Royal Jelly Reduces the Serum Glucose Levels in Healthy Subjects. *J. Med. Food* **2009**, *12*, 1170–1172. [CrossRef] [PubMed]

87. Suzuki, K.M.; Isohama, Y.; Maruyama, H.; Yamada, Y.; Narita, Y.; Ohta, S.; Araki, Y.; Miyata, T.; Mishima, S. Estrogenic activities of Fatty acids and a sterol isolated from royal jelly. *Evid. Based Complement. Altern. Med.* **2008**, *5*, 295–302. [CrossRef] [PubMed]

88. Van Raamsdonk, J.M. Mechanisms underlying longevity: A genetic switch model of aging. *Exp. Gerontol.* **2018**, *107*, 136–139. [CrossRef]

89. Zárate, S.; Stevnsner, T.; Gredilla, R. Role of Estrogen and Other Sex Hormones in Brain Aging. Neuroprotection and DNA Repair. *Front. Aging Neurosci.* **2017**, *9*, 430. [CrossRef] [PubMed]

90. Mishima, S.; Suzuki, K.-M.; Isohama, Y.; Kuratsu, N.; Araki, Y.; Inoue, M.; Miyata, T. Royal jelly has estrogenic effects in vitro and in vivo. *J. Ethnopharmacol.* **2005**, *101*, 215–220. [CrossRef] [PubMed]

91. Lephart, E.D. A review of the role of estrogen in dermal aging and facial attractiveness in women. *J. Cosmet. Derm.* **2018**, *17*, 282–288. [CrossRef] [PubMed]

92. Lambrinoudaki, I.; Augoulea, A.; Rizos, D.; Politi, M.; Tsoltos, N.; Moros, M.; Chinou, I.; Graikou, K.; Kouskouni, E.; Kambani, S.; et al. Greek-origin royal jelly improves the lipid profile of postmenopausal women. *Gynecol. Endocrinol.* **2016**, *32*, 835–839. [CrossRef] [PubMed]

93. Asama, T.; Matsuzaki, H.; Fukushima, S.; Tatefuji, T.; Hashimoto, K.; Takeda, T. Royal Jelly Supplementation Improves Menopausal Symptoms Such as Backache, Low Back Pain, and Anxiety in Postmenopausal Japanese Women. *Evid. Based Complement. Altern. Med.* **2018**, *2018*, 4868412. [CrossRef] [PubMed]

94. Shimizu, S.; Matsushita, H.; Minami, A.; Kanazawa, H.; Suzuki, T.; Watanabe, K.; Wakatsuki, A. Royal jelly does not prevent bone loss but improves bone strength in ovariectomized rats. *Climacteric* **2018**, *21*, 601–606. [CrossRef] [PubMed]

95. Hidaka, S.; Okamoto, Y.; Uchiyama, S.; Nakatsuma, A.; Hashimoto, K.; Ohnishi, S.T.; Yamaguchi, M. Royal jelly prevents osteoporosis in rats: Beneficial effects in ovariectomy model and in bone tissue culture model. *Evid. Based Complement. Altern. Med.* **2006**, *3*, 339–348. [CrossRef] [PubMed]

96. Narita, Y.; Nomura, J.; Ohta, S.; Inoh, Y.; Suzuki, K.M.; Araki, Y.; Okada, S.; Matsumoto, I.; Isohama, Y.; Abe, K.; et al. Royal jelly stimulates bone formation: Physiologic and nutrigenomic studies with mice and cell lines. *Biosci. Biotechnol. Biochem.* **2006**, *70*, 2508–2514. [CrossRef] [PubMed]

97. Kaku, M.; Rocabado, J.M.R.; Kitami, M.; Ida, T.; Uoshima, K. Royal jelly affects collagen crosslinking in bone of ovariectomized rats. *J. Funct. Foods* **2014**, *7*, 398–406. [CrossRef]

98. Munstedt, K.; Henschel, M.; Hauenschild, A.; von Georgi, R. Royal jelly increases high density lipoprotein levels but in older patients only. *J. Altern. Complement. Med.* **2009**, *15*, 329–330. [CrossRef]

99. Yakoot, M.; Salem, A.; Helmy, S. Effect of Memo®, a natural formula combination, on Mini-Mental State Examination scores in patients with mild cognitive impairment. *Clin. Interv. Aging* **2013**, *8*, 975–981. [CrossRef]

100. Pan, Y.; Xu, J.; Chen, C.; Chen, F.; Jin, P.; Zhu, K.; Hu, C.W.; You, M.; Chen, M.; Hu, F. Royal Jelly Reduces Cholesterol Levels, Ameliorates Aβ Pathology and Enhances Neuronal Metabolic Activities in a Rabbit Model of Alzheimer's Disease. *Front. Aging Neurosci.* **2018**, *10*, 50. [CrossRef]

101. Donlon, T.A.; Morris, B.J.; He, Q.; Chen, R.; Masaki, K.H.; Allsopp, R.C.; Willcox, D.C.; Tranah, G.J.; Parimi, N.; Evans, D.S.; et al. Association of Polymorphisms in Connective Tissue Growth Factor and Epidermal Growth Factor Receptor Genes With Human Longevity. *J. Gerontol. A Biol. Sci. Med. Sci.* **2017**, *72*, 1038–1044. [CrossRef]

102. Ewald, C.Y.; Castillo-Quan, J.I.; Blackwell, T.K. Untangling Longevity, Dauer, and Healthspan in *Caenorhabditis elegans Insulin/IGF-1*-Signalling. *Gerontology* **2018**, *64*, 96–104. [CrossRef] [PubMed]

103. Bartke, A. Healthspan and longevity can be extended by suppression of growth hormone signaling. *Mamm. Genome* **2016**, *27*, 289–299. [CrossRef] [PubMed]

104. Altintas, O.; Park, S.; Lee, S.-J.V. The role of *insulin/IGF-1* signaling in the longevity of model invertebrates, *C. elegans* and *D. melanogaster*. *BMB Rep.* **2016**, *49*, 81–92. [CrossRef] [PubMed]

105. Lechler, M.C.; Crawford, E.D.; Groh, N.; Widmaier, K.; Jung, R.; Kirstein, J.; Trinidad, J.C.; Burlingame, A.L.; David, D.C. Reduced *Insulin/IGF-1* Signaling Restores the Dynamic Properties of Key Stress Granule Proteins during Aging. *Cell Rep.* **2017**, *18*, 454–467. [CrossRef] [PubMed]

106. Albert, V.; Hall, M.N. *mTOR* signaling in cellular and organismal energetics. *Curr. Opin. Cell Biol.* **2015**, *33*, 55–66. [CrossRef] [PubMed]

107. Hindupur, S.K.; González, A.; Hall, M.N. The opposing actions of target of rapamycin and AMP-activated protein kinase in cell growth control. *Cold Spring Harb. Perspect. Biol.* **2015**, *7*, a019141. [CrossRef] [PubMed]

108. Kennedy, B.K.; Lamming, D.W. The Mechanistic Target of Rapamycin: The Grand ConducTOR of Metabolism and Aging. *Cell Metab.* **2016**, *23*, 990–1003. [CrossRef]

109. Nakamura, S.; Yoshimori, T. Autophagy and Longevity. *Mol. Cells* **2018**, *41*, 65–72. [CrossRef]

110. Bok, E.; Jo, M.; Lee, S.; Lee, B.R.; Kim, J.; Kim, H.J. Dietary Restriction and Neuroinflammation: A Potential Mechanistic Link. *Int. J. Mol. Sci.* **2019**, *20*, 464. [CrossRef]

111. Yamada, Y.; Kemnitz, J.W.; Weindruch, R.; Anderson, R.M.; Schoeller, D.A.; Colman, R.J. Caloric Restriction and Healthy Life Span: Frail Phenotype of Nonhuman Primates in the Wisconsin National Primate Research Center Caloric Restriction Study. *J. Gerontol. A Biol. Sci. Med. Sci.* **2018**, *73*, 273–278. [CrossRef]

112. Vasconcelos, A.R.; Santos, N.B.d.; Scavone1, C.; Munhoz, C.D. *Nrf2/ARE* Pathway Modulation by Dietary Energy Regulation in Neurological Disorders. *Front. Pharm.* **2019**, *10*, 33. [CrossRef] [PubMed]

113. Detienne, G.; Walle, P.V.d.; Haes, W.D.; Cockx, B.; Braeckman, B.P.; Schoofs, L.; Temmerman, L. Royalactin induces copious longevity via increased translation and proteasome activity in C. elegans. *bioRxiv* **2018**, *2018*, 421818. [CrossRef]

114. Troemel, E.R.; Chu, S.W.; Reinke, V.; Lee, S.S.; Ausubel, F.M.; Kim, D.H. *p38 MAPK* regulates expression of immune response genes and contributes to longevity in *C. elegans*. *PLoS Genet.* **2006**, *2*, e183. [CrossRef] [PubMed]

115. Yu, S.; Driscoll, M. *EGF* signaling comes of age: Promotion of healthy aging in *C. elegans*. *Exp. Gerontol.* **2011**, *46*, 129–134. [CrossRef] [PubMed]

116. Iwasa, H.; Yu, S.; Xue, J.; Driscoll, M. Novel *EGF* pathway regulators modulate *C. elegans* healthspan and lifespan via *EGF* receptor, *PLC-gamma*, and *IP3R* activation. *Aging Cell* **2010**, *9*, 490–505. [CrossRef] [PubMed]

117. Sadowska-Bartosz, I.; Bartosz, G. Effect of antioxidants supplementation on aging and longevity. *Biomed. Res. Int.* **2014**, *2014*, 404680. [CrossRef] [PubMed]

118. Haddad, L.S.; Kelbert, L.; Hulbert, A.J. Extended longevity of queen honey bees compared to workers is associated with peroxidation-resistant membranes. *Exp. Gerontol.* **2007**, *42*, 601–609. [CrossRef]

119. Anderson, K.E.; Ricigliano, V.A.; Mott, B.M.; Copeland, D.C.; Floyd, A.S.; Maes, P. The queen's gut refines with age: Longevity phenotypes in a social insect model. *Microbiome* **2018**, *6*, 108. [CrossRef]

120. Burdock, G.A. Review of the biological properties and toxicity of bee propolis (propolis). *Food Chem. Toxicol.* **1998**, *36*, 347–363. [CrossRef]

121. Henkel, R.; Sandhu, I.S.; Agarwal, A. The excessive use of antioxidant therapy: A possible cause of male infertility? *Andrologia* **2019**, *51*, e13162. [CrossRef]

122. Codling, G.; Al Naggar, Y.; Giesy, J.P.; Robertson, A.J. Concentrations of neonicotinoid insecticides in honey, pollen and honey bees (*Apis mellifera L.*) in central Saskatchewan, Canada. *Chemosphere* **2016**, *144*, 2321–2328. [CrossRef] [PubMed]

123. De Oliveira, R.C.; Queiroz, S.; da Luz, C.F.P.; Porto, R.S.; Rath, S. Bee pollen as a bioindicator of environmental pesticide contamination. *Chemosphere* **2016**, *163*, 525–534. [CrossRef] [PubMed]

124. Tosi, S.; Costa, C.; Vesco, U.; Quaglia, G.; Guido, G. A 3-year survey of Italian honey bee-collected pollen reveals widespread contamination by agricultural pesticides. *Sci. Total Env.* **2018**, *615*, 208–218. [CrossRef] [PubMed]

125. Böhme, F.; Bischoff, G.; Zebitz, C.P.W.; Rosenkranz, P.; Wallner, K. From field to food—will pesticide-contaminated pollen diet lead to a contamination of royal jelly? *Apidologie* **2018**, *49*, 112–119. [CrossRef]

126. Giroud, B.; Bruckner, S.; Straub, L.; Neumann, P.; Williams, G.R.; Vulliet, E. Trace-level determination of two neonicotinoid insecticide residues in honey bee royal jelly using ultra-sound assisted salting-out liquid liquid extraction followed by ultra-high-performance liquid chromatography-tandem mass spectrometry. *Microchem. J.* **2019**, *151*, 104249. [CrossRef]

127. Muresan, C.I.; Schierhorn, A.; Buttstedt, A. The Fate of Major Royal Jelly Proteins during Proteolytic Digestion in the Human Gastrointestinal Tract. *J. Agric. Food Chem.* **2018**, *66*, 4164–4170. [CrossRef]

128. Ostan, R.; Monti, D.; Gueresi, P.; Bussolotto, M.; Franceschi, C.; Baggio, G. Gender, aging and longevity in humans: An update of an intriguing/neglected scenario paving the way to a gender-specific medicine. *Clin. Sci.* **2016**, *130*, 1711–1725. [CrossRef]

International Journal of
Molecular Sciences

Article

Nutritional Mushroom Treatment in Meniere's Disease with *Coriolus versicolor*: A Rationale for Therapeutic Intervention in Neuroinflammation and Antineurodegeneration

Maria Scuto [1,†], Paola Di Mauro [2,†], Maria Laura Ontario [1], Chiara Amato [2], Sergio Modafferi [1], Domenico Ciavardelli [3,4], Angela Trovato Salinaro [1], Luigi Maiolino [2,*] and Vittorio Calabrese [2]

[1]   Department of Biomedical and Biotechnological Sciences, University of Catania, Torre Biologica. Via Santa Sofia, 97, 95123 Catania, Italy; mary-amir@hotmail.it (M.S.); marialaura.ontario@ontariosrl.it (M.L.O.); sergio.modafferi@gmil.com (S.M.); trovato@unict.it (A.T.S.)
[2]   Department of Medical and Surgery Sciences, University of Catania, Via Santa Sofia 78, 95123 Catania, Italy; paola_mp86@hotmail.it (P.D.M.); chiaraamato@hotmail.it (C.A.); calabres@unict.it (V.C.)
[3]   School of Human and Scocial Science, "Kore" University of Enna, Via Salvatore Mazza 1, 94100 Enna, Italy; domenico.ciavardelli@unikore.it
[4]   Centro Scienze dell'Invecchiamento e Medicina Traslazionale-CeSI-Met, via Luigi Polacchi 11, 66100 Chieti, Italy
*   Correspondence: maiolino@policlinico.unict.it
†   These authors contributed equally to this work.

Received: 9 December 2019; Accepted: 27 December 2019; Published: 31 December 2019

Abstract: Meniere's disease (MD) represents a clinical syndrome characterized by episodes of spontaneous vertigo, associated with fluctuating, low to medium frequencies sensorineural hearing loss (SNHL), tinnitus, and aural fullness affecting one or both ears. To date, the cause of MD remains substantially unknown, despite increasing evidence suggesting that oxidative stress and neuroinflammation may be central to the development of endolymphatic hydrops and consequent otholitic degeneration and displacement in the reuniting duct, thus originating the otolithic crisis from vestibular otolithic organs utricle or saccule. As a starting point to withstand pathological consequences, cellular pathways conferring protection against oxidative stress, such as vitagenes, are also induced, but at a level not sufficient to prevent full neuroprotection, which can be reinforced by exogenous nutritional approaches. One emerging strategy is supplementation with mushrooms. Mushroom preparations, used in traditional medicine for thousands of years, are endowed with various biological actions, including antioxidant, immunostimulatory, hepatoprotective, anticancer, as well as antiviral effects. For example, therapeutic polysaccharopeptides obtained from *Coriolus versicolor* are commercially well established. In this study, we examined the hypothesis that neurotoxic insult represents a critical primary mediator operating in MD pathogenesis, reflected by quantitative increases of markers of oxidative stress and cellular stress response in the peripheral blood of MD patients. We evaluated systemic oxidative stress and cellular stress response in MD patients in the absence and in the presence of treatment with a biomass preparation from *Coriolus*. Systemic oxidative stress was estimated by measuring, in plasma, protein carbonyls, hydroxynonenals (HNE), and ultraweak luminescence, as well as by lipidomics analysis of active biolipids, such as lipoxin A4 and F2-isoprostanes, whereas in lymphocytes we determined heat shock proteins 70 (Hsp72), heme oxygenase-1 (HO-1), thioredoxin (Trx), and γ-GC liase to evaluate the systemic cellular stress response. Increased levels of carbonyls, HNE, luminescence, and F2-isoprostanes were found in MD patients with respect to the MD plus *Coriolus*-treated group. This was paralleled by a significant ($p < 0.01$) induction, after *Coriolus* treatment, of vitagenes such as HO-1, Hsp70, Trx, sirtuin-1, and γ-GC liase in lymphocyte and by a significant ($p < 0.05$) increase in the plasma ratio-reduced glutathione (GSH) vs. oxidized glutathione (GSSG). In conclusion, patients affected by MD are

Int. J. Mol. Sci. **2020**, *21*, 284

under conditions of systemic oxidative stress, and the induction of vitagenes after mushroom supplementation indicates a maintained response to counteract intracellular pro-oxidant status. The present study also highlights the importance of investigating MD as a convenient model of cochlear neurodegenerative disease. Thus, searching innovative and more potent inducers of the vitagene system can allow the development of pharmacological strategies capable of enhancing the intrinsic reserve of vulnerable neurons, such as ganglion cells to maximize antidegenerative stress responses and thus providing neuroprotection.

Keywords: redoxomics; glutathione; meniere's disease; neurodegenerative diseases

---

1. Introduction

Prosper Meniere more than 150 ago first described the disease named after him, and to date, although many studies have tried to describe the etiology of Meniere's disease (MD), it still represents a matter of scientific debate. [1]. Among the theories considered to explain its pathophysiology, endolymphatic hydrops with disturbed longitudinal endolymph flow is considered central to MD pathology [2], widely recognized as primary cause leading to cochlear degeneration [3]. Anatomical variation in the position or size of sac and duct in the endolymphatic system, and the presence of viral, autoimmune inflammatory or genetic components are all possible contributory factors to the endolymph homeostasis [4]. Recent studies indicate a pattern of similarity between MD and benign paroxysmal positional vertigo, including age of onset, raising the conceivable possibility that detached saccular otoconia, an event which could be promoted by metabolic disturbances associated with oxidative stress, might represent the fundamental cause of MD [5]. As an hydropic ear pathology, MD is characterized by a triad of symptoms, such as episodic vertigo and tinnitus associated with fluctuating hearing loss, and endolymphatic hydrops, as found on post-mortem examination [2]. However emerging evidence has given rise to the conceivable possibility that MD is a systemic oxidant disorder, where excessive production of free radicals and oxidative stress promote microvascular damage, which is involved in the development of endolymphatic hydrops. Consequently, cellular damage and apoptotic cell death-induced cochleovestibular dysfunction ensues with significant reductions in dendritic innervation densities, and ultrastructural abnormalities reflecting the primary neurotoxic insult [6,7].

While reactive oxygen species at a physiological level play an important role in cellular signaling, excess in free-radical species or oxidative stress due to decreased expression and activity of antioxidant proteins becomes a toxic cause of accelerated aging [8–10]. Thus, the cellular capacity to counteract stressful conditions, known as cellular stress response, requires the activation of pro-survival pathways endowed with increased antioxidant, anti-inflammatory, and antiapoptotic potential [11–13].

Consistent with this notion, integrated survival responses exist in the central and peripheral nervous system, which are controlled by redox-dependent genes, termed vitagenes [14,15]. These include gene coding for proteins that actively operate in detecting and controlling diverse forms of stress and neuronal injuries, such as heat shock proteins (Hsps), γ-GC liase, thioredoxin, sirtuins, and Lipoxin A4 [16]. As a metabolic product of arachidonic acid, LXA4 is an endogenous "stop signal" for inflammatory processes, exhibiting its potent anti-inflammatory potential in various inflammatory disorders, such as arthritis, periodontitis, nephritis, or inflammatory bowel disease [17,18]. Chronic inflammation is known to be central to the progression of Alzheimer's disease (AD), although identification of mechanisms capable of restoring an anti-inflammatory environment compromised in AD pathology remains an area of active investigation [19,20]. Treatment with the pro-resolving mediator aspirin-triggered lipoxin A4 (ATL) resulted in improved cognition, reduced Aβ levels, and enhanced microglia phagocytic activity in Tg2576 transgenic AD mice [21]. Furthermore, LXA4 levels are reduced with age, a pattern significantly more impacted in 3xTg-AD mice [22]. Moreover, in

3xTg-AD mice, up-regulation of lipoxin A4 was induced by aspirin-enhanced cognitive performance while reducing Aβ and phosphorylated-tau (*p*-tau) levels, an effect associated with astrocyte and microglia reactivity [18]. LXA4 action is mediated by LXA4 receptor (ALX) on the cellular membrane, which is known as formyl-peptide receptor-like 1 (FPRL1) [23], and activation of LXA4 signaling can well serve as a robust therapeutic target for mitigating AD-related inflammation and consequential cognitive dysfunction.

Vitagene cellular stress response confers a cytoprotective state not only during aging but also in a variety of human diseases, including cancer, inflammation, and neurodegenerative disorders [24]. Given the broad cytoprotective potential of vitagenes there is now increasing interest in discovering and developing pharmacological agents able to induce stress responses [25]. When appropriately activated, cellular stress response restores redox equilibrium by activating antioxidant and anti-inflammatory pathways, which is of particular importance for brain cells with relatively weak endogenous antioxidant defenses, such as spiral ganglion neurons, centrally involved in the pathogenesis of MD and a preferential site for accumulation of lipoperoxidative hydroxynonenals and protein oxidation carbonyls product, which can disrupt redox homeostasis [26].

Mushrooms, which have been used in traditional medicine for thousands of years [27,28], are emerging as an important nutritional component in the diet capable of modulating the immunity system and inflammatory status. In Asian countries, for instance, modern clinical practice continues to rely on mushroom-derived preparations. According to this, many controlled studies have investigated a long list of mushroom extracts, showing various immunomodulatory biological actions, associated with antioxidant, antiviral, anticancer, and hepatoprotective activities [29,30]. As a result, many traditionally employed mushrooms, including extracts of *Agaricus campestris*, *Pleurotus ostreatus* and *Coriolus versicolor* have shown medicinal effects [31]. In particular, the active principle from *Coriolus versicolor* represents a new class of elements termed biological response modifiers (BRM) [32], which characterize several agents capable of stimulating the immune system, therefore exhibiting various therapeutic effects. Consistent with the neuroinflammatory pathogenesis of neurodegenerative damage occurring in AD, a recent study from our laboratory has provided convincing experimental evidence into the neuroprotective role of *Coriolus* biomass preparation against the neuroinflammatory process, evaluating also the impact of this nutritional intervention on cellular stress response mechanism operating in the central nervous system [33,34].

In the present study we examined the hypothesis that neurotoxic insult represents a critical primary mediator operating in MD pathogenesis, reflected by quantitative increases of markers of oxidative stress and cellular stress response in the peripheral blood of MD patients. We also explore the hypothesis that changes in lipidomics, as well as redox glutathione status associated with increased expression of neuroprotective vitagenes induced through supplementation with mushrooms biomass preparation from Mycology Research Laboratories Ltd., Luton, UK, *Coriolus v.* can provide a novel target for innovative therapeutic approaches aimed at minimizing oxidative stress, neuroinflammation, and neurodegeneration occurring not only in MD, but also in major neurodegenerative disorders such as AD or Parkinson's disease.

2. Results

2.1. Auditory Function Analysis

Profile of Mood States (POMS) analysis (Table 1) revealed in Group A subjects, the group treated with mushroom preparation, a significant improvement of subjective parameters related to the psycho-emotional status of the patients, as compared to untreated MD patients (Group B), where we did not observe particular changes. Table 2 shows homogeneity between the two groups regarding the number of crises, their duration, and the frequency of symptoms. Notably, data in Table 3 illustrates the Tinnitus Handicap Inventory (THI) questionnaire, performed to define the clinical grading of tinnitus severity, showing a statistically significant improvement in the group of patients receiving *Coriolus* mushroom biomass treatment, as compared to the untreated group.

Table 1. Profile of Mood States (POMS).

	Pre-Therapy (T0)		Post-Therapy (T1)	
	Score		Score	
Group	A	B	A	B
Anger (0–48)	28	29	22	29
Confusion (0–28)	17	17	10	16
Depression (0–60)	41	39	25	37
Fatigue (0–28)	16	19	10	19
Tension (0–36)	31	29	13	28
Vigor (0–32)	19	17	19	16
Total Mood Disturbance (−32 to 200)	114 ± 9.8	116 ± 8.6	61 ± 6.11	113 ± 8.1

Table 2. Crisis frequency.

T0	Group A	Group B
Vertigo Attack Frequency		
<2 crisis/year	4 (18.1%)	3 (16.6%)
From 3 to 5 crisis/year	10 (45.4%)	8 (44.4%)
From 6 to 8 crisis/year	8 (36.3%)	7 (38.8%)
Crisis Duration		
<1 h	4 (18.1%)	4 (22.2%)
From 1 to 7 h	12 (54.5%)	9 (50%)
>24 h	6 (27.2%)	5 (27.7%)
Duration of Symptoms		
A few days	15 (68.1%)	11 (61.1%)
Some weeks	6 (27.2%)	6 (33.3%)
A month	1 (4.5%)	1 (5.5%)

Table 3. Tinnitus Handicap Inventory (THI).

Tinnitus Handicap		Inventory	
Pre-Therapy Score	(T0)	Post-Therapy Score	(T1)
Group A	Group B	Group A	Group B
74 ± 2.46	78 ± 2.73	52 ± 1.73 *	74 ± 2.65

* significantly different vs. control untreated MD patients ($p < 0.05$).

To document SNHL, we performed in all subjects, at the initial (T0) phase, tonal audiometry analysis (Figure 1). For both experimental groups, the tonal interest was centered on medium-high frequencies, with an average intensity of 55 dB loss. All subjects in the group A reported in the T1 phase, after treatment, significant changes, both in the frequency range, and in the average loss in dB, as compared to the initial T0 phase. Similarly, speech audiometry analysis revealed in the same subjects receiving mushrooms a significant improvement of intellection threshold, i.e., the ability of verbal discrimination, with respect to the initial T0 phase, where the threshold of intellection and perception that is 100% of the given words was assumed to be 75 db. In contrast to the *Coriolus* biomass-treated group, in patients of Group B, however, we did not detect any significant change compared to thresholds measured at T0 initial phase. This finding was consistent with impedenzometric measures at examination, which revealed in all subjects either at T0 initial phase or at the T1 phase, an average increase in the threshold of stapedial reflexes and the positivity of the Metz test, indicative of cochlear suffering, with no significant differences between the two groups.

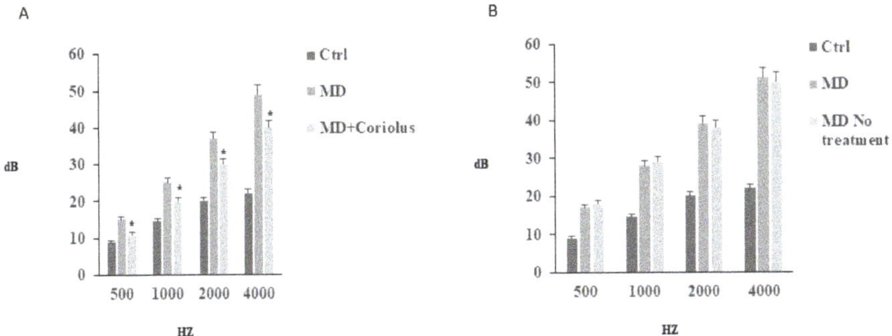

Figure 1. Tonal audiometry analysis. Tonal interest was centered on medium–high frequencies, with an average intensity of 55 dB loss. All subjects reported in both T0 (**B**) and T1 (**A**) phases no significant changes, either in the frequency range, or in the average loss in dB. Speech audiometry analysis, revealed in subjects of group A, who received mushrooms, a significant improvement of intellection threshold, i.e., the ability of verbal discrimination, respect to the initial T0 phase, where the threshold of intellection and perception that is 100% of the given words, was assumed to be 75 db.

2.2. Redoxomics

Modulation of Hsp72, HO-1, Thioredoxin, Sirtuins and γ-GC Liase, in MD Patients after Coriolus Mushroom Supplementation

Oxidative stress plays a role in the pathogenesis of a wide variety of pathological states [35–37]. ROS can oxidize membrane lipids generating lipid hydroperoxides and many aldehydes such as HNE, luminescent by-products, and isoprostanes. HNE can accumulate in cells in relatively high concentrations and cause cell toxicity. Recent studies have also shown that in response to environmental changes and other stressful conditions promoting proteotoxicity [38–40], cells adaptively activate synthesis and accumulation of several members of stress proteins, primarily Hsp70 and HO-1. As reported in Figures 2 and 3, mushroom supplementation with *Coriolus* biomass preparation resulted in up-regulation of the inducible isoforms of both Hsp70 and heme oxygenase-1 (HO-1), in lymphocytes (Figures 2a and 3a), a finding observed also in plasma, (Figures 2b and 3b), as compared to untreated group of MD patients. A representative Western blot obtained probing tissue samples with an antibody specific for the inducible isoform of heat shock proteins 70 (Hsp72) or Heme oxygenase are shown in Figure 2c,d and Figure 3c,d, respectively. Western blot analysis of the Thioredoxin protein also revealed a significant increase in the group of patients treated with *Coriolus* compared to control group, in lymphocyte and plasma (Figure 4a,b). A representative blot of thioredoxin protein is reported in Figure 4c,d. Similar results were also obtained analyzing sirtuin-1 expression. As shown in Figure 5a, Sirtuin-1 immunoreactivity was higher in lymphocytes of a group of MD patients treated for 2 months with *Coriolus* than in the untreated MD group. Consistent with this, plasma sirtuin-1 levels were higher in MD patients supplemented with mushrooms, as compared to the MD group of patients alone (Figure 5b). Representative blots of sirtuin-1 protein are reported in Figure 5c,d, respectively. Another important redoxomic component of vitagene network is γ-GC liase, the rate-limiting enzyme for intracellular glutathione (GSH) synthesis. Notably, GSH concentration and γ-GC liase activity are declining with age in the central nervous system (CNS), a condition associated with increased oxidative stress [41]. Here we report that lymphocyte γ-GC liase levels were higher in MD patients supplemented with mushrooms as compared to the MD group of patients alone (Figure 6a). A representative blot of γ-GC protein is reported in Figure 6b.

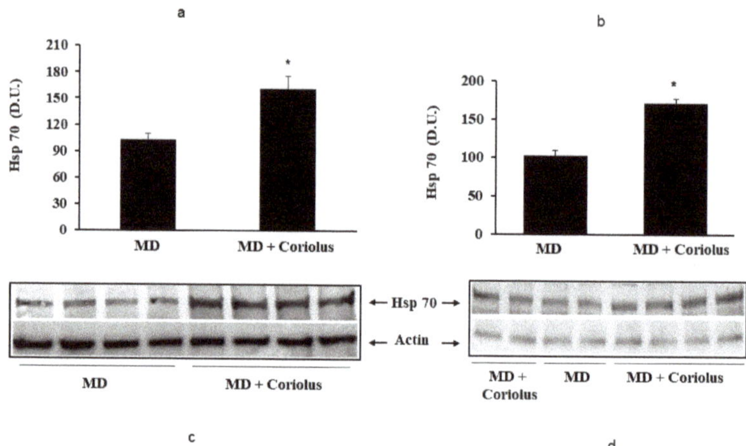

Figure 2. Heat shock protein 70 levels in lymphocytes and in plasma from MD patients. Samples from MD patients were assayed for heat shock protein 70 (Hsp70) by western blot as described in Materials and Methods. A representative immunoblot is shown in (**c**,**d**). β-actin has been used as loading control. The bar graphs (**a**,**b**) show the densitometric evaluation and values are expressed as mean ± SEM of independent analyses on 22 patients (MD plus *Coriolus* biomass) and, respectively, on 18 patients (MD alone), per group. * $p < 0.05$ vs. MD alone. D.U., densitometric units.

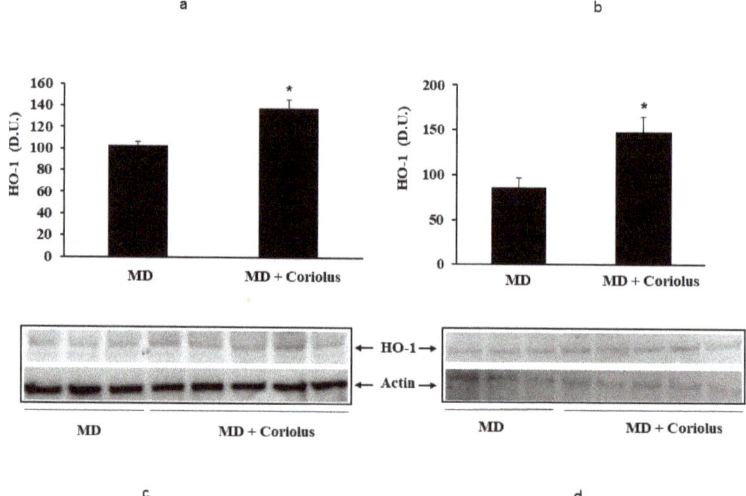

Figure 3. Heme oxygenase-1 levels in lymphocytes and in plasma from MD patients. Samples from MD patients were assayed for heme oxygenase-1 (HO-1) by western blot as described in Materials and Methods. A representative immunoblot is shown. β-actin has been used as loading control (**c**,**d**). The bar graph shows the densitometric evaluation (**a**,**b**) and values are expressed as mean ± SEM of independent analyses on 22 patients (MD plus *Coriolus* biomass) and, respectively, on 18 patients (MD alone), per group. * $p < 0.05$ vs. MD alone. D.U., densitometric units.

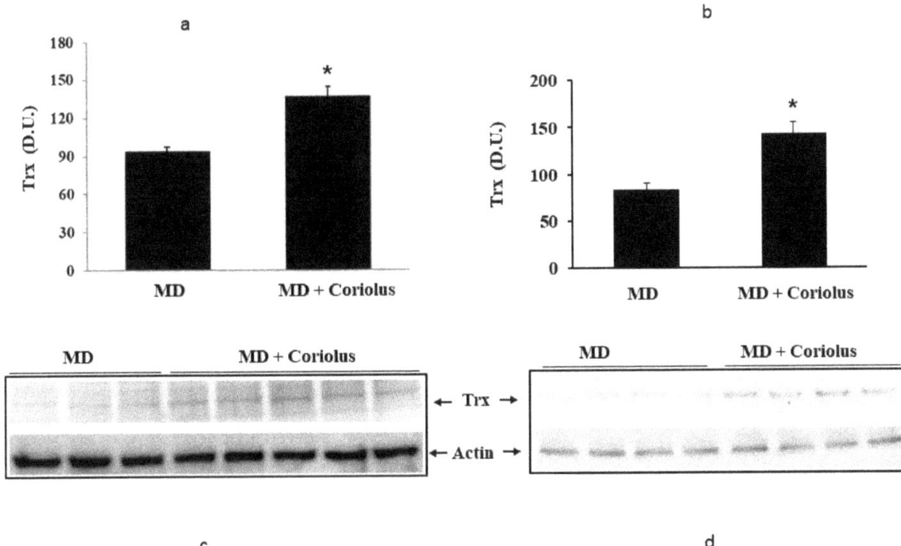

Figure 4. Thioredoxin levels in lymphocytes and in plasma from MD patients. Lymphocyte samples (**a**) and plasma samples (**b**) from MD patients were assayed for thioredoxin (Trx) by western blot as described in Materials and Methods. A representative immunoblot is shown (**c,d**). β-actin has been used as loading control. * $p < 0.05$ vs. MD alone. D.U., densitometric units.

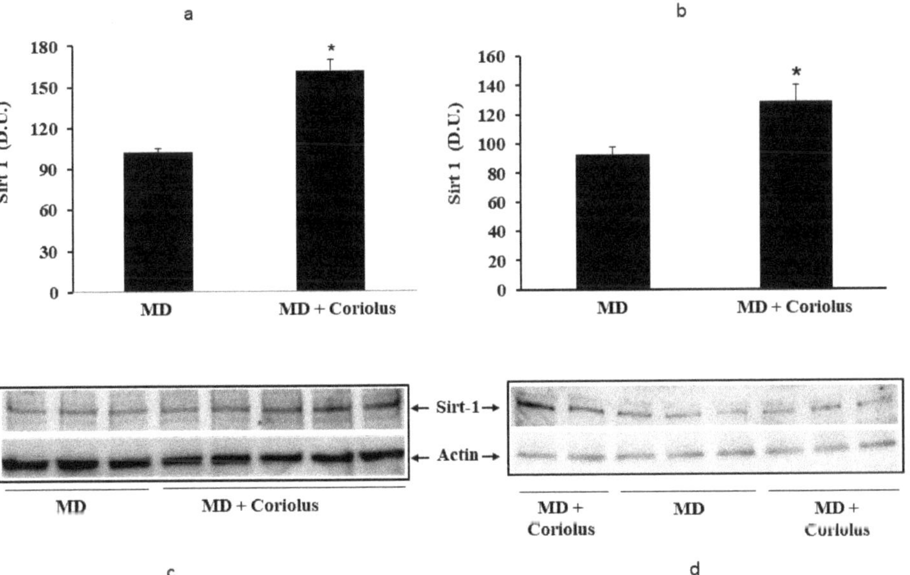

Figure 5. Levels of sirtuin-1 in lymphocytes (**a**) and plasma (**b**) from MD patients. Samples from MD patients were assayed for sirtuin-1 by Western blot as described in Materials and Methods. Representative immunoblots are shown in the same figure (**c,d**). β-actin has been used as loading control. The bar graph shows the densitometric evaluation and values are expressed as mean ± SEM of independent analyses on 22 patients (MD plus *Coriolus* biomass) and, respectively, on 18 patients (MD alone), per group. * $p < 0.05$ vs. MD alone. D.U., densitometric units.

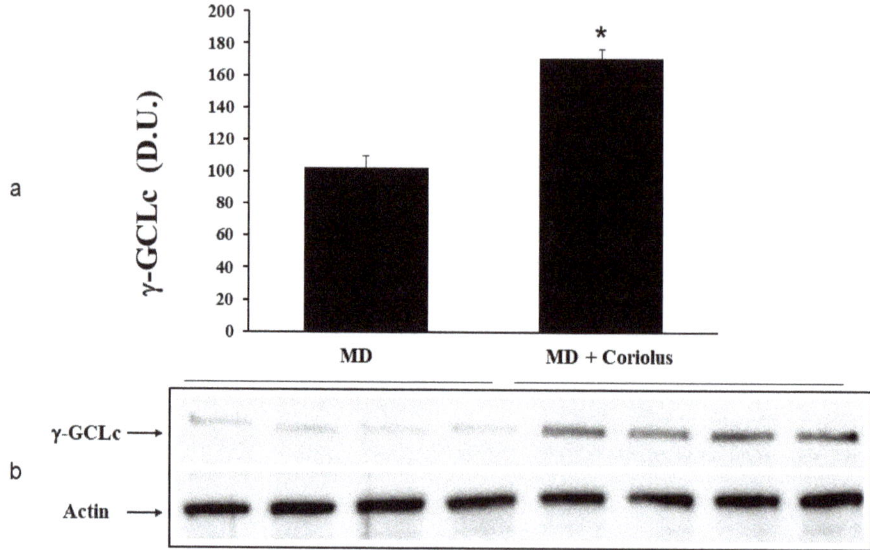

Figure 6. γ-GC liase levels in lymphocytes from MD patients. Plasma samples from MD patients were assayed for γ-GC liase by western blot as described in Materials and Methods. A representative immunoblot is shown (**b**). β-actin has been used as loading control. The bar graph shows the densitometric evaluation and values are expressed as mean ± SEM of independent analyses on 22 patients (MD plus *Coriolus* biomass) and, respectively, on 18 patients (MD alone), per group (**a**). * $p < 0.05$ vs. MD alone. D.U., densitometric units.

2.3. Assessment of Systemic Oxidative Status

Protein and lipid oxidation occurring because of oxidative stress in tissues and organs leads to the formation of carbonyl groups in amino acid residues [41] and, respectively, to 4-hydroxynonenal (HNE) formation from arachidonic acid or other unsaturated fatty acids [42]. As a hallmark for oxidative damage to proteins by free-radical attack, protein carbonylation, by binding via Michael addition to proteins, particularly at cysteine, hystidine, or lysine residues [36], exerts deleterious effects on cell function and viability, being generally unrepairable and leading to production of potentially harmful protein aggregates and to cellular dysfunction. Under conditions of oxidative stress, protein oxidation products measured as protein carbonyls, as well as lipid oxidation products, measured by HNE or ultraweak luminescence, accumulate [6,29,30]. Examination of plasma protein carbonyls (Figure 7a) and HNE (Figure 7b), as well as plasma or lymphocyte ultraweak luminescence levels (Figure 7c) revealed a significant elevation in MD patients respect to *Coriolus*-treated group of MD patients.

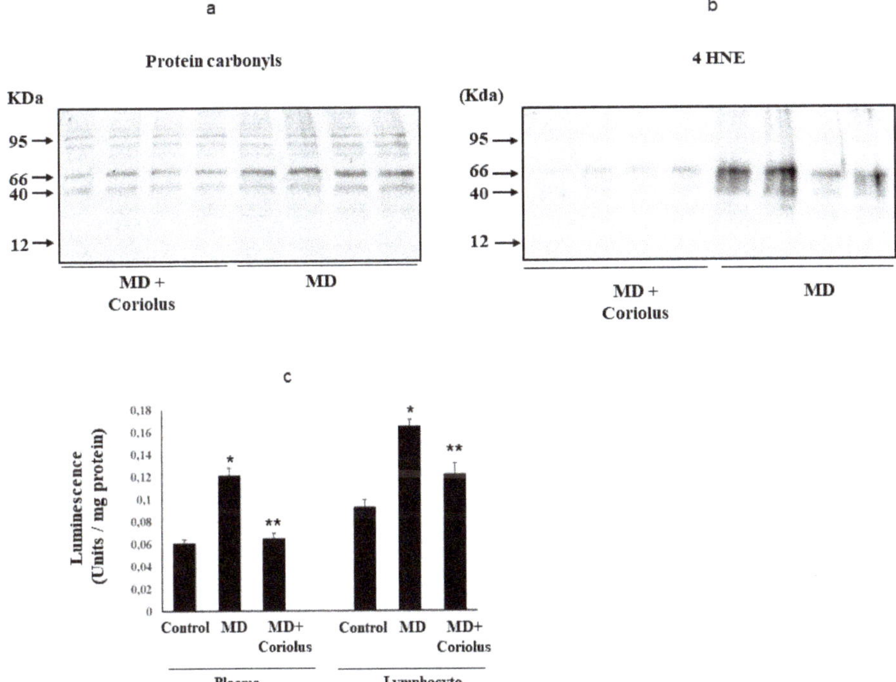

Figure 7. Protein carbonyls, 4-hydroxy-2-nonenals and Spontaneous ultraweak chemiluminescence (UCL) levels in MD patients. Plasma samples from MD patients (**a**,**b**) were assayed for protein carbonyls (DNPH) and 4-hydroxy-2-nonenals (HNE) by Western blot as described in Materials and Methods. Values are expressed as mean ± SEM of independent analyses on 22 patients (MD plus *Coriolus* biomass) and, respectively, on 18 patients (MD alone), per group. * $p < 0.05$ vs. MD alone. D.U., densitometric units. UCL in plasma and lymphocytes of control healthy volunteers and Meniere Diseased (MD) patients, in the absence and presence of Coriolus biomass treatment is shown in (**c**). UCL was measured as described in methods. CTRL: control; MD: Meniere disease patients. (*) $p < 0.05$ vs. control; (**) $p < 0.05$ vs. MD alone.

2.4. Lipidomics Analysis

Oxidation of polyunsaturated fatty acid arachidonic, eicosapentaenoic, docosahexaenoic, linoleic, and dihomo-γ-linolenic generate bioactive lipids. The development of mass spectrometry platforms enabling quantification of diverse lipid species in human urine is of crucial importance to understand metabolic redox homeostasis in normal as well as pathophysiological conditions. Here we demonstrate clearly how administration of *Coriolus* to MD patients increases significantly the powerful anti-inflammatory eicosanoid LXA4 in plasma and lymphocytes as compared to untreated MD patients (Figure 8a,b). The same results were observed in urine, where a large increase in LXA4 was measured after *Coriolus* supplementation (Figure 8c). Consistently, analysis of urine levels of pro-inflammatory eicosanoids 11-dehydro TXB2, isoprostane PGF2α, and isoprostane iPF2α-VI showed the opposite results with significantly higher levels of these bioactive lipids in MD subjects than the levels found in *Coriolus* administered MD patients (Figure 9a–c).

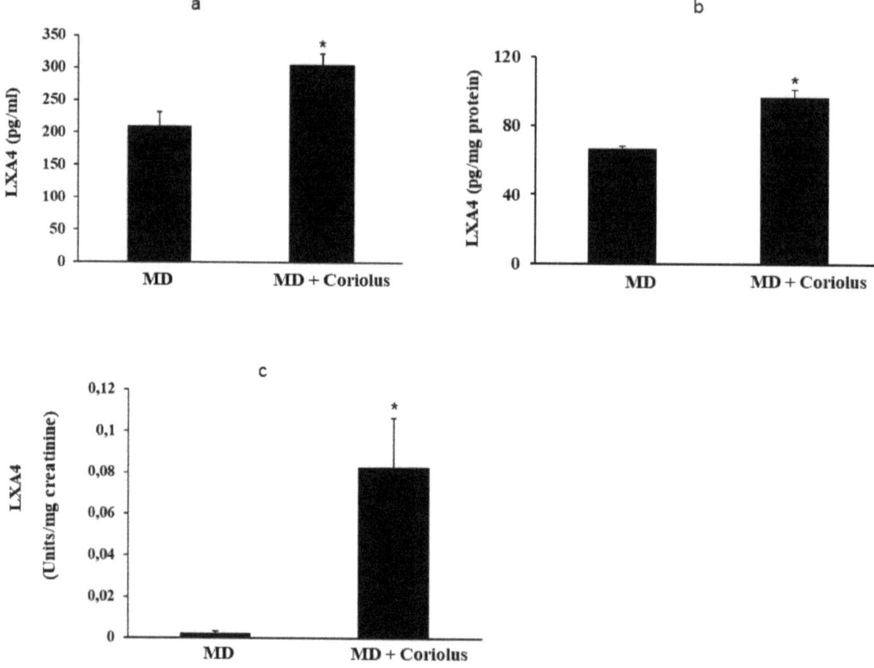

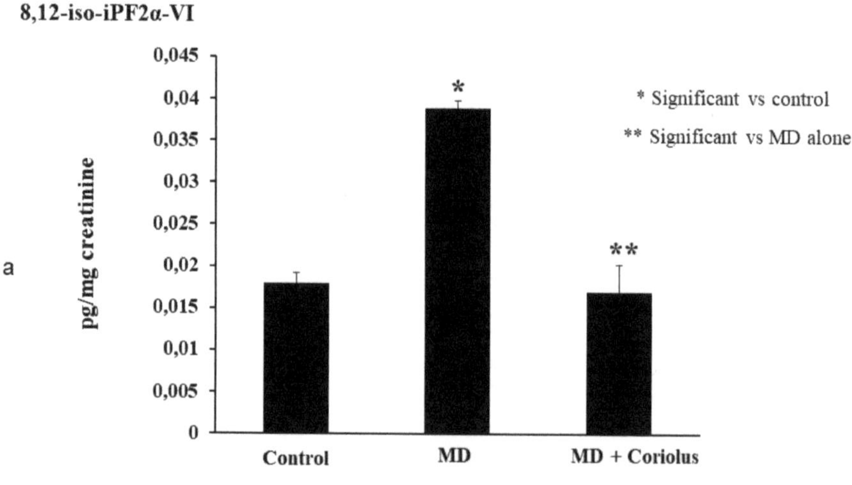

Figure 8. Lipidomic analysis of bioactive lipids. Biolipids are synthesized by oxidation of polyunsaturated fatty acids, arachidonic acid, eicosapentaenoic acid, docosahexaenoic acid, linoleic acid, and dihomo-γ-linolenic acid. The development of enabling mass spectrometry platforms for the quantification of diverse lipid species in human urine is of paramount importance for understanding metabolic redox homeostasis in normal and pathophysiological conditions. Anti-inflammatory eicosanoid LXA4 were measured in plasma, lymphocytes (**a**,**b**) and in urine (**c**), as compared to untreated MD patients.

8,12-iso-iPF2α-VI

Figure 9. *Cont.*

2,3-dinor-8-iso PGF2alpha

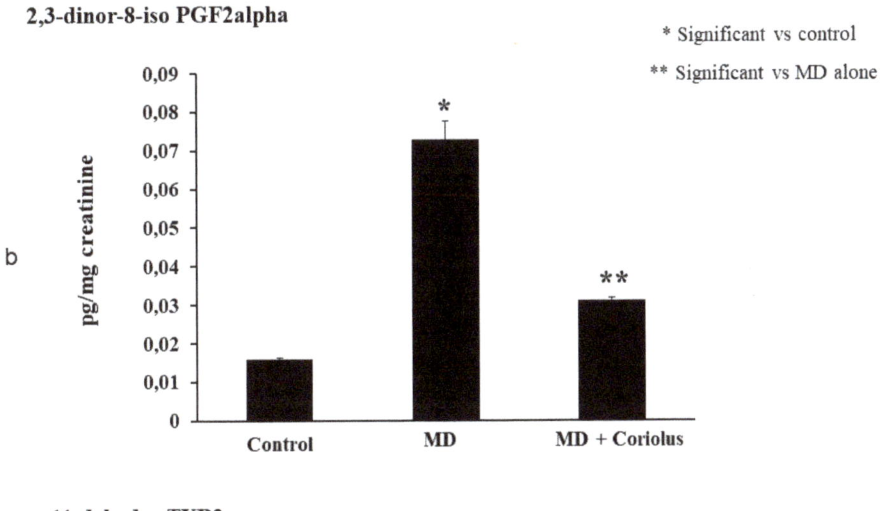

11-dehydro TXB2

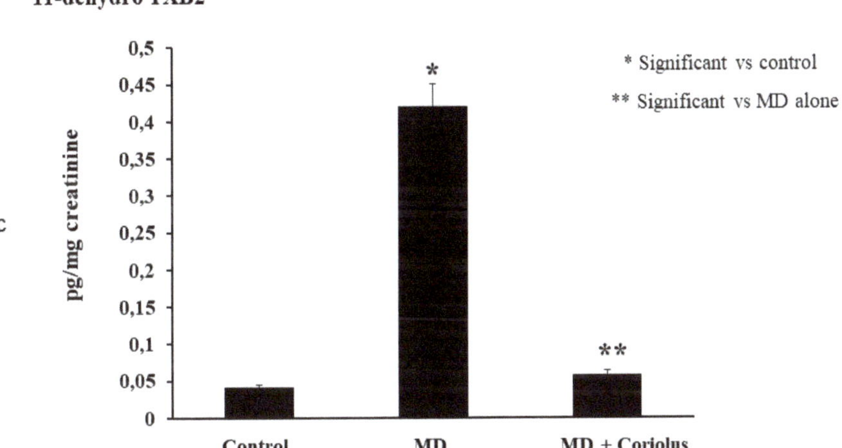

Figure 9. Lipidomic analysis of bioactive lipids. Analysis of urinay pro-inflammatory eicosanoids, 11-dehydro TXB2, isoprostane PGF2α, isoprostane iPF2α-VI, showing opposite results with significant higher levels of these bioactive lipids in MD subjects than the levels found in *Coriolus* administered MD patients are reported in (**a–c**).

Consistent with other findings showing that oxidative stress and altered thiol status in degenerating brain diseases correlates with systemic redox imbalance and oxidative stress, as in AD [31–34], the content of total GSH, reduced and oxidized glutathione and the GSH/GSSG ratio, was determined in the plasma of MD patients as a measure of the antioxidant status and compared with the levels of *Coriolus*-treated MD group (Table 4). We report the plasma from MD patients contained significantly lower levels of GSH as compared to *Coriolus*-supplemented patients, which paralleled to corresponding significantly higher GSSG levels ($p < 0.05$) (Table 4). These changes resulted in a plasma GSH/GSSG ratio which was significantly higher in the group of MD plus *Coriolus* subjects then the ratio found in the MD group alone (Table 4).

Table 4. Plasma and lymphocyte content of total, reduced (GSH) and oxidized (GSSG) glutathione in control and MD patients treated with Coriolus.

	Plasma (nmol/mL)			Lymphocyte (nmol/mg Protein)		
	Control	MD	MD + Coriolus	Control	MD	MD + Coriolus
Total GSH	16.7 ± 2.1	8.33 ± 3.0 *	14.23 ± 2.4 **	9.81 ± 0.8	5.3 ± 0.7 *	7.3 ± 0.5 **
GSH	15.62 ± 2.0	8.44 ± 1.7 *	13.44 ± 1.7 **	9.58 ± 0.6	4.27 ± 0.4 *	7.20 ± 0.5 **
GSSG	0.138 ± 0.01	0.169 ± 0.01 *	0.146 ± 0.01 **	0.093 ± 0.01	0.118 ± 0.01 **	0.096 ± 0.006 **
Ratio GSH/GSSG	113.2 ± 11	56.9 ± 15*	92.05 ± 13 **	96.5 ± 10	42.6 ± 7.9 *	75.0 ± 9.6 **

* Significantly different from control ($p < 0.05$). ** Significantly different from MD alone ($p < 0.05$).

3. Discussion

MD is a chronic illness derived from combined neurodegenerative events occurring at level of spiral ganglion as well as hair cells of the inner ear, associated with a negative impact on the quality of life of individuals, presenting various symptoms, such as temporary hearing loss, dizziness, and tinnitus [7].

After its initial description by Prosper Meniere more than 150 years ago, the disease named after him is still at the center of scientific debate [1]. MD is a hydropic ear pathology, where episodic vertigo, tinnitus, and fluctuating hearing loss coexist with endolymphatic hydrops. [3]. Recent evidence indicates the involvement of oxidative stress in the development of endolymphatic hydrops associated with neuronal ganglion damage with apoptotic neuronal cell death as a prominent factor contributing to SNHL found in the later stages of MD [2]. Thus, it is conceivable that MD, as a systemic oxidant disorder [5], can be also considered, owing to its demonstrated neurodegenerative nature of the neuronal cochlear ganglion component, involved in its pathogenesis, a pursuable investigative model of neurodegeneration. Consistent with this possibility, the present study was undertaken to explore the hypothesis that changes in the redox status of glutathione, stress-responsive vitagenes, and lipidomics, the major determinants in the disruption of redox homeostasis affecting spiral ganglion neurons, may be positively impacted by nutritional intervention with *Coriolus*-MRL biomass supplementation.

Mushrooms have been present in traditional medicine for thousands of years, and are reportedly endowed with immunomodulatory actions, associated with antioxidant, anticancer, antiviral, bacteriostatic, and hepatoprotective properties [43]. Mushroom-derived therapeutics, mainly polysaccharopeptides isolated from *Coriolus versicolor*, are well characterized and commercially available. Here we tested the hypothesis that neurotoxicity is an important causative factor involved in MD pathogenesis, which can be evaluated by measuring markers of oxidative stress and cellular stress response proteins in the peripheral blood of patients with MD. We evaluated in the present study systemic oxidative stress and cellular stress response in 40 patients suffering from MD in the absence and in the presence of treatment with mushroom biomass preparation from *Coriolus*. Systemic oxidative stress was estimated in plasma and urines of patients with MD or MD plus *Coriolus*, by measuring protein carbonyls, HNE, and ultraweak luminescence, as well as active biolipids such as lipoxin A4 and F2- isoprostanes, whereas in the lymphocyte heat shock proteins (HSP) heme oxygenase-1 (HO-1), Hsp70 and thioredoxin (Trx) levels were measured to evaluate the systemic cellular stress response. Increased levels of DNPH, HNE, ultraweak luminescence, and F2-isoprostanes were found in all the samples from MD patients with respect to the MD plus *Coriolus*-treated group. This was paralleled by a significant induction of lymphocyte HO-1, Hsp70, TrxR-1 as well as Sirtuin-1 and by a significant increase in the plasma ratio-reduced glutathione GSH) vs. oxidized glutathione (GSSG).

It is suggested that genetic factors may contribute partly to the etiologies of MD, as some associations have been reported for polymorphisms related to gene coding for protein involved in inflammation, circulation, and blood vessels, such as interleukin 1A (−889C/T), interleukin 6−572C/G), protein kinase C beta type (1425G/A), matrix metalloproteinase-1 (−1607G/2G), methylenetetrahydrofolate reductase (MTHFR) (C677T), prothrombin (G20210A), and complement factor H [44–51], and genes involved in free-radical processes. Although the initial causative factors triggering the disease have not been clarified, various genes and variants have been confirmed to be related to MD, which also suggests a specific family of genetic predisposition and implies genetic factors as key players in the initiation and progression of MD [52]. Consistent with this scenario, inflammation and oxidative stress-induced endolymphatic hydrops have been identified as a secondary pathogenesis of the disease [53]. Thus, MD etiology and pathogenesis appears to be an aberrant response of the adaptive or innate immune system, ultimately mediated by pro-inflammatory and oxidative processes underlying its physiopathological determinism [54]. Several mechanisms are involved in the development of immune-mediated inner-ear pathology, including (a) similarity with potentially harmful component of virus or bacteria, such as cross-reactive epitope inducing inner-ear damage; and (b) generation of pro-inflammatory Interleukin 1 β (IL-1B) or Tumor necrosis factor α (TNF) cytokines and transcriptional nuclear factor kB (NF-kB) [55]. Toll-like receptor coding genes, including TLR3, TLR7, TLR8, and TLR10, are widely reported to contribute to the disease, being directly related to the initiation and progression of MD, thus implying a specific role for the immune system during the pathological processes [56]. This is confirmed by recent findings highlighting the relationships between increased serum levels of IL6 and IL1 with vertigo, a specific complication of MD.

To survive different types of injuries and adapt to environmental changes, neuronal cells have evolved networks of responses capable of detecting and controlling different forms of stress [26–62]. As such, integrated survival mechanisms exist in the brain based on the activity of redox-dependent genes, called vitagenes, capable of sensing stress and including HSP (Hsps), thioredoxin, γ-GC ligase, and sirtuin family proteins, which together with bioactive lipids represent the last step in the "omic" cascade starting from genome, through transcriptome, proteome, and finally to metabolome. Lipid mediators as signaling factors play a fundamental role in the initiation, amplification, and resolution of inflammation [33,34]. Thus, use of urine sample for lipidomic analysis enables reproducible quantification of several lipid metabolites generated by lipoxygenase, cyclooxygenase, and cytochrome P450 activities, such as octadecanoids, eicosanoids, and docosanoids. Lipidomic analysis of urine reveals quantitative data that reflects the alterations in in eicosanoids levels seen in MD patients as compared to normal controls. Lipoxin A4, in particular, is a metabolic product of arachidonic acid, acting as an endogenous "breaking signal" towards inflammatory processes, actively operating in the detection and control of diverse forms of stress in the brain. Owing to its potent anti-inflammatory properties LXA4 positively influences the outcome in many inflammatory disorders, such as nephritis, periodontitis, arthritis, and inflammatory bowel disease [63,64]. Chronic inflammation sustains the progression of neurodegenerative pathologies, including AD and Parkinson's disease, but also in specific neuronal districts, as in the cochleovestibular apparatus and the spiral ganglion neural cells. Thus, identification of mechanisms capable of favorably impacting the pro-inflammatory environment generated in the MD pathology represents an area of active investigation. Consistent with this notion, the activation of the LXA4 pathway could therefore serve as a potential therapeutic target to treat MD associated inflammation and cochleovestibular dysfunction. As LXA4 action is mediated by LXA4 receptor (ALX), a formyl-peptide receptor-like 1 (FPRL1) present on cellular membrane [33], the discovery of agents with the potential of increasing Lipoxin A4 (LXA4), and consequently of reducing inflammatory-mediated endolymphatic hydrops, can be relevant to therapeutics of this disease. Eicosanoid lipoxin A4 (LXA4) decreases toxic compounds such as ROS, inhibits recruitment of activated neutrophils and blocks accumulation of pro-inflammatory cytokines, thereby promoting resolution of inflammation [65].

Our results obtained with a nutritional approach based on a *Coriolus versicolor* biomass supplementation are relevant to innovative therapeutic anti-inflammatory strategies aimed to minimize consequences associated with neurodegeneration and oxidative stress of cochleovestibular system pathologies including not only MD but also sudden sensorineural hear loss (SSNHL) where it has recently demonstrated a critical role played by NLRP3 inflammasome [41]. NLRP3 is a sensor of the intracellular innate immune response expressed in immune cells, including monocytes and macrophages. Activation of the NLRP3 inflammasome results in augmentation of IL-1β secretion and cochlear autoinflammation. [66,67].

Due to different biological routes of actions, ranging from anticancer, antiviral, bacteriostatic, and regulation of of immune function, as well as antioxidant and protectant of hepatocytes [29,68], relevant to the inflammatory disease pathogenesis, mushrooms in the past have been diffusely applied for therapeutic use in traditional medicine [27,28]. It has been shown, in fact, that cytokine response triggered by activated immune cells occurs after stimulation with immunostimulatory molecules derived from mushroom preparations, which are mainly β-glucans [30,57,69–71]. Despite this, however, the active ingredients are not fully characterized, which makes mushroom extracts very difficult to reconcile with current pharmaceutical practices involving highly purified compounds and, therefore, difficult to patent, as they are complex mixtures of molecules of unknown concentrations to be administered for therapeutic purposes. In addition, mushroom-derived polysaccharides are complex molecules that cannot be synthesized, as the mass production of these compounds would require timely and costly extraction processes. Consequently, most research efforts have focused on low molecular weight compounds, such as cordycepin [72], which is a cytotoxic nucleoside analog inhibitor of cell proliferation. However, polysaccharopeptides isolated from *Coriolus versicolor* are well characterized and their commercial diffusion well established. In addition to its medical applications, *Coriolus versicolor* is widely used to degrade organic pollutants such as pentachlorophenol (PCP) [73]. Notably, as previously mentioned, several studies have demonstrated significant ultrastructural reductions in dendritic innervation densities, at level of cochlear ganglion neurons, pointing the possibility that neurotoxicity plays an important role in the pathology of MD [6]. Interestingly, a recent study in mice has found that *Coriolus versicolor* biomass promotes significant increases in dendritic length and branching and total dendritic volume of immature neurons, suggesting a positive effect of oral *Coriolus versicolor* administration on hippocampal neurogenic reserve [74]. Taking all this into account and given the inflammatory pathogenesis of MD degenerative damage, our findings of reduction in oxidative stress and inflammatory mediators associated with increased anti-inflammatory metabolites in mushroom-treated patients has innovative therapeutic potential.

Moreover, increasing evidence suggests that alteration of redox status, overloading of peroxidative product hydroxynonenals (HNE) or protein carbonyls can severely alter redox homeostasis [1]. Thus, the ensuing oxidative stress is a primary causative factor underlying endolymphatic hydrops pathogenesis, associated with cellular degenerative damage and apoptotic cell death affecting vulnerable cells of cochleovestibular apparatus, and thus contributing to the SNHL and vestibular dysfunction found in later stages of MD.

Moreover, it is known that normal auditory function depends on maintenance of the unique ion composition in the endolymph. Hence, reduction of microvascular alterations due to decreased oxidative stress in the inner ear after mushroom treatment has relevant implications [34]. Our data on the modulation of the stress-responsive protein involved in stress tolerance and cell survival are relevant as a potential target of mushrooms therapeutics and nutritional redox approaches, as the ability of neurons to cope with stressful conditions relies upon the capability to activate stress-responsive pro-survival pathways that normally function at a very low level and that result generally in increased synthesis of antioxidant and anti-apoptotic molecules. Among the cellular pathways conferring protection against oxidative stress, a key role is played by vitagenes, which include HSP (Hsps) Hsp70, heme oxygenase-1, and small Hsps, together with thioredoxin, enzymes of Meister cycle for the synthesis of glutathione and sirtuins [10,24]. Given the broad cytoprotective properties of the heat shock response there is now emerging interest in developing pharmacological agents able to potentiate neuroprotective stress responses [26]. When appropriately activated, cellular stress response can restore redox equilibrium and neuronal homeostasis.

4. Materials and Methods

4.1. Chemicals

5,5′-Dithiobis-(2-nitrobenzoic acid) (DTNB), 1,1,3,3-tetraethoxypropane, purified bovine blood SOD, NADH, glutathione (GSH), glutathione disulfide (GSSG), nicotinamide adenine dinucleotide phosphate (β-NADPH, type 1, tetrasodium salt), and glutathione reductase (GR; Type II from Baker's Yeast), were from Sigma Chemicals Co, St. Louis (USA). All other chemicals were from Merck (Darmstadt, Germany) and of the highest grade available.

4.2. Coriolus Versicolor Biomass Preparation

Coriolus versicolor is found almost worldwide; however, its bioactivity varies depending on the habitat in which it grows. To eliminate these variations, established CV-OH1 strain was used which demonstrates rapid and aggressive colonization. According to the manufacturer procedure (Mycology Research Laboratories Ltd., Luton, UK) *Coriolus versicolor* containing both mycelium and primordia (young fruit body) biomass, obtained cultivating the biomass that is grown on a sterilized (autoclaved) substrate. The production process involves the inoculation of sterile organic edible grain with spawn from the mother culture. The fungus is allowed to completely colonize the growth medium aseptically. At the correct stage of development, corresponding to the maximum bioavailability the living biomass is aseptically air-dried, granulated, tested microbiologically, and reduced in powder for tablet preparation. In comparison to *Coriolus* extracts, biomass has the advantage of preserving all nutraceutical potential which is usually reduced with extracts or concentrates, including lyophilization, and thus the activity of the product corresponds with the source mushroom, while being further intensified by using the entire mycelium. Tablets of 500 mg each of the *Coriolus* biomass containing mycelium and primordia of the respective mushroom, kindly provided by Mycology Research Laboratories Ltd. (MRL, Luton, UK), as the product commercially available, were used for experiments. Optimal dosage (200 mg/kg) was chosen according to the dose used in clinical trials with cancer or Human papilloma virus (HPV) patients (3 g/day) [57], a regimen also confirmed by studies in rat [33].

4.3. Ethical Permission

The study was approved by the local Ethics Committee (prot. N. 76/2018/PO, 16 April 2018) and informed consent was obtained from all patients.

4.4. Patients

We enrolled 40 patients (22 males and 18 females, with an average age of 49.5 +/− 14.6 years; range 29–60 years) with MD according to the diagnostic scale of the Committee on Hearing and Equilibrium of the American Academy of Otolaryngology—Head and Neck Surgery published in 1995 for MD [13,58] (two or more definitive spontaneous episodes of vertigo 20 min or longer, audiometrically documented hearing loss on at least one occasion, tinnitus or aural fullness in the treated ear). Patients were divided into two groups, A and B. Group A consisted of 22 patients suffering from cochlear sensorineural hearing loss (SNHL) that was been subjected to treatment with biomass preparation from *Coriolus versicolor* mushroom (MRLs), administered orally in tablets of 500 mg (3 tablets every 12 h, morning and evening, for 2 consecutive months), while Group B, formed of 18 patients, also suffering from cochlear SNHL, was not subjected to any treatment. Constituted exclusion criteria: (i) older than 60 years; (ii) presence of cardiovascular diseases; (iii) presence of metabolic disorders and/or parts; (iv) the presence of external ear pathologies and/or medium; (v) presence of alterations of state-acoustic nerve; (vi) prior learning and/or recent treatment with antioxidant drugs or otherwise active in the compartment cochlear.

All patients, after targeted anamnestic investigation, underwent the T0 initial phase, where the Profile of Mood States (POMS) questionnaire was administered, to assess the emotional and degree of psychological stress status, indexed on the basis of specific elements, such as: Tension–Anxiety (TA), Depression–Discouragement (D), Anger–Hostility (AH), Vigor–Activity (V), Fatigue (F), Confusion–Loss (C), in relation to the impairment caused in each subject from hearing impairment. The POMS original scale contains 65 self-report items using the 5-point Likert Scale. Participants can choose from 0 (not at all) to 4 (extremely). In addition, all subjects were given a tinnitus questionnaire consisting of 40 multiple choice questions to define the impact of symptoms on the patient life. In the groups the grade of severity for each patient was established on the basis of the vertigo attack frequency over a year (from 2 to 8 crisis), the intensity and the duration of symptoms (from a few days, to some weeks, to a month in the most severe case). In addition, the hearing loss degree was assessed instrumentally, allowing staging of the disease in MD patients.

Enrolled patients were also examined to define the qualitative and quantitative characteristics of auditory function: (a) examination ENT; (b) test tone audiometry; (c) speech audiometry; (d) impedenzometry examination. Such instrumental examinations were aimed at defining not only the extent of hearing, but also the location of the SNHL, to define it whether cochlear or retrocochlear. Each patient, either in Group A or Group B, was subjected to blood and urine sampling for biochemical analysis in plasma and lymphocytes, and in urines of specific markers of cellular oxidative stress, lipid and protein oxidative metabolism, cellular stress response (vitagenes), glutathione status (reduced glutathione (GSH), oxidized glutathione (GSSG), and GSH/GSSG ratio and lipoxin A4. Phase T1 in Group A, the mushroom-treated group of patients, was accomplished by oral administration of MRL *Coriolus versicolor* biomass compound for 2 consecutive months, to assess its neuroprotective, anti-neuroinflammatory potential, and thus test the possible protection against the cellular degeneration in general and, particularly, in the inner ear. At 2 months from the beginning of treatment (T1 phase) we evaluated in all patients the degree of evolutive trend of auditory function as well as cellular oxidative stresses, redox status, cellular stress response, and Lipoxin A4, in the blood. The correlative analysis, aimed at highlighting the antioxidant properties of the compound administered and its effects at the cellular level as well as at the clinical level, audiological function, will define a neurobiological clinical model to assess the effectiveness of pharmacological compounds in counteracting oxidative stress and neuroinflammatory damage associated with MD.

4.5. Sampling

Blood (6 mL) was collected after an overnight fast by venopuncture from an antecubital vein into tubes containing ethylenediamine tetraacetic acid (EDTA) as anticoagulant. Immediately after sampling, two blood aliquots were separated: first 2 mL were centrifuged at 10.000× g for 1 min at 4 °C to separate plasma from red blood cells; the remaining aliquot (4 mL) was used for lymphocytes purification. All samples were stored at −80 °C until analysis.

4.6. Lymphocytes Purification

Lymphocytes from peripheral blood were purified by using the Ficoll Paque System following the procedure as suggested by the manufacturer (GE Healthcare, Piscataway, NJ, USA).

4.7. Western Blot Analysis

Plasma samples were processed as such, while the isolated lymphocyte pellet was homogenized and centrifuged at 10,000× g for 10 min. The supernatant was then used for analysis after determination of protein content. Proteins extracted for each sample, at equal concentration (50 μg), were boiled for 3 min in sample buffer (containing 40 mM Tris-HCl pH 7.4, 2.5% SDS, 5% 2-mercaptoethanol, 5% glycerol, 0.025 mg/mL of bromophenol blue) and then separated by SDS-polyacrylamide gel electrophoresis (SDS-PAGE). Separated proteins were transferred onto nitrocellulose membrane (BIO-RAD Hercules, CA, USA) in transfer buffer containing 0.05% of SDS, 25 mM di Tris, 192 mM glycine and 20% v/v methanol. The transfer of the proteins on the nitrocellulose membrane was confirmed by staining with Ponceau Red which was then removed by 3 washes in PBS (phosphate buffered saline) for 5 min/each. Membranes were then incubated for 1 h at room temperature in 20 mM Tris pH 7.4, 150 mM NaCl and Tween 20 (TBS-T) containing 2% milk powder and incubated with appropriate primary antibodies, namely anti-γ-GC liase anti-Hsp70, anti-HO-1, anti-Sirt-1, anti-Trx and anti-HNE polyclonal antibody (Santa Cruz Biotech. Inc.), overnight at 4 °C in TBS-T. The same membrane was incubated with a goat polyclonal antibody anti-beta-actin (SC 1615 Santa Cruz Biotech. Inc., CA, USA, dilution 1:1000) to verify that the concentration of protein loaded in the gel was the same in each sample. The excess of unbound antibodies was removed by 3 washes with TBS-T for 5 min. After incubation with primary antibody, the membranes were washed 3 times for 5 min in TBS-T and then incubated for 1 h at room temperature with the secondary polyclonal antibody conjugated with horseradish peroxidase (dilution 1:500). The membranes were then washed 3 times with TBS-T for 5 min. Finally, the membranes were incubated for 3 min with SuperSignal chemiluminescence detection system kit (Cod 34080 Pierce Chemical Co, Rockford, USA) to display the specific protein bands for each antibody. The immunoreactive bands were quantified by capturing the luminescence signal emitted from the membranes with the Gel Logic 2200 PRO (Bioscience) and analyzed with Molecular Imaging software for the complete analysis of regions of interest for measuring expression ratios. The molecular weight of proteins analyzed was determined using a standard curve prepared with protein molecular weight.

4.8. Glutathione and Glutathione Disulfide Assay

GSH and GSSG were measured by the NADPH-dependent GSSG reductase method as previously reported in Calabrese et al. 2010. Lymphocytes were homogenized on ice for 10 s in 100 mM potassium phosphate, pH 7.5, which contained 12 mM disodium EDTA. For total glutathione, an aliquots (0.1 mL) of homogenates were immediately added to 0.1 mL of a cold solution containing 10 mM DTNB and 5 mM EDTA in 100 mM potassium phosphate, pH 7.5. The samples were then mixed by tilting and centrifuged at 12,000× *g* for 2 min at 4 °C. An aliquot (50 µL) of the supernatant was added to a cuvette containing 0.5 U of GSSG reductase in 100 mM potassium phosphate and 5 mM EDTA, pH 7.5 (buffer 1). After 1 min of equilibration, the reaction was initiated with 220 nmol of NADPH in buffer 1 for a final reaction volume of 1 mL. The formation of a GSH-DTNB conjugate was then measured at 412 nm. The reference cuvette contained equal concentrations of DTNB, NADPH, and enzyme, but not sample. For assay of GSSG, aliquots (0.5 mL) of homogenate were immediately added to 0.5 mL of a solution containing 10 mM N-ethylmaleimide (NEM) and 5 mM EDTA in 100 mM potassium phosphate, pH 7.5. The sample was mixed by tilting and centrifuged at 12,000× *g* for 2 min at 4 °C. An aliquot (500 µL) of the supernatant was passed at one drop/s through a SEP-PAK C18 Column (Waters, Framingham, MA) that had been washed with methanol followed by water. The column was then washed with 1 mL of buffer 1. Aliquots (865 µL) of the combined eluates were added to a cuvette with 250 nmol of DTNB and 0.5 U of GSSG reductase. The assay then proceeded as in the measurement of total GSH. GSH and GSSG standards in the ranges between 0 to 10 nmol and 0.010 to 10 nmol, respectively, added to control samples were used to obtain the relative standard curves, and the results were expressed in nmol of GSH or GSSG, respectively, per mL or mg protein.

4.9. Spontaneous Ultraweak Chemiluminescence Assay

Measurement of chemiluminescence in blood samples was accomplished according to the method of Flecha et al. 1991 [59]. Briefly, aliquots (0.5 mL) of plasma were diluted 1:1 with 30 mM phosphate buffer (pH 7.4), whereas lymphocyte pellet was homogenized and centrifuged at 10,000× *g* for 10 min. Before aliquots (0.5 mL) of the supernatant were taken and diluted 1:1 with 30 mM phosphate buffer (pH 7.4) at 0–4 °C and centrifuged at 10,000 *g* for 3 min at 0–4 °C. Then spontaneous ultraweak chemiluminescence (UCL) was measured in the supernatant at 30 °C with a Turner TD

20/20 luminometer. The sensitivity was adjusted to 50%, and results were expressed as luminescence units/mg protein.

4.10. Lipidomic Analysis

For oxylipins determination plasma and urine samples were extracted essentially as described by Wolfer et al. 2015 [60]. Briefly plasma samples were prepared by transferring 100 µL to the preparation plate after thawing and brief vortexing. A volume of 20 µL of IS working solution and 30 µL of 2% formic acid solution in water were added, and the plate was capped and gently mixed. The SPE plate was conditioned using 200 µL of MeOH and the sorbent equilibrated with 200 µL of H_2O. Samples were transferred from the preparation plate to the SPE, the preparation plate was further washed with 50 µL of MeOH/H_2O 1:1, and the rinsing solution will be added to the SPE plate. Following aspiration, the SPE plate will washed with 200 µL of H_2O + 2% NH_4OH and 200 µL of H_2O/ACN 1:1. Oxylipins will be eluted with 4×25 µL of MeOH + 2% formic acid. The elution fraction was evaporated under N_2 and the residues reconstituted in 120 µL of MeOH/H_2O. Urine samples were prepared by mixing 50 µL of urine with 50 µL of MeOH and 20 µL of IS working solution following thawing and brief vortexing.

Oxylipins determination was carried out with ultrahigh performance liquid chromatography (UHPLC) coupled with mass spectrometry (triple quadrupole, q-tof or orbitrap instrument) the methods and conditions of ionization were chosen to achieve the best results in order of quantification and reproducibility.

4.11. Lipoxin A4 Assay

LXA quantification was performed using an enzyme-linked immunosorbent assay (ELISA) kit following the protocol provided by the company. Biological fluid rates were used, and the measurement performed by a spectrophotometer at a wavelength of 450 nm.

4.12. Determination of Protein

Proteins were estimated by the bicinchoninic acid (BCA) protein assay method [61], using bicinchoninic acid reagent.

4.13. Statistical Analysis

Results were expressed as means ± SEM of $n = 18$ experiments (MD alone) or $n = 22$ experiments (MD plus *Coriolus*), each of which were performed, unless otherwise specified, in triplicate. Data were analyzed by one-way Analysis of Variance (ANOVA), followed by inspection of all differences by Duncan's new multiple-range test. Differences were considered significant at $p < 0.05$.

5. Conclusions

Brain cells, such as spiral ganglion neurons, possessing relatively weak endogenous antioxidant potential, show a particular need for activation of antioxidant pathways, which becomes a central prerequisite under conditions of oxidant insults, such those underlying not only the pathogenesis of MD but also acting in a broad range of age-associated diseases. Aging, in fact, is based on complex mechanisms and systemic processes, whose major gap remains insufficient knowledge about the proactive pathway shift from normal "healthy" aging to disease-associated pathological aging [75]. As a major complication of normal "healthy" aging, the increased risk of age-related diseases, such as cancer, diabetes mellitus, cardiovascular and neurodegenerative diseases, including cochleovestibular dysfunctions that can adversely affect the quality of life in general, with enhanced incidence of co-morbidities and mortality, should be considered.

We evaluated systemic oxidative stress and cellular stress response in MD patients in the absence and in the presence of treatment with a biomass preparation from *Coriolus versicolor*. It was concluded that systemic oxidative stress was reduced in MD patients treated with *Coriolus versicolor*, which was paralleled by a significant induction of vitagenes and by an increased plasma GSH vs. GSSG ratio. Vitagene up-regulation after *Coriolus versicolor* supplementation indicates a maintained response to counteract intracellular pro-oxidant status. In a contextualized global "omics" approach, the combination of redoxomics and lipidomics, being more stable than the metabolome and closer to the phenotype than the transcriptome, represents the most promising "omics" field enabling dissection and perhaps full comprehension at molecular and cellular level, of aging mechanisms and age-related processes.

Approaching the redox biology of the aging inner-ear system, as exploited in the present study, together with broadening of the potential of lipidomic analysis represents an innovative tool for monitoring at the omic level to the extent of oxidative insult and related modifications, allowing the identification of targeted antioxidative cytoprotective vitagene system proteins. The present study also highlights the importance of investigating MD as a convenient model of cochlear neurodegenerative disease.

Author Contributions: M.S. and P.D.M. equally contributed to this work. They were involved in experiment design, work execution and preparation of figures; M.L.O., C.A., D.C., A.T.S., and S.M. were involved in the drafting of this article; L.M. and V.C. designed, executed, and supervised the final version of the manuscript. All authors have read and agreed to the published version of the manuscript.

Funding: V.C. acknowledges support from Piano Ricerca Triennale-linea Intervento 2, University of Catania.

Conflicts of Interest: The authors declare no conflicts of interest.

Abbreviations

MD	Meniere's disease
HNE	4-hydroxynonenal
HSP72	Heat Shock Proteins 70
HO-1	Heme Oxygenase-1
TRX	Thioredoxin
γ-GC	gamma-glutamylcysteine liase
GSH	Reduced Glutathione
GSSG	Oxidized Glutathione
LXA4	Lipoxin A4
AD	Alzheimer's disease
FPRL1	Formyl-Peptide Receptor-Like 1
BRM	Biological Response Modifiers
TA	Tension–Anxiety
D	Depression–Discouragement
AH	Anger–Hostility
V	Vigor–Activity
F	Fatigue
C	Confusion–Loss
POMS	Profile of Mood States
MTHFR	Methylenetetrahydrofolate reductase
TLR	Toll-like receptor coding genes
ROS	Reactive Oxygen Species

References

1. Hallpike, S.C.; Cairns, H. Observations on the pathology of Meniere's syndrome. *J. Laryngol. Otol.* **1938**, *53*, 625–655. [CrossRef]
2. Megerian, C.A.; Cliff, A. Diameter of the cochlear nerve in endolymphatic hydrops: Implications for the etiology of hearing loss in Meniere's disease. *Laryngoscope* **2005**, *9*, 1525–1535. [CrossRef]

3. Capaccio, P.; Pignataro, L.; Gaini, L.M.; Sigismund, P.E.; Novembrino, C.; De Giuseppe, R. Unbalanced oxidative status in idiopathic sudden sensorineural hearing los. *Eur. Arch. Otorhinolaryngol.* **2012**, *269*, 449–453. [CrossRef]
4. Schreiber, B.E.; Agrup, C.; Haskard, D.O.; Luxon, L.M. Sudden sensorineural hearing loss. *Lancet* **2010**, *375*, 1203–1211. [CrossRef]
5. Chau, J.K.; Lin, J.R.; Atashband, S.; Irvine, R.A.; Westerberg, B.D. Systematic review of the evidence for the etiology of adult sudden sensorineural hearing loss. *Laryngoscope* **2010**, *120*, 1011–1021. [CrossRef]
6. Melki, S.J.; Heddon, C.M.; Frankel, J.K.; Levitt, A.H.; Momin, S.R.; Alagramam, K.N.; Megerian, C.A. Pharmacological protection of hearing loss in the mouse model of endolymphatic hydrops. *Laryngoscope* **2010**, *120*, 1637–1645. [CrossRef]
7. Merchant, S.N.; Adams, J.C.; Nadol, J.B. Pathology and pathophysiology of idiopathic sudden sensorineural hearing loss. *Otol. Neurotol.* **2005**, *26*, 151–160. [CrossRef]
8. Calabrese, V.; Dattilo, S.; Petralia, A.; Parenti, R.; Pennisi, M.; Koverech, G.; Calabrese, V.; Graziano, A.; Monte, I.; Maiolino, L.; et al. Analytical approaches to the diagnosis and treatment of aging and aging-related disease: Redox status and proteomics. *Free Radic. Res.* **2015**, *49*, 511–524. [CrossRef]
9. Pilipenko, V.; Narbute, K.; Amara, I.; Trovato Salinaro, A.; Scuto, M.; Pupure, J.; Jansone, B.; Poikans, J.; Bisenieks, E.; Klusa, V.; et al. GABA-containing compound gammapyrone protects against brain impairments in Alzheimer's disease model male rats and prevents mitochondrial dysfunction in cell culture. *J. Neurosci. Res.* **2019**, *97*, 708–726. [CrossRef]
10. Calabrese, V.; Santoro, A.; Trovato Salinaro, A.; Modafferi, S.; Scuto, M.; Albouchi, F.; Monti, D.; Giordano, J.; Zappia, M.; Franceschi, C.; et al. Hormetic approaches to the treatment of Parkinson's disease: Perspectives and possibilities. *J. Neurosci. Res.* **2018**, *96*, 1641–1662. [CrossRef]
11. Scuto, M.C.; Mancuso, C.; Tomasello, B.; Ontario, M.L.; Cavallaro, A.; Frasca, F.; Maiolino, L.; Trovato Salinaro, A.; Calabrese, E.J.; Calabrese, V. Curcumin, Hormesis and the Nervous System. *Nutrients* **2019**, *11*, 2417. [CrossRef]
12. Dattilo, S.; Mancuso, C.; Koverech, G.; Di Mauro, P.; Ontario, M.L.; Petralia, C.C.; Petralia, A.; Maiolino, L.; Serra, A.; Calabrese, E.J.; et al. Heat shock proteins and hormesis in the diagnosis and treatment of neurodegenerative diseases. *Immun. Ageing* **2015**, *12*, 20. [CrossRef]
13. Calabrese, V.; Cornelius, C.; Dinkova-Kostova, A.T.; Iavicoli, I.; Di Paola, R.; Koverech, A.; Cuzzocrea, S.; Rizzarelli, E.; Calabrese, E.J. Cellular stress responses, hormetic phytochemicals and vitagenes in aging and longevity. *Biochim. Biophys. Acta* **2012**, *13*, 86–103. [CrossRef]
14. Cornelius, C.; Trovato Salinaro, A.; Scuto, M.; Fronte, V.; Cambria, M.T.; Pennisi, M.; Bella, R.; Milone, P.; Graziano, A.; Crupi, R.; et al. Cellular stress response, sirtuins and UCP proteins in Alzheimer disease: Role of vitagenes. *Immun. Ageing* **2013**, *10*, 41. [CrossRef]
15. Calabrese, V.; Scapagnini, G.; Davinelli, S.; Koverech, G.; Koverech, A.; De Pasquale, C.; Trovato Salinaro, A.; Scuto, M.; Calabrese, E.J.; Genazzani, A.R. Sex hormonal regulation and hormesis in aging and longevity: Role of vitagenes. *J. Cell Commun. Signal.* **2014**, *8*, 369–384. [CrossRef]
16. Calabrese, V.; Santoro, A.; Monti, D.; Crupi, R.; Di Paola, R.; Latteri, S.; Cuzzocrea, S.; Zappia, M.; Giordano, J.; Calabrese, E.J.; et al. Aging and Parkinson's Disease: Inflammaging, neuroinflammation and biological remodeling as key factors in pathogenesis. *Free Radic. Biol. Med.* **2017**, *115*, 80–91. [CrossRef]
17. Wu, J.; Wang, A.; Min, Z.; Xiong, Y.; Yan, Q. Lipoxin A4 inhibits the production of proinflammatory cytokines induced by beta-amyloid in vitro and in vivo. *Biochem. Biophys. Res. Commun.* **2011**, *408*, 382–387. [CrossRef]
18. Medeiros, R.; Kitazawa, M.; Passos, G.F.; Baglietto-Vargas, D.; Cheng, D.; Cribbs, D.H.; La Ferla, F.M. Aspirin-triggered lipoxin A4 stimulates alternative activation of microglia and reduces Alzheimer disease-like pathology in mice. *Am. J. Pathol.* **2013**, *182*, 1780–1789. [CrossRef]
19. Jean-Louis, T.; Rockwell, P.; Figueiredo-Pereira, M.E. Prostaglandin J2 promotes O-GlcNAcylation raising APP processing by α- and β-secretases: Relevance to Alzheimer's disease. *Neurobiol. Aging* **2018**, *62*, 130–145. [CrossRef]
20. Joshi, Y.B.; Praticò, D. The 5-lipoxygenase pathway: Oxidative and inflammatory contributions to the Alzheimer's disease phenotype. *Front. Cell. Neurosci.* **2015**, *8*, 436. [CrossRef]
21. Dunn, H.C.; Ager, R.R.; Baglietto-Vargas, D.; Cheng, D.; Kitazawa, M.; Cribbs, D.H.; Medeiros, R. Restoration of lipoxin A4 signaling reduces Alzheimer's disease-like pathology in the 3xTg-AD mouse model. *J. Alzheimers Dis.* **2015**, *43*, 893–903. [CrossRef]

22. Gangemi, S.; Pescara, L.; D'Urbano, E.; Basile, G.; Nicita-Mauro, V.; Davì, G.; Romano, M. Aging is characterized by a profound reduction in anti-inflammatory lipoxin A4 levels. *Exp. Gerontol.* **2005**, *40*, 612–614. [CrossRef]

23. Chen, X.Q.; Wu, S.H.; Zhou, Y.; Tang, Y.R. Lipoxin A4-induced heme oxygenase-1 protects cardiomyocytes against hypoxia/reoxygenation injury via p38 MAPK activation and Nrf2/ARE complex. *PLoS ONE* **2013**, *8*, e67120. [CrossRef]

24. Amara, I.; Timoumi, R.; Annabi, E.; Di Rosa, G.; Scuto, M.; Najjar, M.F.; Calabrese, V.; Abid-Essefi, S. Di (2-ethylhexyl) phthalate targets the thioredoxin system and the oxidative branch of the pentose phosphate pathway in liver of Balb/c mice. *Environ. Toxicol.* **2019**, *5*, 78–86. [CrossRef]

25. Trovato Salinaro, A.; Cornelius, C.; Koverech, G.; Koverech, A.; Scuto, M.; Lodato, F.; Fronte, V.; Muccilli, V.; Reibaldi, M.; Longo, A.; et al. Cellular stress response, redox status, and vitagenes in glaucoma: A systemic oxidant disorder linked to Alzheimer's disease. *Front. Pharmacol.* **2014**, *5*, 129. [CrossRef]

26. Calabrese, V.; Cornelius, C.; Mancuso, C.; Ientile, R.; Stella, A.M.; Butterfield, D.A. Redox homeostasis and cellular stress response in aging and neurodegeneration. *Methods Mol. Biol.* **2010**, *610*, 285–308.

27. Elsayed, E.A.; El Enshasy, H.; Wadaan, M.A.; Aziz, R. Mushrooms a potential natural source of anti-inflammatory compounds for medical applications. *Mediat. Inflamm.* **2014**, *2014*, 805841. [CrossRef]

28. El Elsayed, H.; Elsayed, E.A.; Aziz, R.; Wadaan, M.A. Mushrooms and truffles: Historical biofactories for complementary medicine in Africa and in the Middle East. *Evid. Based Complement. Altern. Med.* **2013**, *2013*, 620451. [CrossRef]

29. Paterson, R.R.; Lima, N. Biomedical effects of mushrooms with emphasis on pure compounds. *Biomed, J.* **2014**, *37*, 357–368. [CrossRef]

30. Komura, D.L.; Ruthes, A.C.; Carbonero, E.R. Water-soluble polysaccharides from Pleurotus ostreatus var florida mycelial biomass. *Int. J. Biol. Macromol.* **2014**, *70*, 354–359. [CrossRef]

31. Cui, J.; Goh, K.K.; Archer, R. Characterisation and bioactivity of protein-bound polysaccharides from submerged-culture fermentation of Coriolus versicolor Wr-74 and ATCC-20545 strains. *J. Ind. Microbiol. Biotechnol.* **2007**, *34*, 393–402. [CrossRef] [PubMed]

32. Cheng, K.F.; Leung, P.C. General review of polysaccharopeptides PSP from C versicolor Pharmacological and clinical studies. *Cancer Ther.* **2008**, *6*, 117–130.

33. Trovato Salinaro, A.; Siracusa, R.; Di Paola, R.; Scuto, M.; Fronte, V.; Koverech, G.; Luca, M.; Serra, A.; Toscano, M.A.; Petralia, A.; et al. Redox modulation of cellular stress response and lipoxin A4 expression by Coriolus versicolor in rat brain: Relevance to Alzheimer's disease pathogenesis. *NeuroToxicology* **2016**, *53*, 350–358. [CrossRef] [PubMed]

34. Trovato Salinaro, A.; Pennisi, M.; Crupi, R.; Di Paola, R.; Alario, A.; Modafferi, S.; Di Rosa, G.; Fernandes, T.; Signorile, A.; Maiolino, L.; et al. Neuroinflammation and Mitochondrial Dysfunction in the Pathogenesis of Alzheimer's Disease: Modulation by Coriolus Versicolor (Yun-Zhi) Nutritional Mushroom. *J. Neurol. Neuromed.* **2017**, *2*, 19–28.

35. Butterfield, D.A.; Castegna, A.; Pocernich, C.; Drake, J.; Scapagnini, G.; Calabrese, V. Nutritional approaches to combat oxidative stress in Alzheimer's disease. *J. Nutr. Biochem.* **2002**, *13*, 444–461. [CrossRef]

36. Calabrese, V.; Mancuso, C.; Calvani, M.; Rizzarelli, E.; Butterfield, D.A.; Giuffrida, A. Nitric oxide in the CNS: Neuroprotection vs. *Neurotoxicity. Nat. Neurosci.* **2007**, *8*, 766–775. [CrossRef]

37. Calabrese, V.; Cornelius, C.; Maiolino, L.; Luca, M.; Chiaramonte, R.; Toscano, M.A.; Serra, A. Oxidative stress, redox homeostasis and cellular stress response in Ménière's disease: Role of vitagenes. *Neurochem. Res.* **2010**, *35*, 2208–2217. [CrossRef]

38. Kawaguchi, S.; Hagiwara, A.; Suzuki, M. Polymorphic analysis of the heat-shock protein 70 gene (HSPA1A) in Me'nie're's disease. *Acta Otolaryngol.* **2008**, *128*, 1173–1177. [CrossRef]

39. Calabrese, V.; Cornelius, C.; Trovato Salinaro, A.; Cambria, M.T.; Lo Cascio, M.S.; Di Rienzo, L.; Condorelli, D.; De Lorenzo, A.; Calabrese, E.J. The hormetic role of dietary antioxidants in free radical-related diseases. *Curr. Pharm. Des.* **2010**, *16*, 8778–8783. [CrossRef]

40. Suslu, N.; Yilmaz, T.; Gursel, B. Utility of anti-HSP 70, TNF alpha, ESR, antinuclear antibody, and antiphospholipid antibodies in the diagnosis and treatment of sudden sensorineural hearing loss. *Laryngoscope* **2009**, *119*, 341–346. [CrossRef]

41. Feng, W.; Rosca, M.; Fan, Y.; Hu, Y.; Feng, P.; Lee, H.G.; Monnier, V.M.; Fan, X. Gclc deficiency in mouse CNS causes mitochondrial damage and neurodegeneration. *Hum. Mol. Genet.* **2017**, *26*, 1376–1390. [CrossRef] [PubMed]

42. Esterbauer, H.; Schaur, R.J.; Zollner, H. Chemistry and biochemistry of 4-hydroxynonenal, malonaldehyde and related aldehydes. *Free Radic. Biol. Med.* **1991**, *11*, 81–128. [CrossRef]

43. Trovato Salinaro, A.; Pennisi, M.; Di Paola, R.; Scuto, M.; Crupi, R.; Cambria, M.T.; Ontario, M.L.; Tomasello, M.; Uva, M.; Maiolino, L.; et al. Neuroinflammation and neurohormesis in the pathogenesis of Alzheimer's disease and Alzheimer-linked pathologies: Modulation by nutritional mushrooms. *Immun. Ageing* **2018**, *15*, 8. [CrossRef] [PubMed]

44. Capaccio, P.; Ottaviani, F.; Cuccarini, V.; Ambrosetti, U.; Fagnani, E.; Bottero, A. Methylenetetrahydrofolate reductase gene mutations as risk factors for sudden hearing loss. *Am. J. Otolaryngol.* **2005**, *26*, 383–387. [CrossRef] [PubMed]

45. Capaccio, P.; Ottaviani Cuccarini, V.; Bottero, A.; Schindler, A.; Cesana, B.M. Genetic and acquired prothrombotic risk factors and sudden hearing loss. *Laryngoscope* **2007**, *117*, 547–551. [CrossRef]

46. Capaccio, P.; Cuccarini, V.; Ottaviani, F.; Fracchiolla, N.S.; Bossi, A.; Pignataro, L. Prothrombotic gene mutations in patients with sudden sensorineural hearing loss and cardiovascular thrombotic disease. *Ann. Otol. Rhinol. Laryngol.* **2009**, *118*, 205–210. [CrossRef]

47. Furuta, T.; Teranishi, M.; Uchida, Y.; Nishio, N.; Kato, K.; Otake, H. Association of interleukin-1 gene polymorphisms with sudden sensorineural hearing loss and Meniere's disease. *Int. J. Immunogenet.* **2011**, *38*, 249–254. [CrossRef]

48. Hiramatsu, M.; Teranishi, M.; Uchida, Y.; Nishio, N.; Suzuki, H.; Kato, K. Polymorphisms in genes involved in inflammatory pathways in patients with sudden sensorineural hearing loss. *J. Neurogenet.* **2012**, *26*, 387–396. [CrossRef]

49. Uchida, Y.; Sugiura, S.; Ando, F.; Shimokata, H.; Nakashima, T. Association of the C677T polymorphism in the methylenetetrahydrofolate reductase gene with sudden sensorineural hearing loss. *Laryngoscope* **2010**, *120*, 791–795. [CrossRef]

50. Uchida, Y.; Sugiura, S.; Nakashima, T.; Ando, F.; Shimokata, H. Contribution of 1425G/A polymorphism in protein kinase C-Eta (PRKCH) gene and brain white matter lesions to the risk of sudden sensorineural hearing loss in a Japanese nested case-control study. *J. Neurogenet.* **2011**, *25*, 82–87. [CrossRef]

51. Nishio, N.; Teranishi, M.; Uchida, Y.; Sugiura, S.; Ando, F.; Shimokata, H. Contribution of complement factor H Y402H polymorphism to sudden sensorineural hearing loss risk and possible interaction with diabetes. *Gene* **2012**, *499*, 226–230. [CrossRef] [PubMed]

52. Agrawal, Y.; Minor, L.B. Physiologic effects on the vestibular system in Meniere's disease. *Otolaryngol. Clin. N. Am.* **2010**, *43*, 985–993. [CrossRef] [PubMed]

53. Gürkov, R.; Jerin, C.; Flatz, W.; Maxwell, R. Clinical manifestations of hydropic ear disease (Menière's). *Eur. Arch. Otorhinolaryngol.* **2019**, *276*, 27–40, Erratum in *Eur. Arch. Otorhinolaryngol.* **2019**, *276*, 619–620. [CrossRef] [PubMed]

54. Derebery, M.J. Allergic and immunologic features of Meniere's disease. *Otolaryngol. Clin. N. Am.* **2011**, *44*, 655–666. [CrossRef]

55. Di Renzo, L.; Bianchi, A.; Saraceno, R.; Calabrese, V.; Cornelius, C.; Iacopino, L.; Chimenti, S.; De Lorenzo, A. 174G/C IL-6 gene promoter polymorphism predicts therapeutic response to TNF-α blockers. *Pharmacogenet. Genom.* **2012**, *22*, 134–142. [CrossRef]

56. Requena, T.; Gazquez, I.; Moreno, A.; Batuecas, A.; Aran, I.; Soto-Varela, A.; Santos-Perez, S.; Perez, N.; Perez-Garrigues, H.; Lopez-Nevot, A.; et al. Allelic variants in TLR10 gene may influence bilateral affectation and clinical course of Meniere's disease. *Immunogenetics* **2013**, *65*, 345–355. [CrossRef]

57. Monro, J.A. Treatment of cancer with mushroom products. *Arch. Environ. Health* **2003**, *58*, 533–537. [CrossRef]

58. Hoffer, M. Annual Meeting of the American Academy of Otolaryngology-Head and Neck Surgery. *Arch. Otolaryngol. Head Neck Surg.* **1996**, *122*, 202–203. [CrossRef]

59. Flecha, B.; Llesuy, S.; Boveris, A. Hydroperoxideinitiated chemiluminescence: An assay for oxidative stress in biopsies of heart, liver, and muscle. *Free Radic. Biol.* **1991**, *10*, 93–100. [CrossRef]

60. Wolfer, A.M.; Gaudin, M.; Taylor-Robinson, S.D.; Holmes, E.; Nicholson, J.K. Development and Validation of a High-hroughput Ultrahigh-Performance Liquid Chromatography-Mass Spectrometry Approach for Screening of Oxylipins and Their Precursors. *Anal. Chem.* **2015**, *87*, 11721–11731. [CrossRef]

61. Smith, P.K.; Krohn, R.I.; Hermanson, G.T.; Mallia, A.K.; Gartner, F.H.; Provenzano, M.D.; Fujimoto, E.K.; Goeke, N.M.; Olson, B.J.; Klenk, D.C. Measurement of protein using bicinchoninic acid. *Anal. Biochem.* **1985**, *150*, 76–85. [CrossRef]

62. Calabrese, V.; Giordano, J.; Crupi, R.; Di Paola, R.; Ruggieri, M.; Bianchini, R.; Ontario, M.L.; Cuzzocrea, S.; Calabrese, E.J. Hormesis, cellular stress response and neuroinflammation in schizophrenia: Early onset versus late onset state. *J. Neurosci. Res.* **2017**, *95*, 1182–1193. [CrossRef] [PubMed]

63. Serhan, C.N.; Chiang, N.; Van Dyke, T.E. Resolving inflammation: Dual anti-inflammatory and pro-resolution lipid mediators. *Nat. Rev. Immunol.* **2008**, *8*, 349–361. [CrossRef] [PubMed]

64. Serhan, C.N. Pro-resolving lipid mediators are leads for resolution physiology. *Nature* **2014**, *510*, 92–101. [CrossRef]

65. Wu, S.H.; Liao, P.Y.; Dong, L.; Chen, Z.Q. Signal pathway involved in inhibition by lipoxin A4 of production of interleukins in endothelial cells by lipopolysaccharide. *Inflamm. Res.* **2008**, *57*, 430–437. [CrossRef]

66. Pennisi, M.; Crupi, R.; Di Paola, R.; Ontario, M.L.; Bella, R.; Calabrese, E.J.; Crea, R.; Cuzzocrea, S.; Calabrese, V. Inflammasomes, hormesis, and antioxidants in neuroinflammation: Role of NRLP3 in Alzheimer disease. *J. Neurosci. Res.* **2017**, *95*, 1360–1372. [CrossRef]

67. Abais, J.M.; Xia, M.; Zhang, Y.; Boini, K.M.; Li, P.L. Redox regulation of NLRP3 inflammasomes: ROS as trigger or effector? *Antioxid. Redox Signal.* **2015**, *22*, 1111–1129. [CrossRef]

68. Xu, T.; Beelman, R.B.; Lambert, J.D. The cancer preventive effects of edible mushrooms. *Anticancer Agents Med. Chem.* **2012**, *12*, 1255–1263. [CrossRef]

69. Wasser, S.P. Medicinal mushroom science current perspectives advances evidences and challenges. *Biomed. J.* **2014**, *37*, 345–356. [CrossRef]

70. Lindequist, U.; Kim, H.W.; Tiralongo, E.; van Griensven, L. Medicinal Mushrooms. *Evid Based Complement. Altern. Med.* **2014**, *2014*, 806180. [CrossRef]

71. Da Silva, A.F.; Sartori, D.; Macedo, F.C., Jr.; Ribeiro, L.R.; Fungaro, M.H.; Mantovani, M.S. Effects of glucan extracted from Agaricus blazei on the expression of ERCC5 CASP9 and CYP1A1 genes and metabolic profile in HepG2 cells. *Hum. Exp. Toxicol.* **2013**, *32*, 647–654. [CrossRef] [PubMed]

72. Tuli, H.S.; Sharma, A.K.; Sandhu, S.S.; Kashyap, D. Cordycepin: A bioactive metabolite with therapeutic potential. *Life Sci.* **2013**, *93*, 863–869. [CrossRef] [PubMed]

73. Cui, J.; Chisti, Y. Polysaccharopeptides of Coriolus versicolor physiological activity uses and production. *Biotechnol. Adv.* **2003**, *21*, 109–122. [CrossRef]

74. Ferreiro, E.; Pita, I.; Mota, S.; Valero, J.; Ferreira, N.; Fernandes, T.; Calabrase, V.; Fontes-Ribeiro, C.; Pereira, F.; Rego, A. Coriolus versicolor biomass increases dendritic arboraization of newly-generated neurons in mouse hippocampal dentate gyrus. *Oncotarget* **2018**, *8*, 32929–32942.

75. Nam, S.I.; Yu, G.I.; Kim, H.J.; Park, K.O.; Chung, J.H.; Ha, E.; Shin, D.H. A polymorphism at-1607 2G in the matrix metalloproteinase-1 (MMP-1) increased risk of sudden deafness in Korean population but not at—519A/G in MMP-1. *Laryngoscope* **2011**, *121*, 171–175. [CrossRef] [PubMed]

International Journal of
Molecular Sciences

Review

Food Allergies and Ageing

Massimo De Martinis [1,2,*], Maria Maddalena Sirufo [1,2], Angelo Viscido [3] and Lia Ginaldi [1,2]

[1]   Department of Life, Health and Environmental Sciences, University of L'Aquila, 67100 L'Aquila, Italy;
     maddalena.sirufo@gmail.com (M.M.S.); lia.ginaldi@cc.univaq.it (L.G.)
[2]   Allergy and Clinical Immunology Unit, AUSL 04 Teramo, Italy
[3]   Gastroenterology Unit, Department of Life, Health and Environmental Sciences, University of L'Aquila,
     67100 L'Aquila, Italy; angelo.viscido@univaq.it
*   Correspondence: demartinis@cc.univaq.it; Tel.: +39-0861-429548; Fax: +39-0861-211395

Received: 16 October 2019; Accepted: 5 November 2019; Published: 8 November 2019

Abstract: All over the world, there is an increase in the overall survival of the population and the number of elderly people. The incidence of allergic reactions is also rising worldwide. Until recently, allergies, and in particular food allergies (FAs), was regarded as a pediatric problem, since some of them start in early childhood and may spontaneously disappear in adulthood. It is being discovered that, on the contrary, these problems are increasingly affecting even the elderly. Along with other diseases that are considered characteristics of advanced age, such as cardiovascular, dysmetabolic, autoimmune, neurodegenerative, and oncological diseases, even FAs are increasingly frequent in the elderly. An FA is a pleiomorphic and multifactorial disease, characterized by an abnormal immune response and an impaired gut barrier function. The elderly exhibit distinct FA phenotypes, and diagnosis is difficult due to frequent co-morbidities and uncertainty in the interpretation of in vitro and in vivo tests. Several factors render the elderly susceptible to FAs, including the physiological changes of aging, a decline in gut barrier function, the skewing of adaptive immunity to a Th2 response, dysregulation of innate immune cells, and age-related changes of gut microbiota. Aging is accompanied by a progressive remodeling of immune system functions, leading to an increased pro-inflammatory status where type 1 cytokines are quantitatively dominant. However, serum Immunoglobulin E (IgE) levels and T helper type 2 (Th2 cytokine production have also been found to be increased in the elderly, suggesting that the type 2 cytokine pattern is not necessarily defective in older age. Dysfunctional dendritic cells in the gut, defects in secretory IgA, and decreased T regulatory function in the elderly also play important roles in FA development. We address herein the main immunologic aspects of aging according to the presence of FAs.

Keywords: food allergy; elderly; aging; hypersensitivity; immunosenescence; gut; allergy; inflammation

1. Introduction

Food allergies (FAs) are becoming a relevant public health concern, affecting over 200 million people worldwide and its prevalence is increasing, mainly in developed countries [1]. FAs are characterized by a wide spectrum of manifestations affecting several organs, ranging from mild to severe and life-threatening reactions [2,3]. The diagnosis of a food allergy is complex because different immunologic mechanisms (IgE-mediated, cell-mediated, or mixed) may play a role. As in most immune-mediated diseases, the variability of clinical expression, as well as its growing prevalence, is determined by genotypic, epigenetic, and environmental factors [4–6].

FAs are much more common in children than in adults. Most FAs start in early childhood and usually disappear in adulthood. For this reason, FAs are often considered an almost exclusively pediatric disease. However, although its prevalence is greatest in young children, the occurrence of FA reactions is becoming frequent in the elderly [7,8]. Most studies on epidemiology, immunopathogenesis, and clinical manifestations of FAs have been conducted on children or adolescents rather than in elderly people. However, the demographic distribution of the world population is rapidly changing, with the proportion of older people on the rise and a significant percentage of them haveallergic diseases [9,10]. Parallel to these demographic changes, we can therefore expect that FAs, already increasing in the general population, will also increase in the elderly [11].

It is estimated that the current prevalence of allergic diseases in the elderly reaches 10% but this data is underestimated and also destined to increase [12]. An increasing proportion of children with a FA reaches adulthood and old age, and in some of them, the persistence of the allergic problem occurs. Furthermore, FAs can develop in adulthood and the first symptoms can occur even in elderly subjects [13]. However, to date, there is still not much attention given to FAs in aged people, and symptoms related to FAs, such as vomiting, dyspepsia, diarrhea, pruritus, and skin and respiratory manifestations, often remain undiagnosed in the elderly [14].

Besides the lack of epidemiological data, very little is known about the peculiar immunopathogenetic aspects and the clinical presentation of FAs in old people [15].

The aim of this review is to analyze the pathophysiological mechanisms underlying food allergy in the elderly, emphasizing the most peculiar aspects in this segment of the population, which form the basis of possible intervention measures.

2. Immune System Remodeling in the Elderly

The genetic background controls immunity and inflammation, and influences both the aging process and the development of allergies. Several underlying mechanisms of FAs in the elderly are now recognized, the first of which is immunosenecence, i.e., the peculiar age-related remodeling of the immune system. During senescence, both innate and adaptive immune reactions are deeply changed, favoring the development of FAs [16].

In the elderly, there is an imbalance of lymphocyte sub-populations, characterized by a decrease of naive lymphocytes with an accumulation of memory and senescent lymphocytes. Dysfunctions of immune regulatory cells, thymus involution, hematopoietic stem cell malfunctioning, dysregulation of apoptotic processes, a stress response, and mitochondrial function all contribute to the remodeling of the immune system in the elderly [17]. How the balance between the Th1 and Th2 branches is influenced by the aging process is still a controversial matter. Peripheral T cells from aged subjects are activated, exhibiting higher HLA-DR and CD69 expressions, as well as the increased production of inflammatory cytokines, including IL-1β, IL-6, IL-17, IL-31, and TNFα [18]. Immune responses are skewed toward a proallergenic Th2 profile. In particular, the increased IL-4, IL-5, and other Th2 cytokine production observed in aged subjects suggests a Th2 dominance in the elderly [19]. Such immune profile is the main substrate of the allergic reaction [20]. Moreover, certain Th1 cytokines that are increased in the elderly, such as IL-17, may also contribute to the progression of allergic inflammation. The age-related derangement of the cytokine profile may therefore influence the development of FAs in the elderly [21].

Increased inflammatory cytokines and antigen-presenting cell dysfunction contribute to allergic sensitization and inflammation in the aging. Dendritic cells, the mainstarter of the adaptive responses, exhibit altered costimulatory molecule expression in frail elderly subjects, conditioning dysfunctional antigen processing and presentation, which can elicit allergic responses [22].

Concerning effector cells of allergic reactions, most studies report a reduction in eosinophil degranulation in response to IL-5 stimulation and a decreased mast cell function notwithstanding a normal number of mast cells in the tissues [23].

A compromised T helper function and defects of the isotype switching leading to impaired immunologic memory and lower response to vaccines have been observed in the elderly [24]. Conversely, the IgE isotype is less compromised by aging [25]. In particular, immunosenescence does not influence IgE levels in aged patients with atopy, suggesting the persistence of allergy propensity into advanced age [14,26].

3. The Mucosal Immune System in Aging

Senescence affects not only the systemic immunity but also the local immune responses, especially on the gastrointestinal mucosa. The induction of mucosal tolerance is of paramount importance in mounting protective responses against new dietary antigens, therefore preventing FAs [3]. The gastrointestinal tract is the largest immunologic system with a relevant amount of lymphocytes that are both scattered and aggregated in lymphatic structures (Peyer's patches). This gut immune system exerts a key role in FA development, in particular in the elderly, when the thymic function has almost disappeared [27]. Age-related changes affecting the local immune responses contribute variously to the development of FAs. Mucosal tolerance induction is impaired in the elderly, whereas the effector phase of the allergic reaction is substantially maintained [28]. The mechanism of tolerance to food allergens is an active and ongoing process, and the age-related derangement of regulatory functions mediated by interactions of specific cell types promotes allergic sensitization [29,30]. Changes responsible for the breaking of oral tolerance take place in the gut associated lymphoid tissue. The oral tolerance is generally established in childhood and persists even over 65, unless new unknown allergens are introduced. Usually, new dietary protein intake may induce de novo sensitization in the elderly, whereas oral tolerance established in childhood and young age is generally maintained [31].

Inflammaging, the condition of chronic inflammation that drives senescence [32], increases the tight junction permeability through the effects of proinflammatory cytokines [33]. The epithelial cells in the gastrointestinal tract are themselves responsible for both the production of large amounts of cytokines and the reduction of the proteins of the tight junctions and occludens zonula, leading to an increased gut permeability [34]. This decreased barrier effect results in a rupture of the mechanism of tolerance, which predisposes patients to FAs. The presence of inflammatory cytokines, such as IFN-γ, IL-6, and IL-1β, in the gut mucosa is an important factor in this process [35].

The presence of opsonizing secretory IgA antibodies against food antigens is a central mechanism of mucosal immunity by reducing the attachment, penetration, and invasion of antigens across the mucosal wall. Secretory IgA supplied through breastfeeding protect newborns against harmful antigen penetration, leading to transient tolerance/immunity against oral allergens. This mucosal first-line defense mechanism deteriorates with age and orally induced antigen-specific IgA responses weaken [36]. The immunosenescence itself is associated with a significant reduction in IgA levels in the aged mucosa due to the decreased production by B cells and plasma cells [37]. The reduced IgA levels can reflect both an impaired migration of IgA-secreting plasma cells and their numerical reduction [23]. The decreased production of hyaluronic acid and mucus in the elderly also leads to a reduction in the mechanical protection and transport of antibacterial and defensive proteins to the mucosal surface, including IgA. Moreover, differences in the IgA repertoire between young and old subjects have been described, a difference that probably conditions a decreased efficacy of the IgA mediated defenses in the elderly [38]. IgA deficiency in the elderly is related to the development of FAs and intolerances.

4. Epithelial Barrier and Digestive Function Impairment

In the elderly, the integrity of the gut epithelial barrier is compromised, contributing to the chronic subclinical inflammatory state. Furthermore, the leaky epithelial barrier promotes Th2-immune responses by allowing allergens to penetrate into tissues where they are processed by dendritic cells and macrophages and presented to T cells [39]. Allergen-exposed epithelial cells produce cytokines, including thymic stromal lymphopoietin, which drive Th2 immune responses [40]. Impaired gut permeability therefore contributes to FA development. Derangement of the intestinal barrier integrity associated with aging may arise after gastroenteric mucosa damage [41]. The decreased digestive capacity of the stomach in the elderly, mainly caused by atrophic gastritis, is an additional risk factor for FAs. Gastric atrophy is frequent in the elderly, and depends on underlying diseases, alcohol abuse, or the chronic consumption of drugs, such as proton pump inhibitors or antacids. Long-term use of glucocorticoids determines a variety of serious side effects, including gastrointestinal effects [42]. Chronic alcohol abuse notoriously enhances the gastric mucosa permeability, induces atrophic gastritis, and decreases the gastric secretory capacity [43,44]. The consequent hypoacidity prevents cleavage of the inactive pro-enzyme pepsinogen and the activation of its protease function in the gastric lumen, thus food proteins remain undigested and transit to the intestine. Such intact food proteins can cross the gut mucosa and enter the blood stream, eliciting the production of IgE antibodies. After a consecutive ingestion of the same food protein, the allergen can crosslink IgE on effector cells, namely mast cells, and trigger the release of mediators, including histamine and leukotrienes, which are the elicitors of local and systemic allergic reactions, whose clinical severity is also partly determined by allergen dosage and integrity [45]. Therefore, a physiologically low gastric pH, by allowing an optimal protein digestibility and preventing the sensitizing and eliciting capacity of the allergen, represents an important protective factor against FAs [14].

Food allergens are mostly structurally stable proteins that usually present a greater risk of causing systemic reactions. When digestion is compromised, labile food proteins can also persist partially undigested along the gastrointestinal tract and become food allergens [46]. The gastric protease propepsin is activated only for pH values below 3.0. Furthermore, only acidic chymus entering the duodenum can induce the release of pancreatic enzymes. Thus, because of the decrease in gastric acidity in the elderly population, protein digestion is compromised and harmless proteins are transformed into potentially dangerous allergens [47]. The therapy with proton pump inhibitors in the elderly, through these mechanisms, could thus facilitate the sensitization to food allergens or lower the trigger threshold of the allergic reaction if a FA is already present [48].

Also, age-related changes in organs and systems different from the gut can exert an important role in the development of FAs in the elderly. The skin is one of the main targets of an allergic reaction to food, as well as an important site of primary sensitization. As a result of chronological and environmental factors, the aged skin is characterized by atrophy and dehydration [49]. The progressive loss of structural integrity leads to an impaired immune response and skin barrier function, increased reactive oxygen species and extracellular matrix component, and vascular impairment [50]. Although T-cell-mediated immunity appears decreased, elderly patients can develop contact dermatitis, as well as sensitize themselves through the skin to food allergens [3].

5. Age-Associated Microbial Dysbiosis

In addition to the impaired function of the local immunity and increased gastrointestinal mucosa permeability, age associated alterations of the gut microbiota may also favor FA development in the elderly.

The gut microbiota is a complex ecological system that exerts a central role in several physiological functions, and its composition changes throughout the host's life. It is sensitive to environmental influences and the host's diet, and depends on the host's genetics, gender, and the aging process per se [51].

The system of the secretory IgA plays a critical role not just for the defense against infections but also for the modulation of local immune responses through the maintenance of the intestinal microbiota. Inflammatory processes are associated with dysregulation of the homeostatic interactions between the intestinal microbiota and the aging host [52].

Intestinal microbiota exhibit significant age-associated changes in composition and diversity, as well as in functional features, mainly caused by the immune system remodeling and low-grade chronic inflammation, which respectively characterize immunosenescence and inflammaging [22].

Immunosenescence exerts a key role by modifying the host's response to microbiota, triggering inflammaging, and shifting Th1 versus Th2 responses, thus favoring tolerance dysruption and allergic reaction development [52].

Antibiotics are among the most commonly used drugs in the elderly and are often used improperly. They influence the microbiome composition and function interfering with immune homeostasis. In the geriatric age, antibiotics can further disturb the composition of the microbiota [53,54]. However, even in the elderly, the intestinal flora can be reconstituted by probiotics, but it is not yet known how this can prevent the development of FAs [55].

6. Immune Dysfunctions Due to Nutritional Deficits

Together with the peculiar remodeling of the immune system during senescence, the compromised integrity of epithelial barriers and the sub-clinical chronic inflammatory condition commonly observed in the elderly, a central role in sensitization to food allergens is also played by the lack of micronutrients and vitamins [12].

Micronutrients and antioxidants modulate immune responses and it is suggested that their deficits favor the development of Th2 type responses. For example, deficits of iron, zinc, and vitamin D, which are very common in the elderly, may represent additional risk factors for the onset of allergic reactions during senescence [10].

Zinc is an essential trace element that plays a central role regarding the immune efficiency. Zinc intracellular homeostasis, regulated by metallothioneins and specific transporter proteins, is altered in aging, leading to its decreased availability for immune functions. A reduced zinc level, frequently observed in the elderly, could be responsible for a decreased production of Th1 cytokines, whereas this does not affect the production of Th2 cytokines, thus inducing a cytokine imbalance that promotes the development of allergic diseases [56].

Zinc deficiency contributes to thymic atrophy; immature B cell accumulation; and decreased IgM, Ig-G2a, and IgA subclasses. Stress situations, through pro-inflammatory cytokine production, including IL-6 and TNF-α, are often associated withzinc deficiency. Inflammatory cytokines, permanently increased in the geriatric population, bind zinc ions with a consequent reduced zinc bioavailability and altered immune functions. In particular, decreased levels of zinc induce a reduction of Th1 cytokines, such as IFN-γ, IL-2, and TNF-α, while Th2 cytokines, in particular IL-4, are enhanced. Through this mechanism, zinc deficiency could favor the development of FAs in the elderly [57].

Iron deficiency is also frequent in the elderly [20]. The decreased iron level induces impaired humoral responses, and in particular reduces the production of the IgG4 subclass that physiologically captures the allergens before they can bind to the IgE, thus preventing the activation of effector cells, such as mastocytes and basophils [58].

Several studies suggest that vitamin D deficiency is also very common in the elderly, supporting FA development. Immune dysregulation, in addition to an increased parathyroid hormone level and impaired bone health resulting in enhanced risk of fractures, is a serious consequence of vitamin D deficiency in the elderly [59,60]. The active metabolite of vitamin D, calcitriol, influences T lymphocytes and antigen-presenting cells to induce peripheral tolerance by inhibiting inflammatory responses and promoting regulatory T cells [61]. Vitamin D deficiency is therefore associated with an increased risk of autoimmune and atopic diseases, although the association with IgE levels is not clear [62].

7. Clinical Features of FAs in the Elderly

The variable natural history and the complexity of the possible pathogenetic mechanisms, as well as several age-associated factors, make the diagnosis and management of FAs in the elderly difficult. Nutritional abnormalities and vitamin deficiencies, as well as hormonal imbalances and inflammaging, interacting with genetics, may alter the immune responses, leading to FA development. The age-related decline of physiological functions, in addition to the immune system remodeling, which characterizes senescence, contribute to confer peculiar clinical findings to FA in the elderly [23,63].

Furthermore, despite the normal or increased number of mast cells in the skin of aged subjects with an allergy and a sufficient positive response to prick tests with specific allergens, elderly show weaker cutaneous responses and less intense pomfoid reactions to histamine control [64]. Therefore, since the positive reactions to skin test for an allergy could be partially reduced in the elderly, creating possible risk of false negative skin test, a specific IgE search to diagnose FA is commonly used in older patients [65].

Frailty, comorbidity, and multi-drug intake are conditions commonly found in the elderly and must be taken into account in the management of aged people affected by FAs. Immunologic reactions to foods can be confused with symptoms of other common age-related diseases or be masked by the use of various drugs. Consequently, the characteristic symptoms of FAs often go unnoticed and this contributes to underestimating FA prevalence in old age [7]. Dryness and hyperkeratosis, with consequent itching and increased risk of skin infections, are dermatologic manifestations that often mimic and/or mask the symptoms of an allergy. Cutaneous symptoms, such as atopic dermatitis and urticaria, could also represent manifestations of FAs in the elderly [25]. However, in addition to FAs, even drugs and systemic diseases, mainly hematologic and immune dysfunctions, can also induce urticaria in the elderly. Underlying diseases must, therefore, always be suspected, especially when a new diagnosis of chronic urticaria is made in an elderly person [66,67]. Although aged individuals can respond to immunotherapy for a respiratory allergy, as well as to biological drugs commonly used for the treatment of allergic manifestations, such as urticaria and atopic dermatitis, they are usually excluded from these kinds of therapy [68,69]. This is due to the frequent presence of age-related clinical conditions that are considered contraindications. Moreover, the common occurrence of comorbidities and multidrug intake can affect the therapeutic response and promote the onset of side effects. However, immunotherapy and biologic drugs could also significantly improve the quality of life in the elderly, reducing symptoms and drug consumption [70].

Anaphylaxis is a severe and life-threatening hypersensitivity reaction that can affect allergic patients at any age [45]. Clinical manifestations of anaphylaxis caused by food allergens are less frequent in the elderly compared to young subjects [71]. However, although less common, anaphylaxis exhibits a worst prognosis in older patients [72]. The anaphylaxis mediators released by the mast cells after the binding of the allergen to the IgE anchored to their surface induce profound functional modifications on the cardiocirculatory system, including vasospasms of coronary arteries with reduced myocardial blood flow and arhythmias [10,73]. The age-related susceptibility of the cardiovascular system to mast-cell-derived mediators and underlying comorbidities, such as coronary diseases, contribute to the increased mortality and frequent cardiovascular involvement during anaphylaxis in aged people [74]. In patients with multimorbidities, multidrug prescriptions are important cofactors complicating anaphylactic events in the elderly [75]. Cardiovascular drugs, increasingly prescribed to the elderly, strongly contribute to thegreater probability of a fatal outcome. Beta-blockers and angiotensin-converting enzyme (ACE) inhibitors, commonly used to treat congestive heart failure and hypertension, may in fact contribute to aggravate the impairment of compensatory mechanisms typical of the elderly [76]. Several other drugs may interfere with allergic effector cells of FAs. Nonsteroidal anti-inflammatory drugs, taken for chronic osteoarticular pain, are relevant cofactors in urticaria and anaphylaxis in aged subjects [67]. Tricyclic antidepressants, monoamine oxidase inhibitor, and neuroleptics may increase the cardiac risk of epinephrine administration. All these different drugs could cause hypotension, accelerate and increase exposure to allergens, and mask the symptoms of a possible allergic reaction. Although the advanced age doesnot represent an absolute contraindication to self-injectable adrenaline prescription in those at higher risk of anaphylaxis, impaired neuro-motor coordination, frequent hypomobility, and the common coexistence of osteo-muscular and arthrosic hand pathologies compromise the ability to use auto-injectors, suggesting caution in this prescription [77].

The intake of antiulcer drugs is common in the elderly to cure gastritis, gastroesophageal reflux, gastric ulcers, or in association withcorticosteroids and non-steroidal painkillers to minimize their gastrolesive effects. Gastric hypoacidity and increased permeability of the upper gastrointestinal tract also occur as a result of therapy with acid-suppressive drugs, facilitating the onset of an FA, as well as eosinophilic esophagitis. Elderly patients treated with proton pump inhibitors or H2-receptor blockers are at higher risk for sensitization because dietary proteins both remain incompletely digested and can cross the mucosal barrier more easily due to the increased permeability, thus becoming allergenic [8,49].

8. Conclusions

Adverse food reactions show peculiar characteristics in the elderly that concern both the pathogenesis and the clinic. FAs in the elderly are driven by immunosenescence, as well as the cell aging and tissue modifications that characterize advanced age. The aged gastrointestinal mucosa is central in the development of FAs in the elderly through its compromised digestive properties and structural changes, as well as the alteration of its immune functions linked to immunosenescence and age-related microbiota remodeling. Among the risk factors for the sensitization to food allergens in the elderly, in addition to chronic damage and inflammation of gut epithelia due to the aging process, there are chronic alcohol consumption, chronic infections, multimorbility, polymedication, and drug side effects (Figure 1).

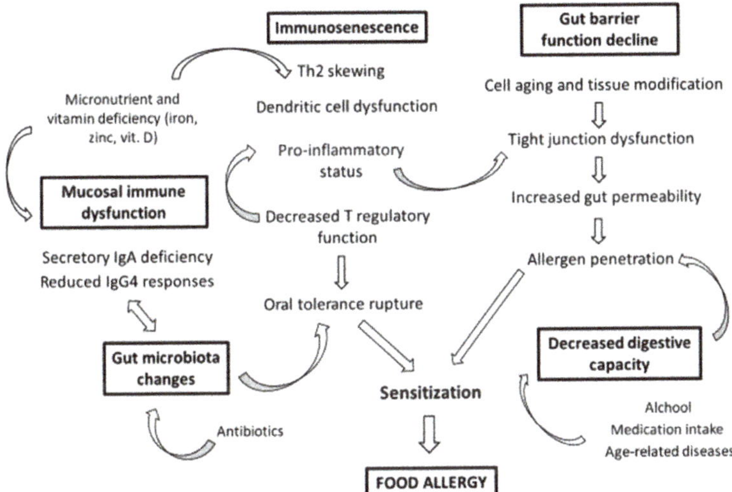

Figure 1. Food allergies in the elderly. The figure shows the main risk factors for the development of a food allergy in the elderly. Immunosenescence and the mucosal immune dysfunction of the gastrointestinal tract are driving forces in the development of food allergies in the elderly. The gut barrier function decline and the compromised digestive properties, as well as the age-related microbiota remodeling, are also central factors for both allergen penetration and sensitization.

Author Contributions: All authors contributed equally to the work.

Funding: The study has not received any funding.

References

1. Tang, M.L.K.; Mullins, R.J. Food allergy: Is prevalence increasing? *Intern. Med. J.* **2017**, *47*, 256–261. [CrossRef] [PubMed]
2. Sicherer, S.H.; Sampson, H.A. Food allergy: A review and update on epidemiology, pathogenesis, diagnosis, prevention, and management. *J. Allergy Clin. Immunol.* **2018**, *141*, 41–58. [CrossRef] [PubMed]
3. De Martinis, M.; Sirufo, M.M.; Viscido, A.; Ginaldi, L. Food allergy insights: A changing landscape. *Arch. Immunol. Ther. Exp.* **2019**, in press.
4. Di Lorenzo, G.; Di Bona, D.; Belluzzo, F.; Macchia, L. Immunological and non-immunological mechanisms of allergic diseases in the elderly: Biological and clinical characteristics. *Immun. Ageing* **2017**, *14*, 23. [CrossRef] [PubMed]

5. Renz, H.; Allen, K.J.; Sicherer, S.H.; Sampson, H.A.; Lack, G.; Beyer, K.; Oettgen, H.C. Food allergy. *Nat. Rev. Dis. Primers* **2018**, *4*, 17098. [CrossRef] [PubMed]

6. De Martinis, M.; Ciccarelli, F.; Sirufo, M.M.; Ginaldi, L. An overview of environmental risk factors in systemic sclerosis. *Expert Rev. Clin. Immunol* **2016**, *12*, 465–478. [CrossRef] [PubMed]

7. Ventura, M.T.; AD'Amato, A.; MGiannini, M.; Carretta, A.; Tummolo, R.A.; Buquicchio, R. Incidence of allergic diseases in an elderly population. *Immunopharmacol. Immunotoxicol.* **2010**, *32*, 165–170. [CrossRef]

8. Möhrenschlager, M.; Ring, J. Food Allergy: An Increasing Problem for the Elderly. *Gerontology* **2011**, *57*, 33–36. [CrossRef]

9. Gavrilov, L.A.; Krut'ko, V.N.; Gavrilova, N.S. The future of human longevity. *Gerontology* **2017**, *63*, 524–526. [CrossRef]

10. Cardona, V.; Guilarte, M.; Luengo, O.; Labrador-Horrillo, M.; Sala-Cunill, A.; Garriga, T. Allergic diseases in the elderly. *Clin. Transl. Allergy* **2011**, *1*, 11. [CrossRef]

11. Nanda, A.; Baptist, A.P.; Divekar, R.; Parikh, N.; Seggev, J.S.; Yusin, J.S.; Nyenhuis, S.M. Asthma in the older adult. *J. Asthma* **2019**. [CrossRef] [PubMed]

12. Willits, E.K.; Park, M.A.; Hartz, M.F.; Schleck, C.D.; Weaver, A.L.; Joshi, A.Y. Food Allergy: A Comprehensive Population-Based Cohort Study. *Mayo Clin. Proc.* **2018**, *93*, 1423–1430. [CrossRef] [PubMed]

13. De Martinis, M.; Sirufo, M.M.; Ginaldi, L. Allergy and Aging: An Old/New Emerging Health Issue. *Aging Dis.* **2017**, *8*, 162–175. [CrossRef] [PubMed]

14. Jensen-Jarolim, E.; Jensen, S.A.F. Food allergies in the elderly: Collecting the evidence. *Ann. Allergy Asthma Immunol.* **2016**, *117*, 472–475. [CrossRef] [PubMed]

15. Milgrom, H.; Huang, H. Allergic Disorders at a Venerable Age: A Mini-Review. *Gerontology* **2014**, *60*, 99–107. [CrossRef] [PubMed]

16. Latza, E.; Duewell, P. NLRP3 inflammasome activation in inflammaging. *Semin. Immunol.* **2018**, *40*, 61–73. [CrossRef]

17. Alberti, S.; Cevenini, E.; Ostan, R.; Capri, M.; Salvioli, S.; Bucci, L.; Ginaldi, L.; De Martinis, M.; Franceschi, C.; Monti, D. Age-Dependent modifications of Type 1 and Type 2 cytokines within virgin and memory CD4+ T cells in humans. *Mech. Ageing Dev.* **2006**, *127*, 560–566. [CrossRef]

18. Ginaldi, L.; De Martinis, M.; Ciccarelli FSaitta, S.; Imbesi, S.; Mannucci, C.; Gangemi, S. Increased levels of interleukin 31 (IL-31) in osteoporosis. *BMC Immunol.* **2015**, *16*, 60. [CrossRef]

19. Ginaldi, L.; De Martinis, M.; Saitta, S.; Sirufo, M.M.; Mannucci, C.; Casciaro, M.; Ciccarelli, F.; Gangemi, S. Interleukin-33 serum levels in postmenopausal women with osteoporosis. *Sci. Rep.* **2019**, *9*, 3786. [CrossRef]

20. Gold, D.R. Allergy: The price paid for longevity and social wealth? *J. Allergy Clin. Immunol.* **2006**, *117*, 148–150. [CrossRef]

21. Diesner, S.C.; Untersmayr, E.; Pietschmann, P.; Jensen-Jarolim, E. Food Allergy: Only a Pediatric Disease? *Gerontology* **2011**, *57*, 28–32. [CrossRef] [PubMed]

22. Accardi, G.; Caruso, C. Immune-inflammatory responses in the elderly: An update. *Immun. Ageing* **2018**, *15*, 11. [CrossRef] [PubMed]

23. Ventura, M.T.; Scichilone, N.; Paganelli, R.; Minciullo, P.L.; Patella, V.; Bonini, M.; Passalacqua, G.; Lombardi, C.; Simioni, L.; Ridolo, E.; et al. Allergic diseases in the elderly: Biological characteristics and main immunological and non-immunological mechanisms. *Clin. Mol. Allergy* **2017**, *15*, 2. [CrossRef] [PubMed]

24. Rosenberg, C.; Bovin, N.V.; Bram, L.V.; Flyvbjerg, E.; Erlandsen, M.; Vorup-Jensen, T.; Petersen, E. Age is an important determinant in humoral and T cell responses to immunization with hepatitis B surface antigen. *Hum. Vaccines Immunother.* **2013**, *9*, 1466–1476. [CrossRef] [PubMed]

25. Tanei, R.; Hasegawa, Y.; Sawabe, M. Abundant immunoglobulin E-positive cells in skin lesions support an allergic etiology of atopic dermatitis in the elderly. *J. Eur. Acad. Dermatol. Venereol.* **2013**, *27*, 952–960. [CrossRef]

26. Gunin, A.G.; Kornilova, N.K.; Vasilieva, O.V.; Petrov, V.V. Age-Related changes in proliferation, the numbers of mast cells, eosinophils, and cd45-positive cells in human dermis. *J. Gerontol. A Biol. Sci. Med. Sci.* **2011**, *66*, 385–392. [CrossRef]

27. Valdiglesiasa, V.; Marcos-Péreza, D.; Lorenzi, M. Immunological alterations in frail older adults: A cross sectional study. *Exp. Gerontol.* **2018**, *112*, 119–126. [CrossRef]

28. Untersmayr, E.; Diesner, S.C.; Bramswig, K.H.; Knittelfelder, R.; Bakos, N.; Gundacker, C.; Lukschal, A.; Wallmann, J.; Szalai, K.; Pali-Schöll, I.; et al. Characterization of intrinsic and extrinsic risk factors for celery allergy in immunosenescence. *Mech. AgeingDev.* **2008**, *129*, 120–128. [CrossRef]

29. Jagger, A.; Shimojima, Y.; Goronzy, J.J.; Weyand, C. Regulatory T cells and immune aging process: A mini-review. *Gerontology* **2014**, *60*, 130–137. [CrossRef]

30. Minciullo, P.L.; Catalano, A.; Mandraffino, G.; Casciaro, M.; Crucitti, A.; Maltese, G.; Morabito, N.; Lasco, A.; Gangemi, S.; Basile, G. Inflammaging and Anti-Inflammaging: The Role of Cytokines in Extreme Longevity. *Arch. Immunol. Ther. Exp.* **2016**, *64*, 111–126. [CrossRef]

31. Anto, J.M.; Bousquet, J.; Akdis, M.; Auffray, C.; Keil, T.; Momas, I.; Postma, D.S.; Valenta, R.; Wickman, M.; Cambon-Thomsen, A.; et al. Mechanisms of the development of allergy: Introducing novel concepts in allergy phenotypes. *J. Allergy Clin. Immunol.* **2017**, *139*, 388–399. [CrossRef] [PubMed]

32. Campisi, G.; Chiappelli, M.; De Martinis, M.; Franco, V.; Ginaldi, L.; Guiglia, R.; Licastro, F.; Lio, D. Pathophysiology of age-related diseases. *Immun. Ageing* **2009**, *6*, 12. [CrossRef] [PubMed]

33. Chung, H.Y.; Kim, D.H.; Lee, E.K.; Chung, K.W.; Chung, S.; Lee, B.; Seo, A.Y.; Chung, J.H.; Jung, Y.S.; Im, E.; et al. Redefining chronic inflammation in aging and age-related diseases: Proposal of the senoinflammation concept. *Aging Dis.* **2019**, *10*, 367–382. [CrossRef] [PubMed]

34. Svoboda, M.; Bilkova, Z.; Muthny, T. Could tight junctions regulate the barrier function of the aged skin? *J. Dermatol. Sci.* **2016**, *81*, 147–152. [CrossRef] [PubMed]

35. Ventura, M.T.; Casciaro, M.; Gangemi, S.; Buquicchio, R. Immunosenescence in aging: Between immune cells depletion and cytokines up-regulation. *Clin. Mol. Allergy* **2017**, *15*, 21. [CrossRef] [PubMed]

36. Mabbott, N.A.; Kobayashi, A.; Sehgal, A.; Bradford, B.M.; Pattison, M.; Donaldson, D.S. Aging and the mucosal immune system in the intestine. *Biogerontol* **2015**, *16*, 133–145. [CrossRef]

37. Frasca, D.; Blomberg, B.B. Effects of aging on B cell function. *Curr. Opin. Immunol.* **2009**, *21*, 425–430. [CrossRef]

38. Ginaldi, L.; Mengoli, L.P.; De Martinis, M. Osteoporosis, inflammation and ageing. In *Handbook on Immunosenescence: Basic Understanding and Clinical Applications*; Springer Nature Switzerland AG: Basel, Switzerland, 2009; pp. 1329–1352.

39. Sato, S.; Kiyono, H.; Fujihashi, K. Mucosal immunosenescence in the gastrointestinal tract: A mini-review. *Gerontology* **2015**, *61*, 336–342. [CrossRef]

40. Al-Sadi, R.; Ye, D.; Said, H.M.; Ma, T.Y. IL-1beta-induced increase in intestinal epithelial tight junction permeability is mediated by MEKK-1 activation of canonical NF-kappaB pathway. *Am. J. Pathol.* **2010**, *177*, 2310–2322. [CrossRef]

41. Tran, L.; Greenwood-Van Meerveld, B. Age-Associated remodeling of the intestinal epithelial barrier. *J. Gerontol. A Biol. Sci. Med. Sci.* **2013**, *68*, 1045–1056. [CrossRef]

42. Ciccarelli, F.; De Martinis, M.; Ginaldi, L. Glucocorticoids in Patients with Rheumatic Diseases: Friends or Enemies of Bone? *Curr. Med. Chem.* **2015**, *22*, 596–603. [CrossRef] [PubMed]

43. Gonzalez-Quintela, A.; Gude, F.; Boquete, O.; Rey, J.; Meijide, L.M.; Suarez, F.; Fernández-Merino, M.C.; Pérez, L.F.; Vidal, C. Association of alcohol consumption with total serum immunoglobulin E levels and allergic sensitization in an adult population-based survey. *Clin. Exp. Allergy* **2003**, *33*, 199–205. [CrossRef] [PubMed]

44. Bakos, N.; Scholl, I.; Szalai, K.; Kundi, M.; Untersmayr, E.; Jensen-Jarolim, E. Risk assessment in elderly for sensitization to food and respiratory allergens. *Immunol. Lett.* **2006**, *107*, 15–21. [CrossRef] [PubMed]

45. De Amici, M.; Ciprandi, G. The age impact on serum total and allergen-specific IgE. *Allergy Asthma Immunol. Res.* **2013**, *5*, 170–174 [CrossRef] [PubMed]

46. Mullin, J.M.; Valenzano, M.C.; Whitby, M.; Lurie, D.; Schmidt, J.D.; Jain, V.; Tully, O.; Kearney, K.; Lazowick, D.; Mercogliano, G.; et al. Esomeprazole induces upper gastrointestinal tract transmucosal permeability increase. *Aliment. Pharmacol. Ther.* **2008**, *28*, 1317–1325. [CrossRef]

47. Untersmayr, E.; Jensen-Jarolim, E. The role of protein digestibility and antiacids on food allergy outcomes. *J. Allergy Clin. Immunol* **2008**, *121*, 1301–1308. [CrossRef]

48. Pali-Scholl, I.; Jensen-Jarolim, E. Anti-acid medication as a risk factor for food allergy. *Allergy* **2011**, *66*, 469–477. [CrossRef]

49. Williamson, S.; Merritt, J.; De Benedetto, A. Atopic dermatitis in the elderly: A review of clinical and pathophysiological hallmarks. *Br. J. Dermatol.* **2019**. [CrossRef]

50. Hänel, K.H.; Cornelissen, C.; Lüscher, B.; Baron, J.M. Cytokines and the skin barrier. *Int. J. Mol. Sci.* **2013**, *14*, 6720–6745. [CrossRef]

51. Thevaranjan, N.; Puchta, A.; Schulz, C.; Naidoo, A.; Szamosi, J.C.; Verschoor, C.P.; Loukov, D.; Schenck, L.P.; Jury, J.; Foley, K.P.; et al. Age-Associated Microbial Dysbiosis Promotes Intestinal Permeability, Systemic Inflammation, and Macrophage Dysfunction. *Cell Host Microbe* **2017**, *21*, 455–466. [CrossRef]

52. Santoro, A.; Ostan, R.; Candela, M.; Biagi, E.; Brigidi, P.; Capri, M.; Franceschi, C. Gut microbiota changes in the extreme decades of human life: A focus on centenarians. *Cell. Mol. Life Sci.* **2018**, *75*, 129–148. [CrossRef] [PubMed]

53. Biagi, E.; Rampelli, S.; Turroni, S.; Quercia, S.; Candela, M.; Brigidi, P. The gut microbiota of centenarians: Signatures of longevity in the gutmicrobiota profile. *Mech. Ageing Dev.* **2017**, *165*, 180–184. [CrossRef] [PubMed]

54. Vaiserman, A.M.; Koliada, A.K.; Marotta, F. Gut microbiota: A player in aging and a target for anti-aging intervention. *Ageing Res. Rev.* **2017**, *35*, 36–45. [CrossRef] [PubMed]

55. Keebaugh, E.S.; Ja, W.W. Breaking Down Walls: Microbiota and the aging gut. *Cell Host Microbe* **2017**, *21*. [CrossRef]

56. Meyer, R.; De Koker, C.; Dziubak, R.; Skrapac, A.K.; Godwin, H.; Reeve, K.; Chebar-Lozinsky, A.; Shah, N. A practical approach to vitamin and mineral supplementation in food allergic children. *Clin. Transl. Allergy* **2015**, *5*, 11. [CrossRef]

57. Maywald, M.; Rink, L. Zinc homeostasis and immunosenescence. *J. Trace Elem. Med. Biol.* **2015**, *29*, 24–30. [CrossRef]

58. Ahluwalia, N.; Sun, J.; Krause, D.; Mastro, A.; Handte, G. Immune function is impaired in iron-deficient, homebound, older women. *Am. J. Clin. Nutr.* **2004**, *79*, 516–521. [CrossRef]

59. Ginaldi, L.; De Martinis, M. Osteoimmunology and Beyond. *Curr. Med. Chem.* **2016**, *23*, 3754–3774. [CrossRef]

60. De Martinis, M.; Sirufo, M.M.; Ginaldi, L. Osteoporosis: Current and emerging therapies targeted to immunological checkpoints. *Curr. Med. Chem.* **2019**. [CrossRef]

61. Chambers, E.S.; Hawrylowicz, C.M. The impact of vitamin D on regulatory T cells. *Curr. Allergy Asthma Rep.* **2011**, *11*, 29–36. [CrossRef]

62. Hyppönen, E.; Berry, D.J.; Wjst, M.; Power, C. Serum 25-hydroxyvitamin D and IgE—A significant but nonlinear relationship. *Allergy* **2009**, *64*, 613–620. [CrossRef] [PubMed]

63. Kamdar, T.A.; Peterson, S.; Lau, C.H.; Saltoun, C.A.; Gupta, R.S.; Bryce, P.J. Prevalence and characteristics of adult-onset food allergy. *J. Allergy Clin. Immunol. Pract.* **2015**, *3*, 114–115. [CrossRef] [PubMed]

64. Scichilone, N.; Callari, A.; Augugliaro, G.; Marcgese, M.; Togias, A.; Bellia, V. The impact of age on prevalence of positive skin prick tests and specific IgE tests. *Respir. Med.* **2011**, *105*, 651–658. [CrossRef] [PubMed]

65. Song, W.J.; Lee, S.M.; Kim, M.H.; Kim, S.H.; Kim, K.W.; Cho, S.H.; Min, K.U.; Chang, Y.S. Histamine and allergen skin reactivity in the elderly population: Results from the Korean Longitudinal Study on Health and Aging. *Ann. Allergy Asthma Immunol.* **2011**, *107*, 344–352. [CrossRef]

66. Magen, E.; Mishal, J.; Schlesinger, M. Clinical and laboratory features of chronic idiopathic urticaria in the elderly. *Int. J. Dermatol.* **2013**, *52*, 1387–1391. [CrossRef]

67. Ventura, M.T.; Cassano, N.; Romita, P.; Vestita, M.; Foti, C.; Vena, G.A. Management of chronic spontaneous urticaria in the elderly. *Drugs Aging* **2015**, *32*, 271–282. [CrossRef]

68. Sirufo, M.M.; Ginaldi, L.; De Martinis, M. Successful Treatment with Omalizumab in a Child with Asthma and Urticaria: A Clinical Case Report. *Front. Pediatr.* **2019**, *7*, 213. [CrossRef]

69. De Martinis, M.; Sirufo, M.M.; Ginaldi, L. Solar Urticaria, a Disease with Many Dark Sides: Is Omalizumab the Right Therapeutic Response? Reflections from a Clinical Case Report. *Open Med.* **2019**, *14*, 403–440. [CrossRef]

70. Sirufo, M.M.; De Martinis, M.; Ginaldi, L. Omalizumab an effective and safe alternative therapy in severe refractory atopic dermatitis: A case report. *Medicine* **2018**, *97*, e10897. [CrossRef]

71. Ridolo, E.; Anti Rogkakou, A.; Ventura, M.T. How to fit allergen immunotherapy in the elderly. *Clin. Mol. Allergy* **2017**, *15*, 17. [CrossRef]

72. Aurich, S.; Dölle-Bierke, S.; Francuzik, W.; Bilo, M.B.; Christoff, G.; Fernandez-Rivas, M.; Hawranek, T.; Pföhler, C.; Poziomkowska-Gęsicka, I.; Renaudin, J.M.; et al. Anaphylaxis in elderly patients: Data from the European Anaphylaxis Registry. *Front. Immunol.* **2019**, *10*. [CrossRef] [PubMed]

73. Worm, M.; Francuzik, W.; Renaudin, J.M.; Bilo, M.B.; Cardona, V.; Scherer Hofmeier, K.; Köhli, A.; Bauer, A.; Christoff, G.; Cichocka-Jarosz, E.; et al. Factors increasing the risk for a severe reaction in anaphylaxis: An analysis of data from The European Anaphylaxis Registry. *Allergy* **2018**, *73*, 1322–1330. [CrossRef] [PubMed]
74. Lee, S.; Hess, E.P.; Nestler, D.M.; Bellamkonda Athmaram, V.R.; Bellolio, M.F.; Decker, W.W.; Li, J.T.; Hagan, J.B.; Manivannan, V.; Vukov, S.C.; et al. Antihypertensive medication use is associated with increased organ system involvement and hospitalization in emergency department patients with anaphylaxis. *J. Allergy Clin. Immunol.* **2013**, *131*, 1103–1108. [CrossRef] [PubMed]
75. González-de-Olano, D.; Lombardo, C.; González-Mancebo, E. The difficult management of anaphylaxis in the elderly. *Curr. Opin. Allergy Clin. Immunol.* **2016**, *16*, 352–360. [CrossRef]
76. Vetrano, D.L.; Foebel, A.D.; Marengoni, A.; Brandi, V.; Collamati, A.; Heckman, G.A.; Hirdes, J.; Bernabei, R.; Onder, G. Chronic diseases and geriatric syndromes: The different weight of comorbidity. *Eur. J. Intern. Med.* **2016**, *27*, 62–67. [CrossRef]
77. Ventura, M.T.; Scichilone, N.; Gelardi, M.; Patella, V.; Ridolo, E. Management of allergic disease in the elderly: Key considerations, recommendations and emerging therapies. *Expert Rev. Clin. Immunol.* **2015**, *11*, 1219–1228. [CrossRef] [PubMed]

International Journal of
Molecular Sciences

Article

Loss of Dystroglycan Drives Cellular Senescence via Defective Mitosis-Mediated Genomic Instability

Guadalupe Elizabeth Jimenez-Gutierrez [1,2,†], Ricardo Mondragon-Gonzalez [1,†],
Luz Adriana Soto-Ponce [1], Wendy Lilián Gómez-Monsiváis [1], Ian García-Aguirre [1],
Ruth Abigail Pacheco-Rivera [2], Rocío Suárez-Sánchez [3], Andrea Brancaccio [4,5],
Jonathan Javier Magaña [3,6,*], Rita C.R. Perlingeiro [7] and Bulmaro Cisneros [1,*]

[1] Departamento de Genética y Biología Molecular, Centro de Investigación y de Estudios Avanzados del Instituto Politécnico Nacional, Ciudad de México 07360, Mexico; gjimenezg@cinvestav.mx (G.E.J.-G.); rmondragon90@gmail.com (R.M.-G.); luz.ponce@cinvestav.mx (L.A.S.-P.); wlgomez@cinvestav.mx (W.L.G.-M.); ian.garcia@cinvestav.mx (I.G.-A.)
[2] Departamento de Bioquímica, Escuela Nacional de Ciencias Biológicas, Instituto Politécnico Nacional, Ciudad de México 11340, Mexico; rpachecor@ipn.mx
[3] Departamento de Genética, Laboratorio de Medicina Genómica, Instituto Nacional de Rehabilitación "Luis Guillermo Ibarra Ibarra", Ciudad de México 14389, Mexico; srossmary@gmail.com
[4] School of Biochemistry, University of Bristol, Bristol BS8 1TD, UK; andrea.brancaccio@icrm.cnr.it
[5] Institute of Chemical Sciences and Technologies "Giulio Natta" (SCITEC), 00168 Roma, Italy
[6] Departamento de Bioingeniería, Escuela de Ingeniería y Ciencias, Instituto Tecnológico y de Estudios Superiores de Monterrey-Campus Ciudad de México, Ciudad de México 14380, Mexico
[7] Department of Medicine, Lillehei Heart Institute, University of Minnesota, Minneapolis, MN 55455, USA; perli032@umn.edu
* Correspondence: magana.jj@tec.mx (J.J.M.); bcisnero@cinvestav.mx (B.C.)
† These authors contribute equally to this work.

Received: 28 May 2020; Accepted: 22 June 2020; Published: 14 July 2020

Abstract: Nuclear β-dystroglycan (β-DG) is involved in the maintenance of nuclear architecture and function. Nonetheless, its relevance in defined nuclear processes remains to be determined. In this study we generated a C2C12 cell-based DG-null model using CRISPR-Cas9 technology to provide insights into the role of β-DG on nuclear processes. Since DG-null cells exhibited decreased levels of lamin B1, we aimed to elucidate the contribution of DG to senescence, owing to the central role of lamin B1 in this pathway. Remarkably, the lack of DG enables C2C12 cells to acquire senescent features, including cell-cycle arrest, increased senescence-associated-β-galactosidase activity, heterochromatin loss, aberrant nuclear morphology and nucleolar disruption. We demonstrated that genomic instability is one driving cause of the senescent phenotype in DG-null cells via the activation of a DNA-damage response associated with mitotic failure, as shown by the presence of multipolar mitotic spindles, which in turn induced the formation of micronuclei and γH2AX foci (DNA-damage marker), telomere shortening and p53/p21 upregulation. Altogether, these events might ultimately lead to premature senescence, impeding the replication of the damaged genome. In summary, we present evidence supporting a role for DG in protecting against senescence, through the maintenance of proper lamin B1 expression/localization and proper mitotic spindle organization.

Keywords: β-Dystroglycan; cellular senescence; lamin B1; DNA-damage response; defective mitosis

1. Introduction

Dystroglycan (DG) is an integral membrane complex that connects the extracellular matrix (ECM) with the intracellular actin-based cytoskeleton, providing structural stability to the plasma membrane (PM) in different tissues and cell types [1–3]. DG is synthesized as a propeptide that separates into α-

and β-DG subunits after proteolytic cleavage [2,4,5]. Both subunits remain together at the PM through the interaction between β-DG's extracellular domain and α-DG's carboxy-terminal globular domain. While α-DG is an extracellular peripheral glycoprotein that binds to various extracellular matrix molecules, including laminin, agrin and perlecan, β-DG is a single-pass transmembrane protein that binds through its cytoplasmic tail to dystrophin, caveolin-3 and other cytoplasmic proteins involved in signal transduction [6–10]. Perturbation of dystroglycan processing is associated with severe congenital disorders and cancer progression [11,12]. In addition, DG has been implicated in cellular processes such as signal transduction and tissue morphogenesis. DG is particularly relevant in skeletal muscle tissue, where it has been classically described to play a key role in stabilizing the sarcolemma of myofibers during the cycles of muscle contraction and relaxation [13]. Upon injury, muscle-specific stem cells (i.e., satellite cells) are activated, proliferate and differentiate into myoblasts that can fuse with pre-existing myofibers, or form new fibers, to overcome the muscle damage. Interestingly, DG is also expressed in satellite cells, where it is essential to enable skeletal muscle regeneration [13,14]. In myoblasts, DG plays an important role in modulating myoblast motility and migration [15]. Therefore, DG has been proven relevant not only in myofibers, but also in myogenic precursors for proper muscle function.

Interestingly, β-DG has the ability to traffic from PM to the nucleus, using the membranous endosome–endoplasmic reticulum network and the importin α2/β1 nuclear import pathway [16–18]. This additional cellular localization suggests potential further roles for β-DG. For instance, nuclear β-DG has been involved in the transcriptional regulation of androgen-responsive transcription factors in prostate cancer [19]. We previously demonstrated that β-DG assembles with the nuclear envelope (NE) proteins emerin and lamins A/C and B1 to maintain nuclear architecture and function in myoblasts [20]. β-DG is subject to nucleocytoplasmic shuttling with an active exportin1/CRM1-mediated nuclear export pathway [21] that together with its nuclear import serves to tightly regulate the nuclear levels of β-DG, thereby allowing effective interactions with binding partners at the NE interface. However, the molecular basis underlying the role of β-DG on NE-associated functions is largely unknown. In this study, we generated DG-null mouse myoblasts (C2C12) using CRISPR-Cas9 technology to analyze in depth the function of β-DG in the nucleus. The initial phenotype noted in DG-null cells was the decrease in lamin B1 levels, which was accompanied by nuclear morphology defects. Therefore, taking the evidence that lamin B1 plays a pivotal role in cellular senescence [22–25], we analyzed the contribution of DG to this cellular process. Cellular senescence is defined as a state of permanent cell cycle arrest that occurs in response to different damaging stimuli, including persistent DNA damage, telomere shortening, oxidative stress and oncogenic signaling [26–28], with silencing of lamin B1 expression being an early and necessary event for senescence to be established [22–25]. We provided evidence showing that DG plays a protective role against senescence, because the lack of DG makes C2C12 cells to acquire senescent features. In addition, we demonstrated that senescence signaling in DG-null cells is triggered by mitotic failure, which in turn elicits a p53-mediated DNA-damage response to arrest the cell cycle, leading to premature senescence.

2. Results

2.1. Generation and Characterization of CRISPR/Cas9-Mediated DG-Null C2C12 Cell Clones

To analyze in depth the functional relationship of β-DG with the NE, we engineered DG knockout cells (DG-KO) on the mouse myogenic cell line C2C12, using CRISPR/Cas9. To silence DG expression, C2C12 cells were transfected with a vector expressing Cas9, the red fluorescence protein (RFP) and one of two different guide RNAs (gRNA1 and gRNA2) targeting the region downstream of the ATG translation initiation codon within the first coding exon of the mouse *Dag1* gene (Figure 1A). After positive selection for RFP and two rounds of negative selection using the IIH6 antibody, which is specific to the αDG laminin binding domain [1,29] fluorescence-activated cell sorting (FACS) and further clonal expansion, two different KO lines (DG-KO1 and DG-KO2) were selected (Figure 1B; see Methods for details). DNA sequencing of the target site was performed to directly identify editing events.

Both DG-KO clones showed indels that generate premature stop codons; thus, only polypeptides with presumably no biological activity are synthesized from DG-KO clones (Figure 1C).

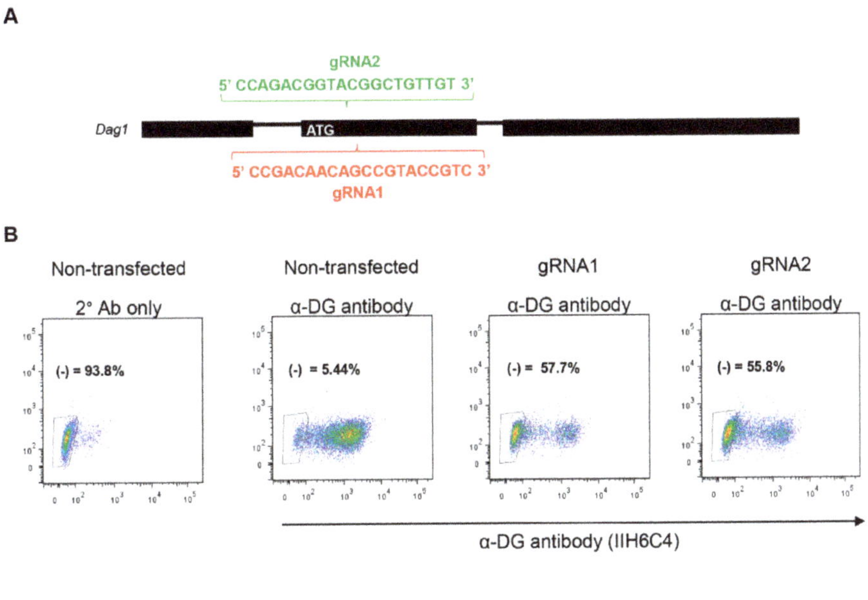

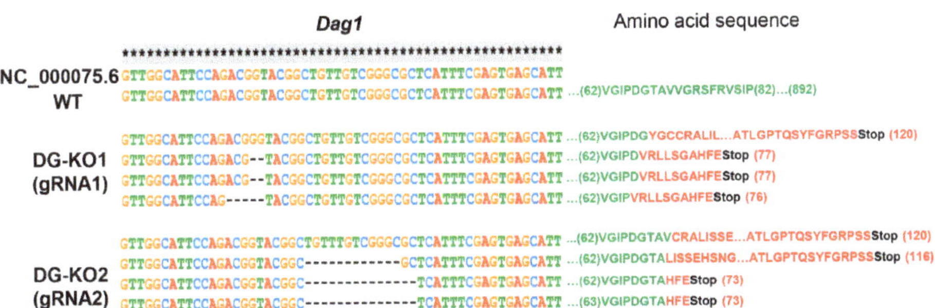

Figure 1. CRISPR/Cas9-engineered dystroglycan knockout (DG-KO) C2C12 cell clones. (**A**) Scheme showing the sequences of guide RNAs (gRNA1 and gRNA2), designed to target *Dag1* gene. (**B**) Fluorescence-activated cell sorting (FACS) analysis on C2C12 cells, gRNA2 or none gRNA (non-transfected cells), and stained with α-DG antibody, IIH6C4. The absence of IIH6C4 reactivity confirmed the lack of functionally glycosylated α-DG. Non-transfected cells incubated only with secondary antibody (2° Ab only) were used to adjust the population negative for α-DG immunostaining (α-DG (-)). Percentages correspond to α-DG (-) population. (**C**) Sequence alignment of mouse *Dag1* gene (annotated) showing the introduction of indels in DG-KO1 and DG-KO2 cell lines compared with WT cells. Amino acid sequence shows the position of the stop codons generated in DG-KO1 and DG-KO2 cell clones.

Owing to the functional relationship of DG with dystrophin-associated proteins (DAPs), DG-KO clones were initially characterized by analyzing the protein levels of various DAPs, namely dystrophin Dp71, α-dystrobrevin and β2-syntrophin. Lysates from both DG-KO1 and DG-KO2 clones showed no β-DG protein expression (Figure 2A; 43 kDa and 26 kDa proteins), and a drastic decrease in the levels of all DAPs analyzed was observed, compared with WT cells (Figure 2B–D). Overall, these data validate DG-KO clones as model for studying DG, including the role of β-DG in NE-associated processes.

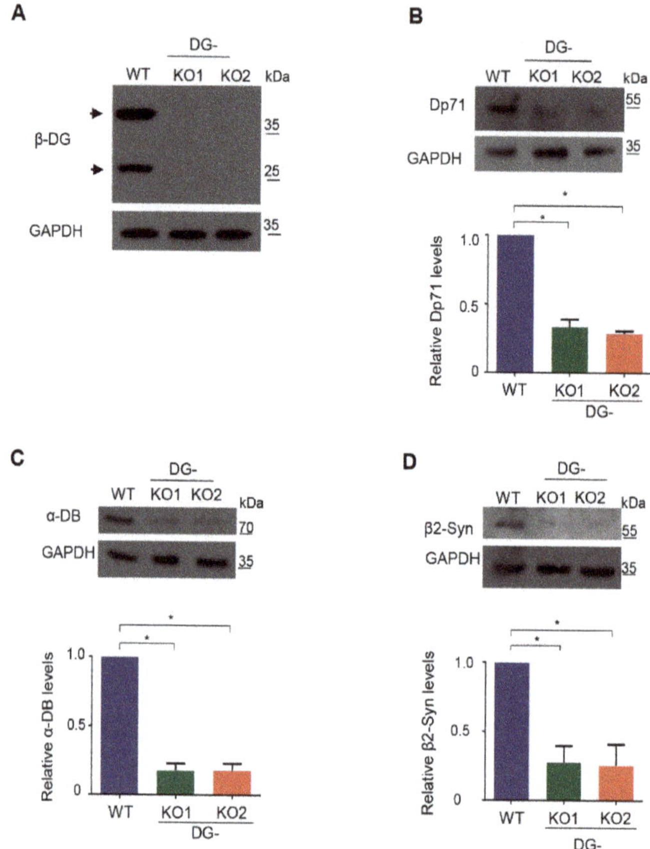

Figure 2. Decreased protein levels of dystrophin associated proteins in DG-KO cells. Lysates from WT, DG-KO1 and DG-KO2 cell cultures were analyzed by SDS-PAGE/WB using specific antibodies against β-DG (**A**), Dp71 (**B**), α-dystrobrevin (α-DB) (**C**), β2-syntrophin (β2-Syn) (**D**) and GAPDH (loading control); representative blots are shown. Bottom graphs: relative protein expression was calculated from three independent experiments and significant differences were calculated using one-way ANOVA and Dunnett´s post hoc test; * $p < 0.05$ in comparison to WT. Data indicate the mean ± SEM.

2.2. DG Deficiency Provokes Altered Localization and Decreased Protein Levels of Lamin B1

Because lamin B1, a critical NE protein, is a β-DG-interacting partner [20], we were prompted to evaluate the impact of the lack of DG on lamin B1 distribution and protein expression. Interestingly, altered immunostaining for lamin B1 and evident nuclear deformities (invaginations) were found in DG-KO1 and DG-KO2 cells (Figure 3A). Consistently, the percentage of cells with aberrant nuclear morphology was clearly higher in DG-KO cell cultures than WT cell culture (right graph). Morphometric analysis of nuclei (nuclear area and circularly index) confirmed significant differences in nuclear shape between WT and DG-KO1 and DG-KO2 cells (Figure 3B). In line with IF/confocal microscopy images, a significant decrease in lamin B1 levels was observed in DG-KO1 and DG-KO2 cells (Figure 3C).

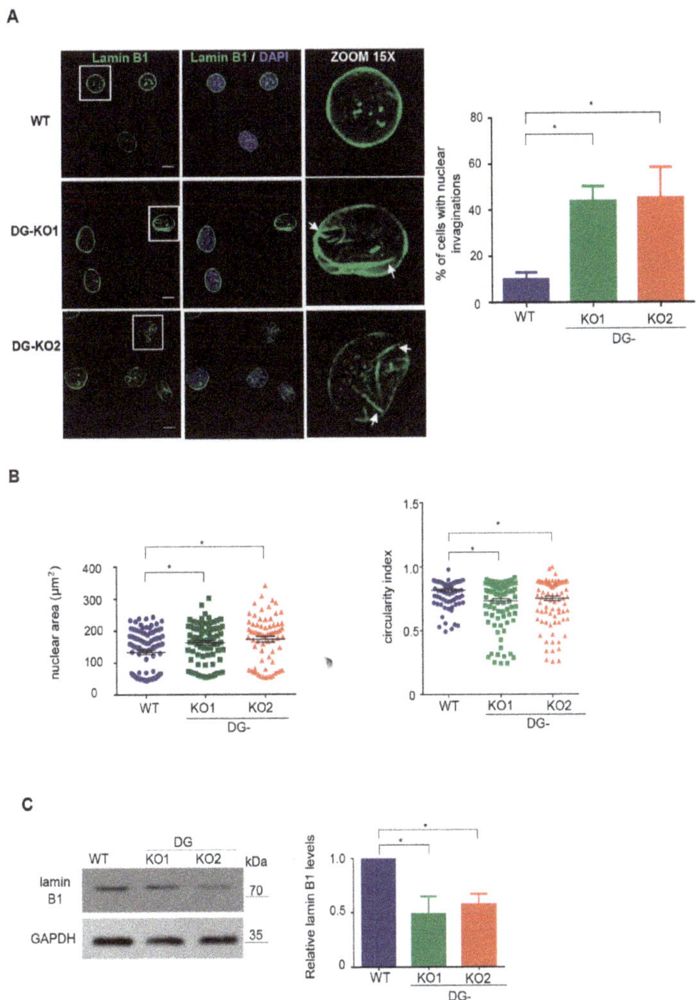

Figure 3. DG-KO cells show altered localization and decreased protein levels of lamin B1. (**A**) WT, DG-KO1 and DG-KO2 cells, seeded on glass coverslips, were fixed and immunostained for lamin B1. Nuclei were stained with diamidino-2-phenylindole (DAPI) prior to confocal laser scanning microscopy analysis (CLSM), and typical images are shown. Bar, 10 μm. Right: the bar graph shows the percentage of nuclei with invaginations, with significant differences calculated in each cell culture from three independent experiments using one-way ANOVA and Dunnett´s post hoc test; data indicate the mean ± SEM (n = 100 nuclei per cell culture; * p < 0.05 compared to WT). (**B**) Nuclear morphometric analysis was carried out on WT, DG-KO1 and DG-KO2 cells, as described in Methods from three separate experiments; significant differences were obtained using a non-parametric Kruskal–Wallis test and post hoc Dunn´s method. Data correspond to the mean ± SEM (n = 100 nuclei per experimental condition; * p < 0.05 compared to WT). (**C**) Lysates from WT and DG-KO1 and DG-KO2 cells were analyzed by SDS-PAGE/WB using antibodies against lamin B1 and GAPDH (loading control). Representative blots from three separate experiments are shown, with significant differences obtained using one-way ANOVA and Dunnett´s post hoc test; * p < 0.05 in comparison to WT. Data indicate the mean ± SEM. Right: densitometric analysis of immunoblot autoradiograms was performed to estimate lamin B1 protein expression.

2.3. The Loss of DG Induces the Expression of Senescence-Associated Features

Lamin B1 downregulation occurs at the onset of cell transition to senescence [23,30]. Thus, we tested whether the reduction in lamin B1 as a consequence of DG loss would lead to a senescent phenotype. To approach this idea, we searched for senescence characteristics in DG-null cells. Decreased proliferative potential and a higher percentage of cells at the G0/G1 phase of the cell cycle were found in DG-KO1 and DG-KO2 cell cultures, compared with WT cells, as shown by MTT and flow cytometry analyses, respectively (Figure 4A,B). G0/G1 arrest might be indicative of premature senescence; thus, the identification of senescent cells in DG-KO cultures was carried out using senescence-associated β-galactosidase activity [SA-β-gal]. Interestingly, increased numbers of senescent cells were found in DG-KO1 and DG-KO2 cultures (20–25%), compared with WT cells (5%) (Figure 4C). In addition to increased β-galactosidase activity, senescent cells display large-flat cell morphology and undergo both nucleolar stress and the loss of perinuclear chromatin [28], among other characteristics. To corroborate the increase in senescent cells in DG-KO cultures, we searched for the presence of these features in DG-null cell cultures, using WT cells induced to senescence by treatment with sodium butyrate (NaBu) for ten days as a positive control for the senescent phenotype. NaBu is a histone deacetylase inhibitor that elicits senescence via irreversible induction of cell cycle arrest [31]. WT cells treated with NaBu showed a flattened and expanded morphology, which markedly contrasts with the typical polygonal morphology of untreated WT cells (Figure 5A). DG-KO2 but not DG-KO1 cells showed a subtle but statistically significant increase in cell surface, compared with WT cells, as determined by phalloidin staining of F-actin and estimation of cell surface area (Figure 5A and right graph). On the other hand, a marked decrease in H3K9me3 foci immunostaining (heterochromatin marker) was observed in DG-null cells in a similar fashion to that observed in NaBu-induced senescent cells (Figure 5B), as determined by confocal laser scanning microscopy (CLSM) and fluorescence intensity quantification (Figure 5B and right graph). Finally, disaggregated nucleoli with a smaller area were found in DG-KO1 and DG-KO2 cell cultures, compared with WT cells, as revealed by immunostaining for the nucleolar protein B23 and nucleolar area quantification (Figure 5C and right graph). However, nucleolar disaggregation was much greater in NaBu-induced senescent cells.

Collectively, the aforementioned data imply that incipient senescence is present in DG-KO cell cultures in the absence of any detectable senescence-inducing stimuli. Thus, we assessed whether the lack of DG sensitizes C2C12 cells to senescence induction. In line with this, the percentage of senescent cells was significantly higher in DG-KO1 cultures (60–70%) than that in WT cells (30%) upon five days of NaBu treatment (Figure 6A). Long-term treatment (10 days) rendered a similar percentage of senescent cells (~75%) between WT and DG-KO cell cultures (Figure 6A).

2.4. Aberrant Multipolar Mitoses in DG-KO Cells Resulted in Micronuclei Formation and Activation of a P53-Mediated DNA Damage Response

A previous study from our group showed that downregulation of DG in C2C12 cells results in an increased number of centrosomes [20]. Therefore, we next assessed whether the lack of DG would lead to aberrant mitosis and, consequently, genomic instability, contributing to the senescent phenotype. WT, DG-KO1 and DG-KO2 cells previously arrested in S phase by double treatment with thymidine were released to allow their progression into mitosis. Cell were immunolabeled for α-tubulin and γ-tubulin to decorate mitotic spindles and centrosomes, respectively, and mitotic cells were visualized by CLSM. Remarkably, a high percentage of DG-KO1 and DG-KO2 cells (80%) showed multipolar mitotic spindles and multidirectional alignment of chromosomes, compared to WT cells (Figure 7A). Because thymidine treatment evokes a DNA damage response by slowing the progression of replication forks [32], we analyzed whether DG-KO cells are more prone to response to DNA damage than WT cells, by monitoring γ-H2AX foci, a DNA repair marker [33]. In line with our hypothesis, a dramatic increase in fluorescence intensity of γ-H2AX foci was observed in thymidine-treated DG-KO cells (Figure 7B) compared with thymidine-treated WT cells. Furthermore, the presence of micronuclei, another faithful indicator of DNA damage and chromosome instability [34], was frequently observed in DG-KO1 (75%) and DG- KO2 (90%) cell cultures upon thymidine treatment, compared with WT culture (20%). To support the latter result, we searched for micronuclei in the absence of any DNA-damage-inducing agent. A significant increase in the percentage of micronuclei-contained cells was observed in DG-KO1 (8%) and DG-KO2 (10%) cultures, compared with WT culture (4%), as shown by lamin B1 immunostaining (Figure S1). Errors in cell division and persistent DNA damage in DG null cells would lead to the activation of the checkpoint proteins p53 and its target proteins p21, which in turn elicits cell cycle arrest and/or senescence. Consistent with this notion, p53 levels were found to be increased in thymidine-treated DG-KO1 and DG-KO2 cells, while augmented levels of p21 were observed only in DG-KO1 cells, compared with WT cells (Figure 7C). Collectively, these data imply that the lack of DG resulted in aberrant multipolar mitosis, which in turn induces a DNA damage response via p53 activation and ultimately cell-cycle arrest.

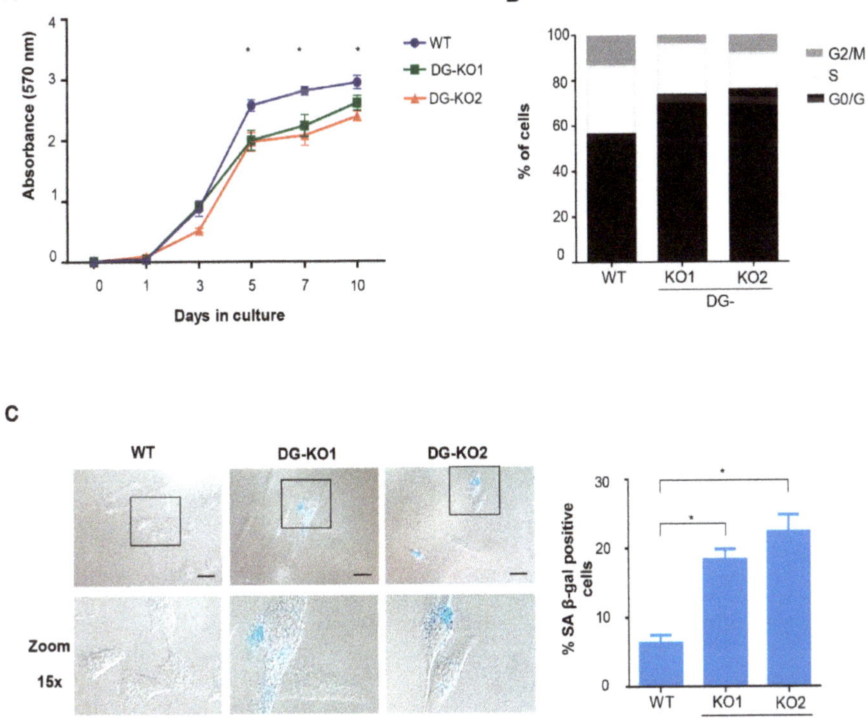

Figure 4. DG-KO cell cultures exhibit decreased proliferation, G0/G1 arrest and senescence. (**A**) MTT-based cell proliferation assays were performed over a 10 days period in WT, DG-KO1 and DG KO2 cell cultures. Data correspond to the mean ± SEM from three independent experiments, with significant differences determined by one-way ANOVA; * $p < 0.05$ compared to WT. (**B**) Cell cycle analysis on WT, DG-KO1 and DG-KO2 asynchronous cell cultures was performed by flow cytometry. A typical graph from three independent experiments is shown. (**C**) Senescent cells were identified in WT, DG-KO1 and DG-KO2 cultures by quantifying SA β-gal activity, and representative images were acquired by light-field microscopy. Bar = 50 μM. Right: the percentage of senescent cells was calculated, and significant differences were obtained from three separate experiments using one-way ANOVA, followed by Dunnett's post hoc test. Data correspond to the mean ± SEM ($n = 200$ cells for each cell culture; * $p < 0.05$ compared to WT).

A **Senescent cell morphology**

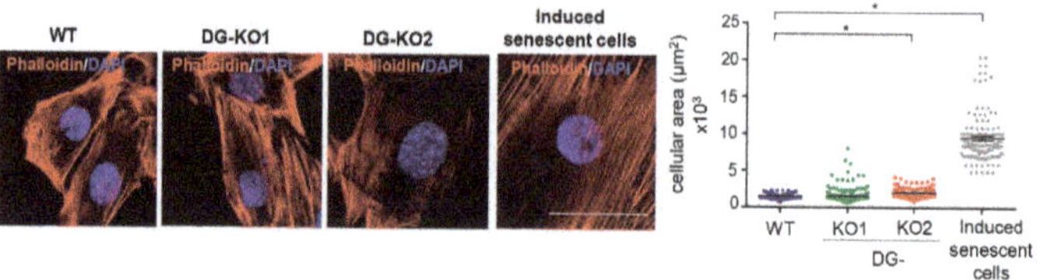

B **Heterochromatin loss**

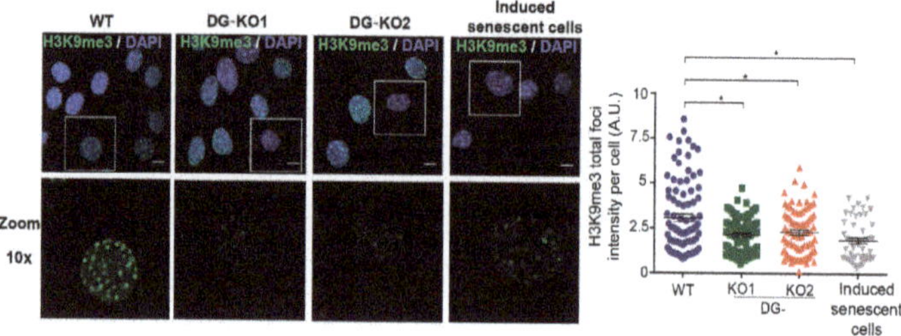

C **Nucleolar disruption**

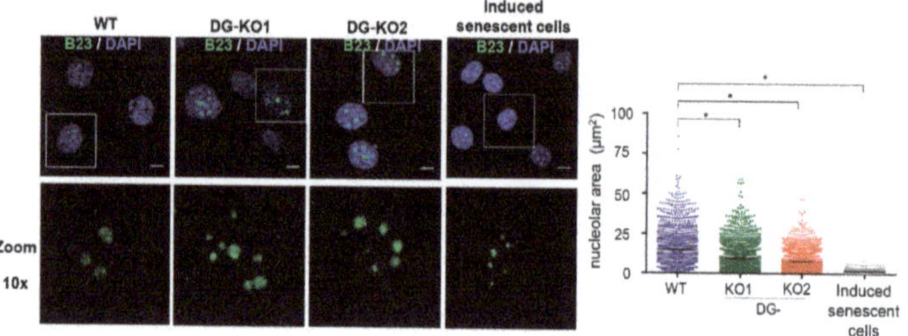

Figure 5. DG-null cells display senescent characteristics. WT, DG-KO1 and DG-KO2 cells were seeded on glass coverslips and fixed prior to CLSM analysis. (**A**) Senescent cell morphology. Cells were labeled with DAPI and phalloidin to visualize nuclei and actin-based cytoskeleton, respectively, and representative images are shown. Bar, 50 μM. Right: the cellular area was estimated using ImageJ software, with significant differences determined by non-parametric Kruskal–Wallis tests and post hoc Dunn´s method. Data correspond to the mean ± SEM (n = 200 cells for each cell culture and from three independent experiments). * $p < 0.05$ in comparison to WT). (**B**) Heterochromatin loss. Cell preparations were immunostained for H3K9me3 followed by DAPI labeling to enable nuclei visualization. Representative confocal microscope images are shown (bar, 10 μM). Right: the fluorescent intensity of H3K9me3 foci was measured using ImageJ software, as described in Methods. Significant differences were determined by non-parametric Kruskal–Wallis tests, followed by post hoc Dunn´s method. Data correspond to the the mean ± SEM (n = 200 cells for each cell culture and from three independent experiments; * $p < 0.05$ in comparison to WT). (**C**) Nucleolar disruption. Cell preparations were immunostained for B23 and labeled with DAPI to decorate nucleoli and nuclei, respectively. Scale bar, 10 μM. Right: nucleolar area was assessed using ImageJ software, as described in Methods (n = 1300 nucleoli per experimental condition). Significant differences were determined by non-parametric Kruskal–Wallis tests and post hoc Dunn´s method; data indicate the mean ± SEM.; * $p < 0.05$ compared to WT.

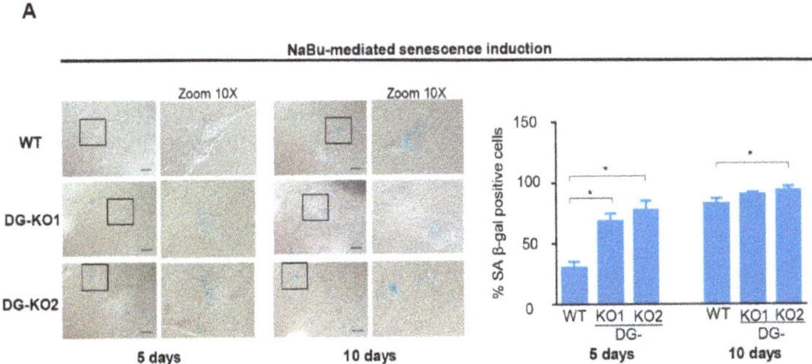

Figure 6. The loss of DG makes C2C12 cells more responsive to senescence induction. (**A**)WT, DG-KO1, and DG-KO2 cell cultures were treated with sodium butyrate (NaBu) for 5 or 10 days to induce senescence, and senescent cells were identified by SA β-gal activity; typical images were acquired by light-field microscopy. Bar = 50 μM. Right: graph shows the percentage of senescent cells obtained from three independent assays ($n = 100$ cells for each cell culture). Significant differences were determined by one-way ANOVA followed by Dunnett´s multiple comparison test; * $p < 0.05$ compared to WT. Data correspond to the mean ± SEM.

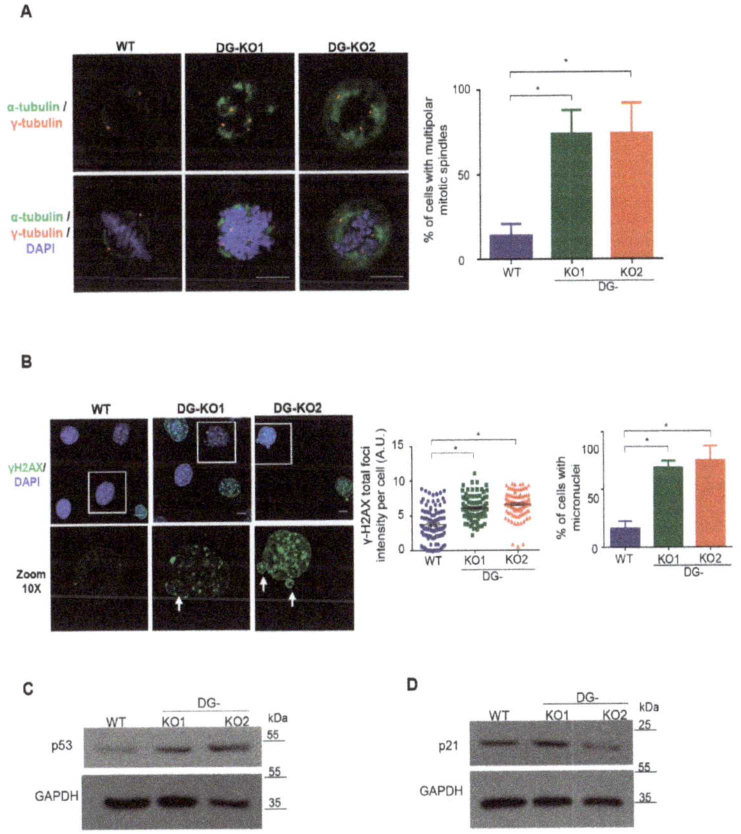

Figure 7. Mitotic failure activates a p53-dependent DNA damage response in DG-null cells. (**A**) WT, DG-KO1 and DG-KO2 cells cultured on coverslips were arrested in S phase by double treatment with

thymidine and further release into cell cycle for 4 h to progress into mitosis. Afterwards, cell preparations were immunolabeled for α-tubulin and γ-tubulin to decorate mitotic spindles and centrosomes, respectively, and counterstained with DAPI to visualized nuclei, prior to CLSM analysis. Right. The percentage of multipolar mitotic spindles was determined from three separate experiments ($n = 50$ cells from each cell culture). Significant differences were calculated using one-way ANOVA and Dunnett´s multiple comparison test; * $p < 0.05$ compared to WT. Data indicate the mean ± SEM. (**B**) WT, DG-KO1 and DG-KO2 cells were cultured on coverslips and treated with thymidine as per panel A. Cell preparations were then immunolabeled for γ2HAX and counterstained with DAPI to decorate nuclei, prior to CLSM analysis. Right: the fluorescence intensity of γ2HAX foci was calculated, and significant differences were determined by non-parametric Kruskal–Wallis tests followed by Dunn´s post hoc analysis. Data correspond to the mean ± SEM from three separate experiments; * $p < 0.05$ compared to WT. Far right: the number of cells with micronuclei was calculated and significant differences were obtained by one-way ANOVA and Dunnett´s post hoc analyses. Data correspond to the mean ± SEM from three independent assays ($n = 100$ cells for each cell culture); * $p < 0.05$ compared to WT. (**C**) and (**D**) Lysates from WT, DG-KO1 and DG-KO2 cells cultures, previously treated with thymidine as per panel A, were analyzed by SDS-PAGE/WB, using specific antibodies against p53 (C), p21 (D) and the loading control, and representative blots from two independent experiments are shown.

Because DNA damage response can induce telomerase shortening irrespective of telomerase activity [35], we were prompted to estimate telomerase length in DG-deficient and WT cells of similar culture passage (6–8 passage) by in situ hybridization (FISH), using a telomere oligonucleotide fluorescein-labeled probe (Figure 8). It is assumed that the probe hybridizes quantitatively to telomeric repeats, and hence the integrated telomere foci fluorescence intensity of a single nucleus is directly related to the length of their telomeres [36,37]. We observed that the fluorescence intensity of telomere foci was significantly less intense in DG-null cells, compared with WT cells (Figure 8A).

A

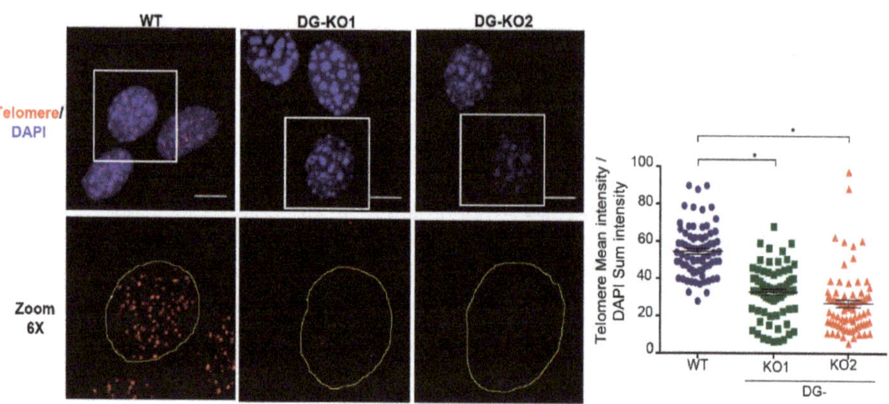

Figure 8. DG-KO cells exhibit telomere shortening. (**A**)WT, DG-KO1 and DG-KO2 cells grown on coverslips were processed for FISH using a specific telomere probe (see Methods), and nuclei were decorated by staining with DAPI. Representative CLSM images are shown; bar, 10 μM. The right graph shows the relative telomere length determined by the telomere mean intensity divided by DAPI sum intensity. Significant differences were determined by non-parametric Kruskal–Wallis tests, followed by post hoc Dunn´s analysis. Data indicate the mean ± SEM from three separate experiments ($n = 75$ cells per cell culture); * $p < 0.05$ compared to WT).

3. Discussion

In this study, we generated a C12C12 myoblasts-based model with no expression of DG (α-DG and β-DG) using CRISPR-Cas9 technology to provide insights into the nuclear function of β-DG. In earlier studies, we showed that β-DG is involved in maintaining the structure and function of the NE [20,21]; nevertheless, the specific mechanisms underlying its role in these nuclear processes remains to be determined. We isolated two DG-knockout clones (DG-KO1 and DG-KO2) that were generated from two different gRNAs. We used this strategy in order to validate that the phenotypes observed were not due to off-target cleavage by the CRISPR-Cas9 system. DG-KO cells were intentionally sorted from the glycosylated α-DG negative population and, accordingly, expanded DG-KO clones showed no expression of β-DG and decreased levels of dystrophin Dp71, α-dystrobrevin and β2-syntrophin, three well characterized partners of β-DG [38], which validated our DG-deficient cell system.

We focused our research on the previously observed interaction between β-DG and lamin B1 [20]. Lamin B1 belongs to a group of type V intermediate filament proteins known as the lamins. These proteins are the main component of the nuclear lamina, which provides stability to the nuclear structure, and regulates nuclear processes such as transcription, chromatin organization, cell cycle, among others [39]. The abnormalities in the nuclear lamina observed in DG-KO cells confirmed previous observations suggesting a key role for β-DG in maintaining the integrity of this compartment [21]. Moreover, the decreased levels of lamin B1 in the absence of DG is particularly relevant as the downregulation of lamin B1 is a key event mediating premature senescence [22,40]. This has been attributed to the plethora of nuclear processes that lamin B1 regulates, such as heterochromatin architecture, cell cycle progression, nuclear morphology, gene expression and splicing [41,42]. Therefore, we assessed whether DG-KO cells would acquire a senescent phenotype. Consistently, DG-KO cells exhibited several senescent marks in the absence of any senescence-inducing stimuli, including reduced cell proliferation with arrest at G0/G1, elevated SA-β-gal activity, nucleolar disaggregation, senescent cell morphology and loss of heterochromatin. The cellular transition to senescence is associated with extensive chromatin reorganization and gene expression changes. Specifically, lamin B1 downregulation occurring during senescence facilitates the spatial relocalization of perinuclear H3K9me3-positive heterochromatin [43]. Furthermore, downregulation of SUV39H1 during the establishment of senescence may promote DNA repair, leading to genome destabilization due to deheterochromatinization of repetitive DNA, which in turn results in cell cycle arrest [44]. Thus, lamin B1 may contribute to senescence by the spatial reorganization of chromatin and through gene repression [43]. In this scenery, characterization of the DG-KO cell gene expression profile, including the genomic DNA methylation pattern, will help to determine the epigenetic regulation occurring in response to the loss of DG. It is worth noting that treatment with the histone deacetylase inhibitor NaBu induced senescence in proliferating myoblasts and that this effect was exacerbated in DG-KO cells. A possible explanation is that β-DG is required to stabilize lamin B1 at the nuclear lamina so that it can attenuate induced senescence. Indeed, perturbation of β-DG nuclear trafficking causes both mistargeting and reduced protein levels of lamin B1, leading ultimately to aberrant nuclear architecture [21]. Thus, tight control of nuclear β-DG content is physiologically relevant to preserve β-DG-lamin B1 interaction, thereby allowing the cell to finely tune nuclear activity in response to cellular stimuli. While increased susceptibility to senescence might be a consequence of lamin B1 alteration in the absence of β-DG, it remains to be explored whether additional mechanisms related to β-DG functions (e.g., signaling) are also contributing to this process.

Cellular senescence is induced by different damaging stimuli, including extended replication, DNA damage, oxidative stress, telomere shortening and oncogenic signaling [45,46]. In an attempt to further understand how the lack of DG results in senescence, we invoked an earlier study that might connect DG with DNA damage. We previously demonstrated that DG downregulation resulted in over-duplicated centrosomes in C2C12 cells [20], an aberrant characteristic associated with multipolar mitosis [47,48]. Supporting our assumption, multipolar mitotic spindles were frequently found in DG-null cells, compared with WT culture. Consistent with mitotic defects driving chromosome instability and DNA damage response [49], DG-KO cells exhibited an increased number of micronuclei-containing cells and apparent shortening of telomeres, compared with WT cells. The formation of micronuclei occurred to a much greater extent when DG-null cells were subjected to thymidine-mediated DNA damage. Thymidine treatment evokes a DNA damage response by slowing the progression of replication forks [32]. Furthermore, numerous intensely stained foci of phosphorylated H2AX histone (γH2AX) were found in DG-KO cells after thymidine exposure. γH2AX orchestrates DNA repair by recruiting repair factors to the surrounding of double-strand break (DSB) sites, including MRE11/NBS1/RAD50, MDC1, 53BP1 and BRCA1 [33,50]. Supporting the idea that DG deficiency makes the cell more prone to DNA-damage response, upregulation of the p53 pathway (p53 and p21 proteins) was found in DG-KO cells upon thymidine treatment. p53 plays a pivotal role for senescence induction; the DNA damage response activates ataxia telangiectasia (ATM) and Rad3-related (ATR) kinases, which in turn activate the p53/p21 axis by phosphorylation of both p53 and its ubiquitin ligase Mdm2, leading to the stabilization of p53 levels [51]. However, differences in p53 pathway activation between DG-KO1 and DG-KO2 cells due to inter-clonal heterogeneity cannot be ruled out. This issue deserves further investigation.

How DG-KO cells acquire multiple centrosomes, a hallmark of cancer cells [52], is unknown. Centrosome amplification could result from altered centrosome replication and/or cytokinesis failure. Numerous proteins that regulate the centrosome duplication cycle have been identified, including Polo-like kinase-4, cyclin-dependent kinase 2, and SPD-2 [53]; however, none of them has been linked with DG so far. It is worth noting that β-DG localized to the cleavage furrow and midbody in cytokinesis [54]; thus, DG deficiency might lead centrosome amplification through impaired cytokinesis. Nonetheless, the fact that no binucleated cells were observed in DG-KO cultures opposes this hypothesis. On the other hand, considering that B-type lamins have been involved in the assembly and maintenance of mitotic spindles in Xenopus [55], it is possible that aberrant multipolar spindles in DG-KO cells emerge, at least in part, due to depleted lamin B1 levels exhibited by DG-null cells. Clearly, further research is required to elucidate a role for DG, if any, on centrosome duplication/mitosis organization. Although CRISPR-Cas9 genome editing ablates the expression of both α- and β-DG, we believe that the senescent phenotype of DG-KO cells is mechanistically linked to the nuclear deficiency of β-DG, because lamin B1, the central hub of cellular senescence, is a β-DG interacting partner [20,21]. Nonetheless, the possibility that the lack of α-DG drives the cell to senescence by perturbing the outside-in signaling pathway across the ECM-cytoskeleton-nucleus axis [56] cannot be ruled out. Furthermore, the rescue of DG expression in DG-KO cells is required to undoubtedly demonstrate the contribution of β-DG to cellular senescence.

In summary, overall our data are consistent with the paradigm that interfering with DG function results somehow in aberrant multipolar mitoses, which in turn evokes a p53-dependent DNA-damage response, arresting cell cycle progression and thereby inducing senescence, to avoid propagation of damaged genomes.

4. Materials and Methods

4.1. Cell Culturing and Treatments

Mouse C2C12 myoblasts were cultured in Dulbecco's modified Eagle's medium (DMEM) (Invitrogen, Carlsbad, CA, USA) supplemented with 10% (*v/v*) fetal bovine serum, 50 U/ml penicillin, 50 µg/ml streptomycin, and 1 mM sodium pyruvate at 37 °C, in a humidified 5% CO2 cell incubator. For senescence induction, cells were treated for 5 or 7 days with sodium butyrate (NaBu 5 mM, Sigma-Aldrich, St Louis) diluted in PBS 1X or vehicle alone. To analyze mitosis, cells were blocked at the S phase using double treatment with thymidine (2 mM) and then released from arrest by washing with PBS and plating in fresh culture medium on glass coverslips for 3–4 h (metaphase-anaphase).

4.2. Generation of DG-KO C2C12 Cell Lines by CRISPR-Cas

Two different single guide RNAs were designed to target the first coding exon of *Dag1* gene, using the crispr.mit.edu online tool: gRNA1 (5′ CCGACAACAGCCGTACCGTC 3′) and gRNA2 (5′ CCAGACGGTACGGCTGTTGT 3′) gRNAs were cloned into pSpCas9(BB)-red fluorescent protein (RFP) plasmid (modified from pSpCas9(BB)-2A-GFP, a gift from Feng Zhan–Addgene plasmid # 48138; http://n2t.net/addgene:48138;RRID:Addgene_48138), Addgene (Watertown, MA, USA). C2C12 cells were transfected with gRNA1- or gRNA2-Cas9-RFP plasmids using lipofectamine (LTX) with plus reagent (Thermo Fisher Scientific, Waltham, MA, USA). Forty-eight hours post transfection, RFP positive cells were sorted using FACSAria (BD Biosciences, Woburn, MA, USA). After expansion, cells were collected with enzyme-free cell dissociation buffer (Gibco–Thermo Fisher Scientific, Waltham, MA, USA) and incubated with anti-CD16/CD32 antibody (Mouse BD Fc Block, 2.4G2; BD Biosciences, Woburn, MA, USA) on ice for 5 minutes, and subsequently with anti-α-DG antibody (IIH6C4) on ice for 30 minutes. Following PBS washes, cells were incubated with goat anti-mouse IgG-Alexa Fluor 488 secondary antibody on ice for 20 minutes. Cells were then washed with PBS and resuspended in FACS buffer (10% FBS in PBS). Cells negative for α-DG staining were sorted by FACS and expanded. WT cells incubated with or without IIH6C4 were used to set the gates for positive or negative α-DG staining, respectively. Upon expansion, sorted cells had a second round of sorting for α-DG negative staining and single cells were collected in a 96-well plate for clonal expansion. DG-KO clones were screened for β-DG by Western blotting using anti-β-DG antibodies (MANDAG2). Two clones, DG-KO1 and DG-KO2, were expanded and characterized by sequencing the DNA region targeted by the gRNAs, to confirm *Dag1* gene disruption. Cell cultures between passage six and twelve were used for all analyses.

4.3. Antibodies

The following primary antibodies were used. Mouse monoclonal antibodies against α-dystrobrevin (α-DB; BD Transduction Laboratories, Becton Dickinson, Franklin Lakes, NJ, USA), α-DG (IIH6C4 (IIH6, 05-593; Millipore, Sigma-Aldrich, St. Louis, MO, USA), β-DG (MANDAG2 [57]), α-tubulin (sc-32293; Santa Cruz Biotechnology, CA, USA), p53 (#2524; Cell Signaling Technology, MA, USA), p21 (#2946; Cell Signaling Thecnology, MA, USA), and GAPDH (sc-32233; Santa Cruz Biotechnology, CA, USA). Rabbit polyclonal antibodies against B23 (sc-6013-R; Santa Cruz Biotechnology, CA, USA), dystrophin Dp71 (+78Dp71; Genemed Synthesis Inc., San Francisco, CA, USA), lamin B1 (Ab16048; Abcam, Cambridge, UK), γ-tubulin (sc10732; Santa Cruz Biotechnology, CA, USA), H3K9me3 (ab8898; Abcam, Cambridge, UK) and γ-H2AX (#07-164; Millipore, Sigma-Aldrich, St. Louis, MO, USA). Goat polyclonal antibody against β2-syntrophin (SC-13766; Santa Cruz Biotechnology, CA, USA) was also used.

4.4. Western Blotting

C2C12 cell culture lysates were electrophoresed on 10% SDS-polyacrylamide gels and transferred onto nitrocellulose membranes (Bio-Rad Laboratories Inc., Berkeley, CA, USA). Membranes were blocked in TBST (100 mM Tris-HCL pH 8.0, 150 mM NaCL, 0.5% (*v/v*) Tween-20) with low fat-dried milk and then incubated overnight at 4 °C with the appropriate primary antibodies. The specific protein signal was developed using the corresponding secondary antibodies and enhanced chemiluminescence western blotting detection system (ECL TM; Amersham Pharmacia, GE Healthcare), according to the manufacturer′s instructions. Images were acquired for densitometric analysis with a Gel Doc EZ System (Bio-Rad Laboratories Inc., Berkeley, CA, USA), using Image Lab 6.0.1 software (Bio-Rad Laboratories Inc., Berkeley, CA, USA). To normalize protein expression from the same sample and on the same blot, the band intensity of the target protein was divided by the band intensity of the loading protein.

4.5. Immunofluorescence and Confocal Microscopy Analysis

Cells cultured on coverslips were fixed with 4% paraformaldehyde in PBS for 10 min, permeabilized with 0.2% Triton X-100-PBS, blocked with 0.5% fetal bovine serum and 3% bovine serum albumin (BSA) in PBS and then incubated overnight at 4 °C with the corresponding primary antibodies. The following day, cells were washed with 0.05% Triton-X-100-PBS for 5 min and then with PBS alone three times, prior to be incubated for 1 h at room temperature with the appropriate fluorochrome-conjugated secondary antibody. For double immunolabeled samples, this was followed by overnight incubation at 4 °C with corresponding primary antibodies and the next day, cells were incubated with secondary fluorochrome-conjugated antibodies. Where indicated, F-actin was labelled using TRITC-conjugated Phalloidin (Sigma-Aldrich St. Louis, MO, USA) diluted 1:500 in PBS for 10 min at room temperature. Finally, coverslip preparations were incubated for 20 min at room temperature with 0.2 µg/mL diamidino-2-phenylindole (DAPI; Sigma Aldrich) for nuclei visualization, mounted on microscope slides with VectaShield (Vector Laboratories Inc., Burlingame, CA, USA) and further analyzed by confocal laser scanning microscopy (CLSM; Eclipse Ti Series, Nikon Corporation Healthcare Business Unit, Japan) using a 63× (NA = 1.2) oil-immersion objective. The analysis of digitized images was carried out using ImageJ, 1.49 software (Wayne Rasband National Institutes of Health, USA. http://imageJ.nih.gov.ij). For morphometric analysis of nuclei, raw images were calibrated and converted to 8-bit gray scale, to set up a threshold for nuclei selection. Then, nuclear area and circularity parameters were calculated, as described previously [58]. The nucleolar area (μm^2) was calculated on maxima projection images, using a 3D objects counter, as described previously [59]. To quantify the fluorescence intensity of γ-H2AX (DNA-damage marker and H3K9me3 (heterochromatin marker) foci, the Find Maxima function from ImageJ was used, as previously described [60]. Data were plotted using Prism6 software.

4.6. Flow Cytometry and Cell Proliferation Assays

Cells were trypsinized and washed twice with PBS prior to being fixed with 80% ethanol for 30 min, stained for DNA with 1 µg/mL DAPI (Sigma-Aldrich) for 20 min and transferred to flow cytometry tubes for cell cycle analysis in a BD LSR-Fortessa flow cytometer (BD Biosciences, San Jose, CA, USA), using the ModFit LT software (Verity Software House, Topsham, ME). For proliferation assays, cells were harvested and plated in triplicate onto 12-well microplates at (Corning, Costar), at a density of 1×10^3 cells/mL. Cell proliferation was assessed for 10 days using the MTT [3-(4,5-dimethylthiazole)-2-5-diphenyl tetrazolium bromide] commercial kit (Sigma-Aldrich) and following the manufacturer's instructions. Absorbance was measured at 570 nm on a Molecular Devices Spectra Max Plus384 microplate reader (Molecular Devices, Sunnyvale, CA, USA).

4.7. Fluorescence in Situ Hybridization (FISH) and Relative Telomere Length Determination

Cells grown on coverslips were fixed with 4% paraformaldehyde in PBS 1× for 10 min, washed three times in PBS 1X and permeabilized with Triton 0.2% in PBS 1X for 12 min. Cell preparations were treated for 20 min at 37 °C with 100 µL of RNase (1 µg/mL), washed three times with PBS 1X and dried. Afterwards, coverslips were incubated in hybridization buffer (20 mM Na2HPO4 [Ph 7.4], 20 mM Tris [pH 7.4], 60% formamide, 10% BSA) with 1ng/µL Cy3-conjugated telomere probe (Cy3 conjugated G-strand probe [5′-GGGTTAGGGTTAGGGTTA-3′]) added. Coverslips were then incubated at 80 °C for 2 h in the dark for denaturation, prior to incubation at room temperature overnight for hybridization. Next day, coverslips were washed twice with SSC 2×/1% Tween 20, for 10 min at 60 °C; twice with SSC 1×/0.1% Tween 20 and once with SSC 0.5×/0.1% Tween 20. Finally, cells preparations were incubated for 10 min at room temperature with DAPI (0.2 µg/µL, Sigma-Aldrich Inc.) for nuclei visualization, washed with PBS 1×, and mounted on microscope slides with VectaShield (Vector Laboratories, Inc., Burlingame, CA, USA) for confocal microscopy analysis. The analysis of the number of telomere foci and its relative length of the Q-FISH technique was performed using the Find Maxima function of

ImageJ software, 1.49 version image analysis (Wayne Rasband National Institutes of Health, USA. http://imageJ.nih.gov.ij). Raw images were converted to 8-bit gray scale to set up a binary mask that allowed the analysis of fluorescence intensity of the foci within a DAPI-positive region. The relative telomere length was calculated as follows: the telomere mean intensity was divided by the sum intensity of the DAPI signal, as described previously [61].

4.8. Senescence-Associated β-Galactosidase (SA-β-Gal) Assay

Cells seeded on coverslips were stained with SA-β-Gal following manufacturer's instructions (Senescent Cell Histochemical Staining Kit, Sigma-Aldrich, St. Louis, MO, USA). Blue stained cells expressing β-galactosidase (senescent cells) were observed under bright-field microscopy using differential interference contrast.

5. Conclusions

In summary, overall our data are consistent with the paradigm that interfering with DG function results somehow in aberrant multipolar mitoses, which in turn evokes a p53-dependent DNA-damage response, arresting the cell cycle progression and thereby inducing senescence, to avoid the propagation of damaged genomes.

Supplementary Materials: Supplementary materials can be found at http://www.mdpi.com/1422-0067/21/14/4961/s1.

Author Contributions: Conceptualization; B.C., G.E.J.-G., R.M.-G. and R.C.R.P.; Methodology; G.E.J.-G., R.M.-G., L.A.S.-P., W.L.G.-M., I.G.-A., R.A.P.-R. and R.S.-S.; Formal analysis; B.C., G.E.J.-G., R.M.-G., L.A.S.-P., W.L.G.-M., I.G.-A. and A.B.; Resources; B.C., J.J.M., R.C.R.P., R.A.P.-R. and A.B.; Writing; B.C., G.E.J.-G., R.M.-G. and A.B.; Funding acquisition and Project Administration; B.C. and J.J.M. All authors have read and agreed to the published version of the manuscript.

Funding: This research was funded by Bilateral Agreement CNR (Italy)/CINVESTAV (Mexico) (Reference CNR2018/4) and SEP-CINVESTAV (Reference 242) grants (BC) and the National Institutes of Health (NHI), grants R01 AR055299 and AR071439 (R.C.R.P.).

Acknowledgments: We are grateful to Jesús Pablo Gómez Islas for technical assistance and to Steve J Winder for critical reading of the manuscript.

Conflicts of Interest: The authors declare no conflict of interest.

Abbreviations

NaBu	Sodium butyrate
PM	Plasma membrane
NE	Nuclear envelope
DAPs	Dystrophin-Associated Proteins
ECM	Extracellular matrix
SA-β-gal	Senescence associated β-galactosidase
FACs	Fluorescence-activated cell sorting
SDS PAGE	Sodium dodecyl sulfate polyacrylamide gel
CLSM	Confocal laser scanning microscopy analysis

References

1. Ervasti, J.M.; Campbell, K.P. Membrane organization of the dystrophin-glycoprotein complex. *Cell* **1991**, *66*, 1121–1131. [CrossRef]

2. Ibraghimov-Beskrovnaya, O.; Ervasti, J.M.; Leveille, C.J.; Slaughter, C.A.; Sernett, S.W.; Campbell, K.P. Primary structure of dystrophin-associated glycoproteins linking dystrophin to the extracellular matrix. *Nature* **1992**, *355*, 696–702. [CrossRef] [PubMed]

3. Suzuki, A.; Yoshida, M.; Yamamoto, H.; Ozawa, E. Glycoprotein-binding site of dystrophin is confined to the cysteine-rich domain and the first half of the carboxy-terminal domain. *FEBS Lett.* **1992**, *308*, 154–160. [CrossRef]

4. Holt, K.H.; Crosbie, R.H.; Venzke, D.P.; Campbell, K.P. Biosynthesis of dystroglycan: Processing of a precursor propeptide. *FEBS Lett.* **2000**, *468*, 79–83. [CrossRef]

5. Smalheiser, N.R.; Kim, E. Purification of cranin, a laminin binding membrane protein. Identity with dystroglycan and reassessment of its carbohydrate moieties. *J. Biol. Chem.* **1995**, *270*, 15425–15433. [CrossRef]

6. Boffi, A.; Bozzi, M.; Sciandra, F.; Woellner, C.; Bigotti, M.G.; Ilari, A.; Brancaccio, A. Plasticity of secondary structure in the N-terminal region of beta-dystroglycan. *Biochim. Biophys. Acta* **2001**, *1546*, 114–121. [CrossRef]

7. Di Stasio, E.; Sciandra, F.; Maras, B.; Di Tommaso, F.; Petrucci, T.C.; Giardina, B.; Brancaccio, A. Structural and functional analysis of the N-terminal extracellular region of beta-dystroglycan. *Biochem. Biophys. Res. Commun.* **1999**, *266*, 274–278. [CrossRef]

8. Huang, X.; Poy, F.; Zhang, R.; Joachimiak, A.; Sudol, M.; Eck, M.J. Structure of a WW domain containing fragment of dystrophin in complex with beta-dystroglycan. *Nat. Struct. Biol.* **2000**, *7*, 634–638. [CrossRef]

9. Chung, W.; Campanelli, J.T. WW and EF hand domains of dystrophin-family proteins mediate dystroglycan binding. *Mol. Cell Biol. Res. Commun.* **1999**, *2*, 162–171. [CrossRef]

10. Ishikawa-Sakurai, M.; Yoshida, M.; Imamura, M.; Davies, K.E.; Ozawa, E. ZZ domain is essentially required for the physiological binding of dystrophin and utrophin to beta-dystroglycan. *Hum. Mol. Genet.* **2004**, *13*, 693–702. [CrossRef]

11. Muntoni, F. Journey into muscular dystrophies caused by abnormal glycosylation. *Acta Myol. Myopathies Cardiomyopathies Off. J. Mediterr. Soc. Myol.* **2004**, *23*, 79–84.

12. Singh, J.; Itahana, Y.; Knight-Krajewski, S.; Kanagawa, M.; Campbell, K.P.; Bissell, M.J.; Muschler, J. Proteolytic enzymes and altered glycosylation modulate dystroglycan function in carcinoma cells. *Cancer Res.* **2004**, *64*, 6152–6159. [CrossRef]

13. Cohn, R.D.; Henry, M.D.; Michele, D.E.; Barresi, R.; Saito, F.; Moore, S.A.; Flanagan, J.D.; Skwarchuk, M.W.; Robbins, M.E.; Mendell, J.R.; et al. Disruption of DAG1 in differentiated skeletal muscle reveals a role for dystroglycan in muscle regeneration. *Cell* **2002**, *110*, 639–648. [CrossRef]

14. Dumont, N.A.; Wang, Y.X.; von Maltzahn, J.; Pasut, A.; Bentzinger, C.F.; Brun, C.E.; Rudnicki, M.A. Dystrophin expression in muscle stem cells regulates their polarity and asymmetric division. *Nat. Med.* **2015**, *21*, 1455–1463. [CrossRef] [PubMed]

15. Thompson, O.; Moore, C.J.; Hussain, S.A.; Kleino, I.; Peckham, M.; Hohenester, E.; Ayscough, K.R.; Saksela, K.; Winder, S.J. Modulation of cell spreading and cell-substrate adhesion dynamics by dystroglycan. *J. Cell Sci.* **2010**, *123*, 118–127. [CrossRef] [PubMed]

16. Oppizzi, M.L.; Akhavan, A.; Singh, M.; Fata, J.E.; Muschler, J.L. Nuclear translocation of beta-dystroglycan reveals a distinctive trafficking pattern of autoproteolyzed mucins. *Traffic (Copenhagen, Denmark)* **2008**, *9*, 2063–2072. [CrossRef]

17. Lara-Chacón, B.; de León, M.B.; Leocadio, D.; Gómez, P.; Fuentes-Mera, L.; Martínez-Vieyra, I.; Ortega, A.; Jans, D.A.; Cisneros, B. Characterization of an Importin alpha/beta-recognized nuclear localization signal in beta-dystroglycan. *J. Cell. Biochem.* **2010**, *110*, 706–717. [CrossRef]

18. Gracida-Jiménez, V.; Mondragón-González, R.; Vélez-Aguilera, G.; Vásquez-Limeta, A.; Laredo-Cisneros, M.S.; Gómez-López, J.D.; Vaca, L. Retrograde trafficking of β-dystroglycan from the plasma membrane to the nucleus. *Sci. Rep.* **2017**, *7*, 9906. [CrossRef]

19. Mathew, G.; Mitchell, A.; Down, J.M.; Jacobs, L.A.; Hamdy, F.C.; Eaton, C.; Rosario, D.J.; Cross, S.S.; Winder, S.J. Nuclear targeting of dystroglycan promotes the expression of androgen regulated transcription factors in prostate cancer. *Sci. Rep.* **2013**, *3*, 2792. [CrossRef]

20. Martínez-Vieyra, I.A.; Vásquez-Limeta, A.; González-Ramírez, R.; Morales-Lázaro, S.L.; Mondragón, M.; Mondragón, R.; Ortega, A.; Winder, S.J.; Cisneros, B. A role for β-dystroglycan in the organization and structure of the nucleus in myoblasts. *Biochim. Biophys. Acta* **2013**, *1833*, 698–711. [CrossRef]

21. Vélez-Aguilera, G.; de Dios Gómez-López, J.; Jiménez-Gutiérrez, G.E.; Vásquez-Limeta, A.; Laredo-Cisneros, M.S.; Gómez, P.; Winder, S.J.; Cisneros, B. Control of nuclear β-dystroglycan content is crucial for the maintenance of nuclear envelope integrity and function. *Biochim. Biophys. Acta* **2018**, *1865*, 406–420. [CrossRef] [PubMed]

22. Shimi, T.; Butin-Israeli, V.; Adam, S.A.; Hamanaka, R.B.; Goldman, A.E.; Lucas, C.A.; Shumaker, D.K.; Kosak, S.T.; Chandel, N.S.; Goldman, R.D. The role of nuclear lamin B1 in cell proliferation and senescence. *Genes Dev.* **2011**, *25*, 2579–2593. [CrossRef] [PubMed]

23. Freund, A.; Laberge, R.M.; Demaria, M.; Campisi, J. Lamin B1 loss is a senescence-associated biomarker. *Mol. Biol. Cell* **2012**, *23*, 2066–2075. [CrossRef] [PubMed]

24. Garvalov, B.K.; Muhammad, S.; Dobreva, G. Lamin B1 in cancer and aging. *Aging* **2019**, *11*, 7336–7338. [CrossRef]

25. Shah, P.P.; Donahue, G.; Otte, G.L.; Capell, B.C.; Nelson, D.M.; Cao, K.; Aggarwala, V.; Cruickshanks, H.A.; Rai, T.S.; McBryan, T.; et al. Lamin B1 depletion in senescent cells triggers large-scale changes in gene expression and the chromatin landscape. *Genes Dev.* **2013**, *27*, 1787–1799. [CrossRef]

26. Campisi, J. Senescent cells, tumor suppression, and organismal aging: Good citizens, bad neighbors. *Cell* **2005**, *120*, 513–522. [CrossRef]

27. Adams, P.D. Healing and hurting: Molecular mechanisms, functions, and pathologies of cellular senescence. *Mol. Cell* **2009**, *36*, 2–14. [CrossRef]

28. Wang, A.S.; Dreesen, O. Biomarkers of Cellular Senescence and Skin Aging. *Front. Genet.* **2018**, *9*, 247. [CrossRef]

29. Ervasti, J.M.; Campbell, K.P. A role for the dystrophin-glycoprotein complex as a transmembrane linker between laminin and actin. *J. Cell Biol.* **1993**, *122*, 809–823. [CrossRef]

30. Dreesen, O.; Chojnowski, A.; Ong, P.F.; Zhao, T.Y.; Common, J.E.; Lunny, D.; Lane, E.B.; Lee, S.J.; Vardy, L.A.; Stewart, C.L.; et al. Lamin B1 fluctuations have differential effects on cellular proliferation and senescence. *J. Cell Biol.* **2013**, *200*, 605–617. [CrossRef]

31. Petrova, N.V.; Velichko, A.K.; Razin, S.V.; Kantidze, O.L. Small molecule compounds that induce cellular senescence. *Aging Cell* **2016**, *15*, 999–1017. [CrossRef]

32. Bolderson, E.; Scorah, J.; Helleday, T.; Smythe, C.; Meuth, M. ATM is required for the cellular response to thymidine induced replication fork stress. *Hum. Mol. Genet.* **2004**, *13*, 2937–2945. [CrossRef] [PubMed]

33. Podhorecka, M.; Skladanowski, A.; Bozko, P. H2AX Phosphorylation: Its Role in DNA Damage Response and Cancer Therapy. *J. Nucleic Acids* **2010**, *2010*. [CrossRef] [PubMed]

34. Maass, K.K.; Rosing, F.; Ronchi, P.; Willmund, K.V.; Devens, F.; Hergt, M.; Herrmann, H.; Lichter, P.; Ernst, A. Altered nuclear envelope structure and proteasome function of micronuclei. *Exp. Cell Res.* **2018**, *371*, 353–363. [CrossRef]

35. Hewitt, G.; Jurk, D.; Marques, F.D.; Correia-Melo, C.; Hardy, T.; Gackowska, A.; Anderson, R.; Taschuk, M.; Mann, J.; Passos, J.F. Telomeres are favoured targets of a persistent DNA damage response in ageing and stress-induced senescence. *Nat. Commun.* **2012**, *3*, 708. [CrossRef] [PubMed]

36. Hultdin, M.; Grönlund, E.; Norrback, K.; Eriksson-Lindström, E.; Just, T.; Roos, G. Telomere analysis by fluorescence in situ hybridization and flow cytometry. *Nucleic Acids Res.* **1998**, *26*, 3651–3656. [CrossRef] [PubMed]

37. O'Sullivan, J.N.; Finley, J.C.; Risques, R.A.; Shen, W.T.; Gollahon, K.A.; Moskovitz, A.H.; Gryaznov, S.; Harley, C.B.; Rabinovitch, P.S. Telomere length assessment in tissue sections by quantitative FISH: Image analysis algorithms. *Cytometry. Part A* **2004**, *58*, 120–131. [CrossRef] [PubMed]

38. Ervasti, J.M.; Sonnemann, K.J. Biology of the striated muscle dystrophin-glycoprotein complex. *Int. Rev. Cytol.* **2008**, *265*, 191–225. [CrossRef]

39. Dechat, T.; Pfleghaar, K.; Sengupta, K.; Shimi, T.; Shumaker, D.K.; Solimando, L.; Goldman, R.D. Nuclear lamins: Major factors in the structural organization and function of the nucleus and chromatin. *Genes Dev.* **2008**, *22*, 832–853. [CrossRef]

40. Lukášová, E.; Kovařík, A.; Kozubek, S. Consequences of Lamin B1 and Lamin B Receptor Downregulation in Senescence. *Cells* **2018**, *7*, 11. [CrossRef] [PubMed]

41. Hutchison, C.J. Do lamins influence disease progression in cancer? *Adv. Exp. Med. Biol.* **2014**, *773*, 593–604. [CrossRef]

42. Camps, J.; Erdos, M.R.; Ried, T. The role of lamin B1 for the maintenance of nuclear structure and function. *Nucleus (Austin, Tex.)* **2015**, *6*, 8–14. [CrossRef]

43. Sadaie, M.; Salama, R.; Carroll, T.; Tomimatsu, K.; Chandra, T.; Young, A.R.; Narita, M.; Pérez-Mancera, P.A.; Bennett, D.C.; Chong, H.; et al. Redistribution of the Lamin B1 genomic binding profile affects rearrangement of heterochromatic domains and SAHF formation during senescence. *Genes Dev.* **2013**, *27*, 1800–1808. [CrossRef] [PubMed]

44. Sidler, C.; Kovalchuk, O.; Kovalchuk, I. Epigenetic Regulation of Cellular Senescence and Aging. *Front. Genet.* **2017**, *8*, 138. [CrossRef]

45. Martins, F.; Sousa, J.; Pereira, C.D.; da Cruz, E.S.O.A.B.; Rebelo, S. Nuclear envelope dysfunction and its contribution to the aging process. *Aging Cell* **2020**, *19*, e13143. [CrossRef] [PubMed]

46. Prieto, L.I.; Graves, S.I.; Baker, D.J. Insights from In Vivo Studies of Cellular Senescence. *Cells* **2020**, *9*, 954. [CrossRef] [PubMed]

47. Gergely, F.; Basto, R. Multiple centrosomes: Together they stand, divided they fall. *Genes Dev.* **2008**, *22*, 2291–2296. [CrossRef] [PubMed]

48. Vitre, B.D.; Cleveland, D.W. Centrosomes, chromosome instability (CIN) and aneuploidy. *Curr. Opin. Cell Biol.* **2012**, *24*, 809–815. [CrossRef]

49. Lerit, D.A.; Poulton, J.S. Centrosomes are multifunctional regulators of genome stability. *Chromosome Res.* **2016**, *24*, 5–17. [CrossRef]

50. Georgoulis, A.; Vorgias, C.E.; Chrousos, G.P.; Rogakou, E.P. Genome Instability and γH2AX. *Int. J. Mol. Sci.* **2017**, *18*. [CrossRef] [PubMed]

51. Mijit, M.; Caracciolo, V.; Melillo, A.; Amicarelli, F.; Giordano, A. Role of p53 in the Regulation of Cellular Senescence. *Biomolecules* **2020**, *10*, 420. [CrossRef] [PubMed]

52. Gönczy, P. Centrosomes and cancer: Revisiting a long-standing relationship. *Nat. Rev.. Cancer* **2015**, *15*, 639–652. [CrossRef] [PubMed]

53. Pihan, G.A. Centrosome dysfunction contributes to chromosome instability, chromoanagenesis, and genome reprograming in cancer. *Front. Oncol.* **2013**, *3*, 277. [CrossRef]

54. Higginson, J.R.; Thompson, O.; Winder, S.J. Targeting of dystroglycan to the cleavage furrow and midbody in cytokinesis. *Int. J. Biochem. Cell Biol.* **2008**, *40*, 892–900. [CrossRef]

55. Tsai, M.Y.; Wang, S.; Heidinger, J.M.; Shumaker, D.K.; Adam, S.A.; Goldman, R.D.; Zheng, Y. A mitotic lamin B matrix induced by RanGTP required for spindle assembly. *Science (New York, N.Y.)* **2006**, *311*, 1887–1893. [CrossRef] [PubMed]

56. Hieda, M. Signal Transduction across the Nuclear Envelope: Role of the LINC Complex in Bidirectional Signaling. *Cells* **2019**, *8*. [CrossRef] [PubMed]

57. Pereboev, A.V.; Ahmed, N.; thi Man, N.; Morris, G.E. Epitopes in the interacting regions of beta-dystroglycan (PPxY motif) and dystrophin (WW domain). *Biochim. Biophys. Acta* **2001**, *1527*, 54–60. [CrossRef]

58. Filippi-Chiela, E.C.; Oliveira, M.M.; Jurkovski, B.; Callegari-Jacques, S.M.; da Silva, V.D.; Lenz, G. Nuclear morphometric analysis (NMA): Screening of senescence, apoptosis and nuclear irregularities. *PLoS ONE* **2012**, *7*, e42522. [CrossRef]

59. García-Aguirre, I.; Alamillo-Iniesta, A.; Rodríguez-Pérez, R.; Vélez-Aguilera, G.; Amaro-Encarnación, E.; Jiménez-Gutiérrez, E.; Vásquez-Limeta, A.; Samuel Laredo-Cisneros, M.; Morales-Lázaro, S.L.; Tiburcio-Félix, R.; et al. Enhanced nuclear protein export in premature aging and rescue of the progeria phenotype by modulation of CRM1 activity. *Aging Cells* **2019**, *18*, e13002. [CrossRef]

60. Zhang, H.; Sun, L.; Wang, K.; Wu, D.; Trappio, M.; Witting, C.; Cao, K. Loss of H3K9me3 Correlates with ATM Activation and Histone H2AX Phosphorylation Deficiencies in Hutchinson-Gilford Progeria Syndrome. *PLoS ONE* **2016**, *11*, e0167454. [CrossRef]

61. Tichy, E.D.; Sidibe, D.K.; Tierney, M.T.; Stec, M.J.; Sharifi-Sanjani, M.; Hosalkar, H.; Mubarak, S.; Johnson, F.B.; Sacco, A.; Mourkioti, F. Single Stem Cell Imaging and Analysis Reveals Telomere Length Differences in Diseased Human and Mouse Skeletal Muscles. *Stem Cell Rep.* **2017**, *9*, 1328–1341. [CrossRef] [PubMed]

Article

Peripheral Circulating Exosomal miRNAs Potentially Contribute to the Regulation of Molecular Signaling Networks in Aging

Hongxia Zhang and Kunlin Jin *

Department of Pharmacology and Neuroscience, University of North Texas Health Science Center, Fort Worth, TX 76107, USA; Hongxia.Zhang@unthsc.edu
* Correspondence: kunlin.jin@unthsc.edu; Tel.: +1-817-735-2579

Received: 4 February 2020; Accepted: 9 March 2020; Published: 11 March 2020

Abstract: People are living longer than ever. Consequently, they have a greater chance for developing a functional impairment or aging-related disease, such as a neurodegenerative disease, later in life. Thus, it is important to identify and understand mechanisms underlying aging as well as the potential for rejuvenation. Therefore, we used next-generation sequencing to identify differentially expressed microRNAs (miRNAs) in serum exosomes isolated from young (three-month-old) and old (22-month-old) rats and then used bioinformatics to explore candidate genes and aging-related pathways. We identified 2844 mRNAs and 68 miRNAs that were differentially expressed with age. TargetScan revealed that 19 of these miRNAs are predicated to target the 766 mRNAs. Pathways analysis revealed signaling components targeted by these miRNAs: mTOR, AMPK, eNOS, IGF, PTEN, p53, integrins, and growth hormone. In addition, the most frequently predicted target genes regulated by these miRNAs were EIF4EBP1, insulin receptor, PDK1, PTEN, paxillin, and IGF-1 receptor. These signaling pathways and target genes may play critical roles in regulating aging and lifespan, thereby validating our analysis. Understanding the causes of aging and the underlying mechanisms may lead to interventions that could reverse certain aging processes and slow development of aging-related diseases.

Keywords: exosomes; aging; serum; functional enrichment analysis; ingenuity pathway analysis; miRNA-mRNA networks; aging-related disease

---

1. Introduction

Aging is a highly complex biological process that is often accompanied by a general decline in tissue function and an increased risk for aging-related diseases, such as cardiovascular disease, stroke, cancer, and neurodegenerative diseases. Indeed, as average lifespan continues to increase, aging-related functional decline, such as cognitive impairment, will likely become a health care priority [1]. For example, the most common form of dementia is Alzheimer's disease (AD), but a large proportion of cognitive impairment cases in the aged population is not due to AD but rather to normal aging process. Thus, it is important to identify ways to maintain functional integrity during aging [2]. Many theories have been proposed to explain why we age [3]. Recently, we proposed a new theory positing that aging is the process of continuous impairment of microcirculation in the body [4]. Indeed, compelling evidence indicates that systemic factors in the blood profoundly reverse aging-related impairments [5–7], which are influenced by specific rejuvenating or deteriorating factors, e.g., proteins, microRNAs (miRNAs), and mRNAs [8]. Thus, many circulating factors have been identified as attractive biomarkers for tissue-specific diseases and aging [9,10]. However, the mechanisms underlying the contributions of blood-derived factors to aging remain unclear.

Research over the last two decades has demonstrated that cells mainly communicate by releasing extracellular vesicles (EV) that can act on nearby cells (paracrine signaling) or end up in circulating body fluids, with possible effects at distant sites (endocrine signaling) [11]. Exosomes are small EVs (approximately 50–150 nm in diameter) of endosomal origin that initially form as intraluminal vesicles inside late endosomal compartments. Indeed, exosomes contain many specific proteins, mRNAs, miRNAs, and long noncoding RNAs [12] and play a vital role in cell communication by transferring their cargo between source and target cells, which is also important in aging and aging-related disease [13]. For example, injection of serum exosomes from young mice into old mice could alter the expression pattern of aging-associated molecules to mimic that of young mice [14]. In addition, studies have documented that exosomes from brain cells can cross the blood-brain barrier (BBB) and serve as peripheral circulating biomarkers of cognitive impairment in AD [15–17], and, blood exosomes can also cross the BBB to target brain cells and affect brain function [18–21]. Thus, peripheral circulating exosomes have diagnostic and therapeutic potential. However, most studies have focused on establishing exosomal protein or miRNA profiles for comparing disease states and matched controls, and few studies have focused on characterizing proteins and miRNAs in peripheral circulating exosomes during normal aging [22]. Therefore, it is critical to define the profiles for exosomal proteins and miRNAs that can be transferred from exosome to recipient cells. Importantly, it has been estimated that miRNAs regulate ~31% of all eukaryotic genes by promoting degradation of their mRNAs or inhibiting their translation [23,24]. Indeed, miRNA-mediated regulation governs metabolism, immunity, lifespan, cell proliferation, apoptosis, and development [25–27], as well as pathological processes such as cancer and cardiovascular and neurodegenerative disease [28–30]. Therefore, among the exosomal cargo that is transferred to recipient cells, miRNAs likely have the greatest downstream impact on cell functions. To explore the role of circulating exosomes in aging processes, exosomal miRNAs must be more broadly characterized. In addition, recent evidence suggests that numerous signaling pathways regulate normal aging processes. However, research is lacking concerning how aging affects co-expression profiles for exosomal miRNAs and mRNAs and how miRNA-mRNA regulatory networks systematically influence aging processes.

To address shortcomings in our knowledge of exosomal miRNA functions, we used next-generation sequencing to establish miRNA and mRNA profiles for circulating exosomes isolated from young and old rats. We also investigated the possible role of exosomal miRNAs in aging by analyzing the biological importance of the miRNA targets and in major signaling pathways associated with aging using bioinformatic tools including Gene Ontology (GO) enrichment, Kyoto Encyclopedia of Genes and Genomes (KEGG) enrichment and pathways, eukaryotic orthologous groups (KOG) function classification, and Ingenuity Pathway Analysis (IPA). Our findings may provide a basis for understanding the physiological consequences of aging-related changes in the makeup of circulating miRNAs and could lead to potential interventions for aging-related diseases.

2. Results

2.1. Characterization of Serum Exosomes

We first characterized the protein content of serum exosomes isolated from young and old rats using Western blotting. Serum exosomes from each of young and old rats were positive for the exosome markers, CD63 and CD9 (Figure 1A). Nanoparticle tracking analysis (NTA) (Figure 1B) verified a strong enrichment of particles in the range 40–120 nm, with mean size of 82 +/− 0.8 nm, supporting a multimodal size distribution of exosomes with a peak diameter of 70–120 nm, consistent with previous reports [31,32]. In addition, transmission electron microscopy (TEM) was used to confirm that the purified particles were membrane bound, round and heterogeneous in size (40–120 nm) (Figure 1C).

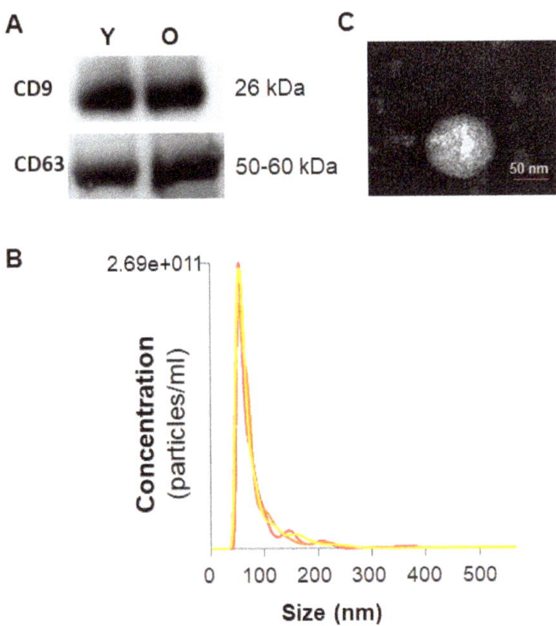

Figure 1. Characterization of serum exosomes. (**A**) Western blotting for CD63 and CD9 in serum exosomes isolated from young and old rats. (**B**) Average overall size distribution of exosomes from serum of old rats using the Nanoparticle Tracking Analysis. (**C**) Representative transmission electron microscopy image showing the typical morphology and size range of exosomes from serum of old rat. Y, serum exosomes from young rats; O, serum exosomes from old rats.

2.2. Differentially Expressed RNAs in Serum Exosomes with Age

To determine whether aging affects the levels of serum exosomal RNAs, RNA profiles were determined by next-generation RNA Sequencing. After quality control and filtering, a total of 35117 RNAs, including mRNA, miRNAs and other type of RNAswere identified in exosomes from serum of young and old rats (Supplementary data). Following application of thresholds for significance, 2736 (17.9%) were down-regulated and 108 (7%) were up-regulated in serum exosomes from old rats ($p < 0.05$, >1.5-fold change; Figure 2A), among identified 15272 mRNA. In addition, 600 miRNAs were identified after quality control, among which 68 were relatively abundant in old rats, including 28 that were down-regulated and 40 that were up-regulated serum exosomes from old rats ($p < 0.05$, >1.5-fold change; Figure 2B). A volcano plot (Figure 2C) and cluster analysis (Figure 2D) revealed the overall distribution of differentially expressed mRNAs and miRNAs of serum exosomes with age after analysis with TargetScan.

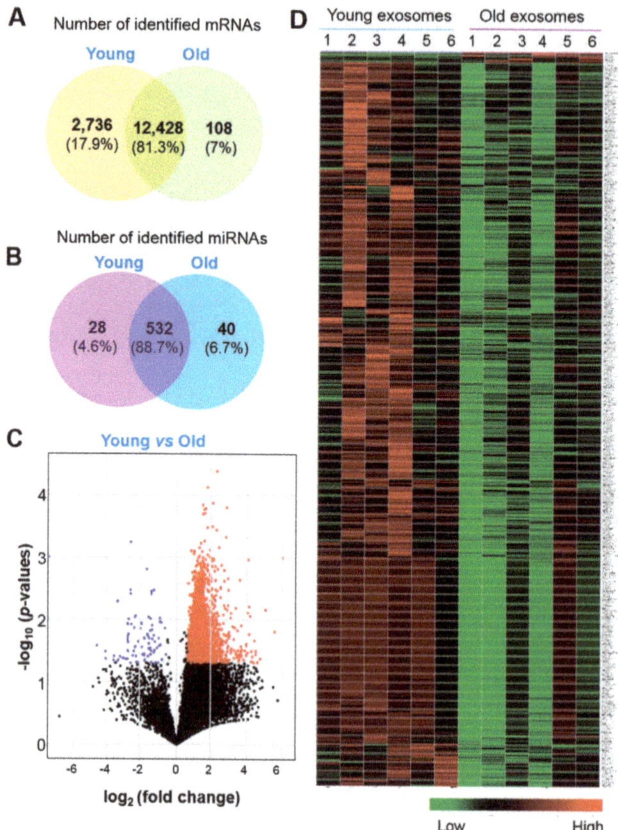

Figure 2. Profiles for mRNAs and miRNAs of serum exosomes from young and old rats. (**A,B**) Venn diagram of all differentially expressed mRNAs (**A**) and miRNAs (**B**) identified in serum exosomes. (**C**) Volcano plot for comparing the differentially expressed exosomal mRNAs and miRNAs in serum from young and old rats after analysis with TargetScan (fold change > 1.5 and $p < 0.05$). (**D**) Heatmap of the differentially expressed mRNAs and miRNAs in serum exosomes from young and old rats ($n = 6$ each group) after TargetScan (fold change > 1.5 and $p < 0.05$). Young, serum exosomes from young rats; Old, serum exosomes from old rats.

2.3. Identification of miRNA-Targeted mRNAs

MiRNAs regulate expression of specific genes via hybridization to mRNAs to promote their degradation in order to inhibit their translation or both [33]. A volcano plot revealed the overall distribution of the exosomal miRNAs we identified in this study (Figure 3A). To study the possible functional roles of the differentially expressed miRNAs, their potential mRNA targets were analyzed with Targetscan. Among the 68 miRNAs, only 19 were associated with 766 of the 2844 mRNAs that were differentially expressed with age (Figure 3B), suggesting that these miRNAs contribute to the age-dependent regulation of specific mRNAs. Among them, 5 mRNAs were down-regulated and 14 were up-regulated in serum exosomes from old rats compared with those from young rats (Figure 3C). MiRNA-483-3p and miRNA-489-3p were detected only in exosomes from young rats, and miRNA-187-3p, miRNA-202-3p, miRNA-450b-5p, miRNA-501-3p, miRNA-511-5p, and miRNA-598-3p were detected only in exosomes from old rats (Table 1). Figure 3D presents results of a cluster analysis of differentially expressed miRNAs for each sample.

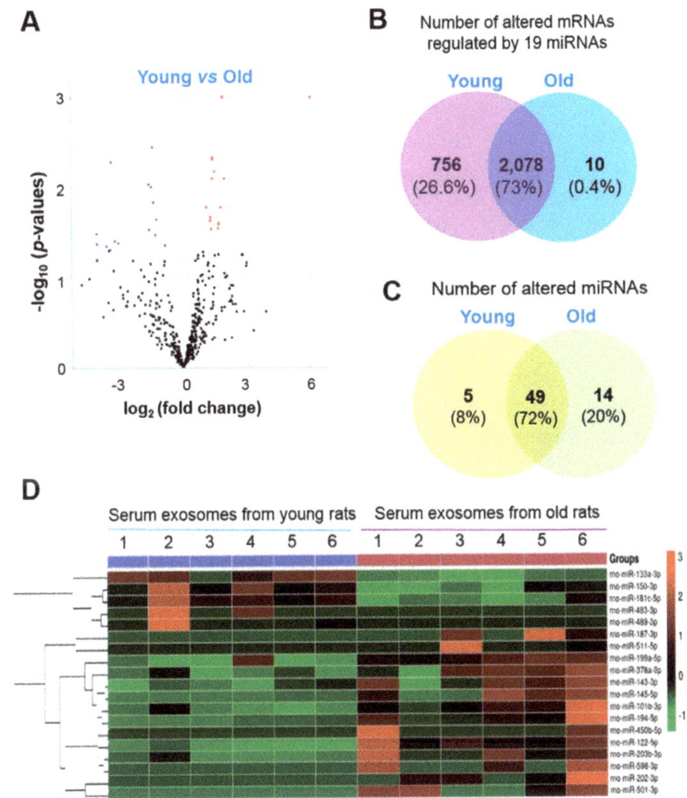

Figure 3. Profiles for differentially expressed miRNAs of serum exosomes from young and old rats. (**A**) Volcano plot showing the differentially expressed miRNAs (fold change > 1.5 and $p < 0.05$). (**B**) Venn diagram showing the differentially expressed mRNAs. (**C**) Venn diagram of the differentially expressed miRNAs. (**D**) Heat map of hierarchical clustering of 19 miRNAs that were identified in serum exosomes from young and old rats ($n = 6$ each group).

Table 1. List of circulating miRNAs for which expression differed with age.

ID	*p*-Value	Fold Change	Expression Level (Old vs. Young)
rno-miR-101b-3p	0.0295919	3.0300921	Up
rno-miR-122-5p	0.005130112	10.50495019	Up
rno-miR-133a-3p	0.000973173	−3.631787956	Down
rno-miR-143-3p	0.033831212	2.469145377	Up
rno-miR-145-5p	0.032740298	17.04524832	Up
rno-miR-150-3p	0.016694482	−3.404719964	Down
rno-miR-181c-5p	0.006634803	−2.79454293	Down
rno-miR-187-3p	0.001	64	Up
rno-miR-194-5p	0.042745612	16.8112814	Up
rno-miR-199a-5p	0.009166599	3.06058214	Up
rno-miR-202-3p	0.001	64	Up
rno-miR-203b-3p	0.03954997	9.395112765	Up
rno-miR-378a-3p	0.043871243	1.797380304	Up
rno-miR-450b-5p	0.001	64	Up
rno-miR-483-3p	0.001	−64	Down
rno-miR-489-3p	0.001	−64	Down
rno-miR-501-3p	0.001	64	Up
rno-miR-511-5p	0.001	64	Up
rno-miR-598-3p	0.001	64	Up

2.4. GO Enrichment Analysis of miRNA-Targeted mRNAs

To gain a better understanding of the potential role of these exosomal miRNAs in aging, we used Blastp to carry out functional annotation and enrichment analysis of their target genes identified in the GO enrichment analysis. Functional annotation was categorized by biological process, cellular component and molecular function, and only the top 10 GO terms having the smallest p-value were considered. These categories represent the annotation of the functional enrichment of targeted genes, and a lower p-value represents a greater functional enrichment of a relative term. Notably, almost all the genes listed under these GO terms were downregulated in serum exosomes from old rats. This analysis revealed several enriched functional categories and target genes, including genes involved in the posttranslational modification of proteins, metabolic processes, cell communication, molecular function, and intracellular signal transduction (Figure 4A).

2.5. KOG and KEGG Enrichment and Analyses

KOG was used to functionally classify mRNAs (766) regulated by the 19 exosomal miRNAs that were differentially expressed between young and old rats. Among the resultant 25 KOG classifications, genes involved in "signal transduction mechanisms" were the ones most commonly targeted (151 genes), followed by "general function prediction only" (119 genes), "transcription" (56 genes), "posttranslational modification and protein turnover" (55 genes), and "intracellular function and secretion and vesicular transport "(40 genes) (Figure 4B).

KEGG is a comprehensive knowledge base for both functional interpretation and practical application of genomic information [34]. KEGG pathway analysis identified 20 pathways that differed significantly ($p < 0.05$) between exosomes of young and old rats (Figure 4C). Among these pathways, the following were found to be involved in aging and lifespan: insulin resistance, mitogen-activated protein kinase (MAPK) signaling, PI3 kinase (PI3K)–Akt signaling, mammalian target of rapamycin (mTOR) signaling, toll-like receptor signaling, FoxO signaling, ErbB signaling, longevity-regulating signaling, and resistance to inhibitors of epidermal growth-factor receptor (EGFR) tyrosine kinase (Figure 4C).

2.6. Analysis of Pathways and Interaction Networks

We then carried out IPA for molecular pathways associated with serum exosomal miRNAs during aging. The results showed that 163 IPA canonical pathways were predicted to be significantly related to the expression of serum exosomal miRNAs, based on $p < 0.05$. The top 22 most strongly aging-associated pathways targeted by miRNAs are shown in Figure 5A. Those discovered aging-related signaling pathways included insulin, integrin, ErbB, neuregulin, mTOR, opioid, telomerase, phosphatase and tensin homolog 10 (PTEN), insulin-like growth factor-1 (IGF-1), adenosine monophosphate-activated protein kinase (AMPK), growth hormone, endothelial nitric oxide synthase (eNOS), nitric oxide, fibroblast growth factor (FGF), cyclic adenosine monophosphate (cAMP), sphingosine, platelet-derived growth factor (PDGF), docosahexaenoic acid (DHA), triggering receptor expressed on myeloid cells 1 (TREM1), and p53, suggesting that miRNAs target multiple biological pathways that modulate aging.

Figure 5B presents the IPA network results, and Table 2 lists the miRNAs involved in the nine pathways. Similar to IPA results, the networks contained genes predicted to be involved in metabolism, growth hormone signaling, and oxidative stress. As shown in Figure 5B, each pathway was linked with several gene transcripts, and individual genes could be regulated by several miRNAs. This suggested that the serum exosomal miRNAs that regulate crosstalk between pathways differ among young and old rats. The most common proteins in the networks were eukaryotic translation initiation factor 4E binding protein 1 (EIF4EBP1), insulin receptor (INSR), phosphoinositide dependent protein kinase 1 (PDPK1), PTEN, paxillin (PXN), and IGF-1 receptor (IGF-1R) that were targeted by the most prominent miRNAs (Figure 6A). Overall, the results establish putative functions between miRNAs and their target mRNAs, molecular networks, and biological pathways that modulate the makeup of serum exosomal miRNAs in young versus old animals. Figure 6B shows one such example of miRNA-mediated regulation.

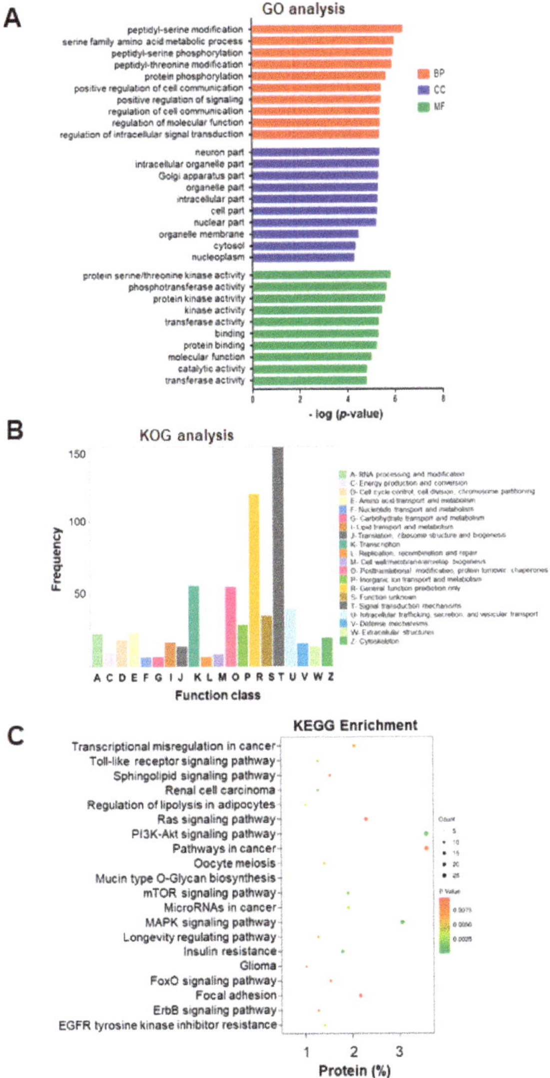

Figure 4. Gene Ontology (GO) analysis, eukaryotic orthologous groups (KOG) functional classification, and Kyoto Encyclopedia of Genes and Genomes (KEGG) pathway analysis of target genes (mRNAs) regulated by the 19 miRNAs that were differentially expressed between young and old rats. (**A**) GO annotation of predicted targets. The top 10 most enriched GO terms are listed in terms for biological process (BP), cellular component (CC), and molecular function (MF) based on p-values. (**B**) KOG functional classification of target genes. The vertical axis represents the frequency of target genes classified into the specific categories, and the horizontal axis represents the KOG functional classification. (**C**) The top 20 most common KEGG pathways of the differentially expressed mRNAs regulated by the 19 miRNAs. Fold change > 1.5 and $p < 0.05$. GO, gene ontology; KOG, eukaryotic orthologous groups; KEGG, Kyoto Encyclopedia of Genes and Genomes.

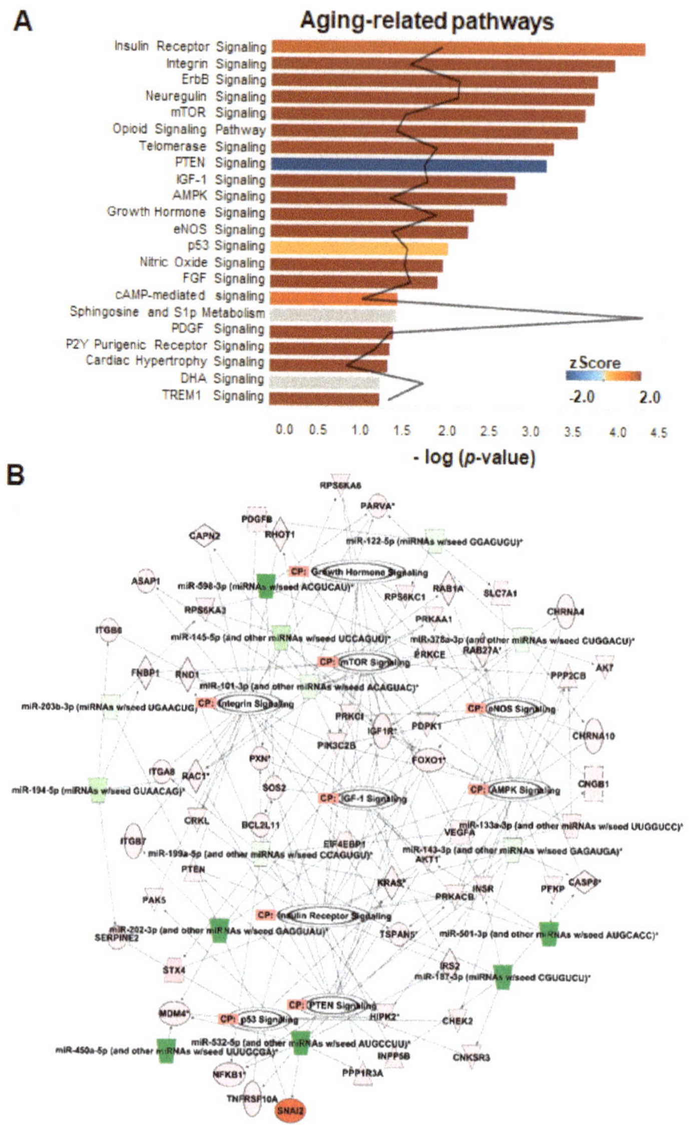

Figure 5. Ingenuity Pathway Analysis (IPA) of the differentially expressed miRNAs in serum exosomes from young and old rats. (**A**) IPA showing the 22 most significant aging-related pathways involving mRNAs, whose expression is regulated by differentially expressed miRNAs in serum exosomes from young and old rats. Each Z score represents the upregulation or downregulation of gene expression based on young vs. old. The black curve denotes the ratio between the number of the differentially expressed target genes and the total number of genes in each of these pathways. (**B**) IPA-predicted network for the differentially expressed miRNAs showing predicted targets and their association with biological functions in aging-related signaling pathways governed by the following factors: growth hormone signaling, mammalian target of rapamycin (mTOR) signaling, endothelial nitric oxide synthase (eNOS) signaling, integrin signaling, insulin-like growth factor-1 (IGF-1) signaling, AMP-activated protein kinase (AMPK) signaling, insulin receptor signaling, p53 signaling, and phosphatase and tensin homolog 10 (PTEN) signaling.

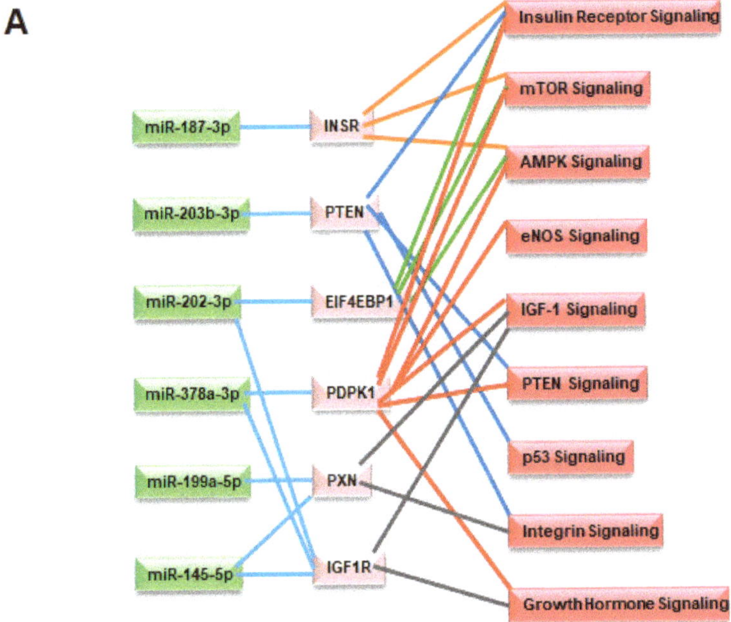

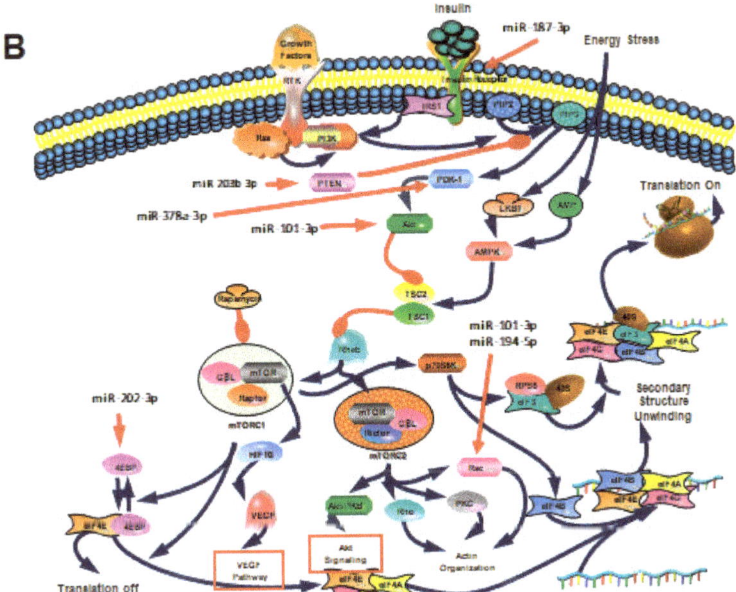

Figure 6. The most common target genes and mTOR pathway regulated by the differentially expressed exosomal miRNAs. (**A**) *EIF4EBP1, INSR, PDPK1, PTEN, PXN,* and *IGF-1R* are the most common network genes targeted by the prominent circulating exosomal miRNAs including miR-187-3p, miR-203b-3p, miR-202-3p, miR-378a-3p, miR-199a-5p, and miR-145-5p. (**B**) IPA networks showing the regulatory effects of the differentially expressed miRNAs from rat serum on mTOR signaling.

Table 2. IPA of genes targeted by 19 miRNAs that were differentially expressed with age.

Ingenuity Canonical Pathways	−log (*p*-Value)	Related miRNA	Target Genes	Full Name
Insulin receptor signaling	4.48	miR-378a-3p	PDK1	3-phosphoinositide-dependent protein kinase 1
		miR-187-3p	INSR	insulin receptor
		miR-202-3p	4E-BP1	Eukaryotic translation initiation factor 4E-binding protein 1
		miR-101b-3p	AKT	RAC-alpha serine/threonine-protein kinase
		miR-199a-5p	STX4	Syntaxin-4
		miR-203b-3p	PTEN	phosphatase and tensin homolog deleted on chromosome
mTOR signaling	3.77	miR-378a-3p	PDK1	3-phosphoinositide-dependent protein kinase 1
		miR-187-3p	INSR	insulin receptor
		miR-202-3p	eIF4E-BP1	Eukaryotic translation initiation factor 4E-binding protein 1
		miR-101b-3p	AKT	RAC-alpha serine/threonine-protein kinase
		miR-194-5p	RAC	Aryl-hydrocarbon-interacting protein-like 1
AMPK signaling	2.84	miR-378a-3p	PDK1	phosphoinositide-dependent kinase-1
		miR-187-3p	INSR	insulin receptor
		miR-202-3p	eIF4E-BP1	Eukaryotic translation initiation factor 4E-binding protein 1
		miR-101b-3p	AKT	RAC-alpha serine/threonine-protein kinase
eNOS signaling	2.37	miR-378a-3p	PDK1	phosphoinositide-dependent kinase-1
		miR-187-3p	CASP8	caspase-8
		miR-101b-3p	AKT	RAC-alpha serine/threonine-protein kinase
		miR-122-5p	CAT1	cationic amino acid transporter 1
		miR-143-3p	CASP8	caspase-8
IGF-1 signaling	2.94	miR-378a-3p	PDK1, IGF-1R	phosphoinositide-dependent kinase-1, Insulin-like growth factor 1 receptor
		miR-202-3p	IGF-1R	Insulin-like growth factor 1 receptor
		miR-101b-3p	AKT	RAC-alpha serine/threonine-protein kinase
		miR-145-5p	IGF-1R, PXN	insulin-like growth factor 1 receptor, Paxillin
		miR-199a-5p	PXN	paxillin
PTEN signaling	3.31	miR-378a-3p	PDK1	phosphoinositide-dependent kinase-1
		miR-101b-3p	BCL2L11, AKT	bcl-2-like protein 11, RAC-alpha serine/threonine-protein kinase
		miR-203b-3p	PTEN	phosphatase and tensin homolog deleted on chromosome
		miR-532-5p	NF-κB	Nuclear factor NF-kappa-B
p53 signaling	2.13	miR-202-3p	MDM4	protein Mdm4
		miR-101b-3p	AKT	RAC-alpha serine/threonine-protein kinase
		miR-203b-3p	PTEN	phosphatase and tensin homolog deleted on chromosome
		miR-532-5p	MDM4, Slug	Protein Mdm4, Zinc finger protein SNAI2
		miR-450b-5p	MDM4	Protein Mdm4
		miR-194-5p	PAI-1	Glia-derived nexin
		miR-199a-5p	HIPK2	Homeodomain-interacting protein kinase 2
		miR-143-3p	HIPK2, Chk2	Homeodomain-interacting protein kinase 2, Serine/threonine-protein kinase Chk2
Integrin signaling	4.13	miR-378a-3p	PARVIN-α	parvin alpha
		miR-101b-3p	ASAP1, PARVIN-α, AKT	arf-GAP with SH3 domain, ANK repeat and PH domain-containing protein 1, parvin alpha, RAC-alpha serine/threonine-protein kinase
		miR-145-5p	PXN, CRKL	Paxillin, Crk-like protein
		miR-122-5p	PDGFβ	Platelet-derived growth factor subunit B
		miR-203b-3p	PTEN	phosphatase and tensin homolog deleted on chromosome
		miR-598-3p	PARVIN-α	parvin alpha
		miR-199a-5p	PXN	Paxillin
Growth hormone signaling	2.44	miR-378a-3p	PDK1	phosphoinositide-dependent kinase-1
		miR-202-3p	IGF-1R	Insulin-like growth factor 1 receptor
		miR-145-5p	IGF-1R	Insulin-like growth factor 1 receptor

3. Discussion

The past two decades have witnessed the use of heterochronic blood exchange techniques, including heterochronic parabiosis, heterochronic blood or plasma transfer, or heterochronic apheresis, as tools for studying the biology of aging. Indeed, heterochronic blood exchange from a young to an old animal resulted in rejuvenation, whereas accelerated aging in a young animal was observed after heterochronic blood exchange from an old animal [5]. To explore the underlying mechanism, we used Exo-NGS analysis to compare the expression profiles for mRNAs and miRNAs in serum exosomes isolated from young and old rats. We Identified 68 miRNAs and 2844 mRNAs in serum exosomes that were differentially expressed between young and old rats. In contrast to mRNAs, little is known about changes in miRNA abundance in the aging process. For this reason, we focused on circulating miRNAs, which serve as potential biomarkers and therapeutic targets for aging-related disease. To determine how these circulating miRNAs affect aging, it is important to identify the targets for each miRNA. Our data revealed that, of the 68 differentially expressed serum exosome miRNAs we identified, 19 were predicated to target 766 differentially expressed mRNAs based on TargetScan analysis. Among the 19 miRNAs, 14 were more abundant in exosomes from old rats than from young rats, and five were less abundant. These results are consistent with reports that the abundance of the majority of these 14 miRNAs including miR-150-3p, miR-378-3p, miR-199a-5p, miR-145-5p, miR-598-3p, miR-122-5p, miR-194-5p, miR-203a-3p, miR-202-3p, miR-145-5p, and miR-532-5p, was elevated in blood or tissue samples from older humans, mice and rats [17,35–41]. Our data also confirmed that miR-181a-5p and miR-133a-3p decreased with age [40,42]. These 14 miRNAs have been linked with aging, and the expression of some of them has been associated with cancer, longevity, inflammatory responses, and aging-related neurodegenerative and cardiac diseases [17,36–46]. Collectively, the abundance of the majority of our differentially expressed miRNAs has been previously reported to be altered with age, suggesting roles for these miRNAs in lifespan. Interestingly, downregulation of miR-181a-5p in serum exosomes from old rats correlates negatively with the expression of pro-inflammatory cytokines IL-6 and TNFα and correlates positively with that of the anti-inflammatory cytokines TGFβ and IL-10 in the serum of rhesus monkeys [42]. Notably, the abundance of IL-6 and TNF-α has been correlated with aging [47]. Therefore, certain exosomal miRNAs may contribute to aging by regulating systemic inflammation, and the makeup of these miRNAs may serve as a biological signature of aging.

We used Blastp and GO to functionally annotate miRNA-regulated genes and, identified biological processes that are altered by changes in exosomal miRNAs abundance changes with age. Among these processes, the most highly represented and enriched terms were protein posttranslational modification, metabolic process, cell communication, molecular function, and intracellular signal transduction, implying that these miRNAs may provide a significant link between aging and multiple biological processes through their regulation of target genes [48]. KEGG pathway analysis revealed that the mRNAs targeted by these miRNA targets were enriched in known aging-related signaling pathways [49–51]. The GO and KEGG analysis also revealed that most of the miRNA-targeted mRNAs are involved in signaling pathways and biological processes, that are critical for aging, suggesting that circulating miRNAs may help regulate the rate of aging and therefore are potential biomarkers for aging. Any individual miRNA may have the potential to act on numerous target genes, and therefore, multiple miRNAs have the potential to modulate numerous biological pathways. Hence, the impact of miRNAs on any particular pathway(s) can be assessed most effectively by examining any synergism between the miRNAs [52]. To further investigate how any single miRNA-mRNA interaction regulates aging-related pathways, we performed IPA and found that the altered circulating miRNAs target the signaling pathways governed by insulin, integrin, mTOR, AMPK, PTEN, IGF-1, growth hormone, eNOS and p53, which are crucial pathways in aging and lifespan [49–51]. For example, we found that miRNA-187-3p can regulate INSR mRNA and that miRNA-378a-3p andmiRNA-202-3p can regulate IGF-1R mRNA. Studies have documented an inverse correlation between cellular miRNA-187 levelss and glucose-stimulated insulin secretion [53] and that miRNA-378a may play a role in insulin resistant and the consequent of obesity [54]. It is well documented that the insulin/IGF-1 pathway plays a critical role in aging and longevity across a wide spectrum of species [55–57]. Evidence includes that either reducing the level of circulating IGF-1 or reducing the expression of IGF-1R increases longevity [57]; moreover, the loss of one allele of the *Igf-1* receptor increases the lifespan of mice by 33% [58]. We also found that miR-187-3p, miR-202-3p, and miR-378a-3p regulate the mRNA levels of *INSR*, *EIF4EBP1* and *PDK1*, the genes for which are targeted by) mTOR signaling pathway. The mTOR pathway integrates both intracellular and extracellular signals and serves as a central regulator of cell metabolism, proliferation and survival, and it also controls lifespan by regulating translation through activation of p70S6K and inhibition of the translation repressor eIF4EBP [59]. For example, knocking down three translational regulators, namely eIF4E, eIF4G, and eIF2B homologs, in *C. elegans* extends worm lifespan [60–62], and modulation of the translation of their mRNAs by a dominant-negative form of TOR extends lifespan [63]. Recent studies have shown that the lifespan of different mouse strains can be extended significantly when mTOR inhibitor of rapamycin is administrated [64,65]. There is no clear explanation how a reduction in signaling via mTOR or insulin/IGF-1 affects lifespan. However, one potential explanation is that global mRNA translation is reduced after inhibiting either of these signaling pathways, which may reduce the burden and energetic demands associated with protein folding, repair, and degradation, thus maintaining better overall protein homeostasis [51]. Our findings support this hypothesis.

In addition to the insulin/IGF-1 and mTOR pathways, many other signaling pathways, such as the PTEN pathway, also modulate lifespan [66]. Indeed, PTEN has significant implications for extending human longevity through its antioxidant activity and contribution to the benefits of caloric restriction as well as its involvement DNA-damage reduction, inhibition of DNA replication, and tumor suppression [67]. We found that miR-203b-3p can target PTEN. Notably, signaling pathways, such as the insulin/IGF-1, mTOR and PTEN pathways, may individually regulate aging and lifespan. However, these signaling networks are not autonomous but connected through some specific mediators. For instance, mTOR has two complexes, namely mTOR complex 1 (mTORC1) and mTOR complex 2 (mTORC2) [59]. mTORC1 is regulated by Akt, and mTORC2 is an Akt activator [68]. PI3 kinase signaling activates mTORC2, which in turn activates a number of other kinases, including PKCα. Consistently, we found that the EIF4EBP1, INSR, PDPK1, PTEN, PXN, and IGF1R overlap and are regulated by at least two circulating miRNAs, and each of these pathways may play a unique role in aging [49–51].

Taken together, our findings suggest that changes in the makeup of circulating exosomal miRNAs with age not only can be considered as a potential predictor of animal age but also may contribute to aging via several key signaling pathways that regulate aging and lifespan. It will be important to identify and understand the mechanisms of rejuvenation and accelerated aging, because the findings concerning rejuvenation can potentially reverse deleterious processes of aging, whereas the findings concerning accelerated aging may pinpoint potential pathways for interventions that may slow the rate of aging and the incidence of aging-related disease. The challenge for the future will be to determine how these mediators map onto the different pathways and interact with each other, and to decipher how they contribute to the molecular mechanisms in aging.

4. Materials and Methods

4.1. Isolation of Serum Exosomes

Whole blood was collected from young (three-month-old) or old (22-month-old) rats ($n = 6$ per group) via cardiac puncture into BD Vacutainer®Plus Glass Serum blood collection tubes (Becton Dickinson, NJ, USA). Whole blood samples were allowed to clot by standing at room temperature for 30 min, and the clots were removed by centrifugation for 10 min at 1000× g at 4 °C. The isolated serum samples were aliquoted and stored at −80 °C.

Serum exosomes from young or old rats were isolated using the ExoQuick Exosome precipitation kit (System Biosciences, CA, USA). Briefly, serum (500 µL) was centrifuged at 3000× g for 15 min at

4 °C to eliminate cells and cell debris. The supernatant was transferred to a sterile micro-tube, and an appropriate volume of exosome precipitation solution from the kit was added, with incubation for 30 min at 4 °C. The mixture was then centrifuged at 1500× g for 30 min at 4 °C, and the exosome pellet was re-suspended in sterile phosphate-buffered saline at 4 °C.

4.2. Characterization of Serum Exosomes

Both the concentration and average size of the isolated serum exosomes were determined by nanoparticle-tracking analysis (NTA) using the Exosome Nanosight Analysis Service of System Biosciences (Palo Alto, CA, USA). The serum exosomes were also observed using transmission electron microscopy (TEM, FEI Tecnai G2 Spirit BioTwin, OR, USA) to determine morphology and the extent of dispersion, this analysis was performed at the Electron Microscopy Core Facility at the University of Texas Southwestern Medical Center, TX, USA. The enrichment of exosomes was determined by Western blotting using antibodies against exosomes components such as CD63 andCD9.

4.3. Western Blotting

Serum exosomes were lysed in RIPA buffer and the protein concentration was determined using the Quick Start Bradford protein assay (Pierce™ BCA Protein Assay kit, Thermo Fisher Scientific, MA, USA). The lysates (10 µg) were electrophoresed through 8–12% SDS-PAGE gels, and the separated proteins were transferred to a nitrocellulose membrane. The membrane was incubated in blocking buffer (5% milk in Tris-buffered saline with 0.05% *w/v* Tween-20) for 1 h at room temperature and then incubated overnight at 4 °C with mouse antibody against rat CD63 (1:1000, BD Pharmingen, CA, USA) and CD9 (1:1000, BD Pharmingen). Immunopositivity was detected with a horseradish peroxidase (HRP)—conjugated secondary antibodies and the Pierce enhanced chemiluminescence (ECL) substrate (Thermo Fisher Scientific, MA, USA). The data were recorded and analyzed using the ChemiDoc Imaging System (Bio-Rad).

4.4. Isolation of Total RNA fromEexosomes and Next-Generation RNA Sequencing

Total RNA was isolated from serum exosomes using the SeraMir Exosome RNA Purification Column kit (System Biosciences, CA, USA). For each sample, 1 µL of the final RNA eluate was used for measurement of RNA concentration with the Agilent Bioanalyzer Small RNA Assay using the Bioanalyzer 2100 Expert instrument (Agilent Technologies, Santa Clara, CA, USA). Serum exosomal RNAs (N = 6 each group) were sent to the Exo-NGS™ (Exosomal RNA-Seq) services for next-generation RNA sequencing (System Biosciences, CA, USA) using small RNA libraries. Next-generation RNA sequencing was performed on an Illumina NextSeq instrument (Illumina, CA, USA) with 1 × 75 bp single-end reads at an approximate depth of 10–15 million reads per sample.

4.5. Data Processing

Raw data were analyzed using an integrated UCSC genome browser on the Banana Slug analytics platform (UCSC, CA, USA). Briefly, the exosome Small RNA-seq Analysis kit was initiated with a data quality check of each input sequence using FasQC (Wellcome Sanger Institute, UK) an open-source quality control tool for analyzing high-throughput sequence data. Following the quality-control step, the RNA-seq reads were processed to detect and remove unknown nucleotides at the ends of reads, trim sequencing adaptors, and filter reads for quality and length, using FastqMcf, which is part of the EA-utils package (ExpressionAnalysis, NC, USA) and PRINSEQ (http://prinseq.sourceforge.net/, USA). FastQC was then repeated to analyze the trimmed reads, thus allowing a before and after comparison. Sequence reads in the improved set were mapped to the reference genome using Bowtie, an ultrafast, memory-efficient short-read aligner. Expression analyses, including computation of read coverage and noncoding RNA abundance, were performed using the open-source software SAMtoolsand Picard (Github, CA, USA).

4.6. Bioinformatics Analysis

After data processing, expression statistics for the normalized reads were evaluated using analysis of variance to identify differentially expressed genes. Differentially expressed genes were selected if the fold changes (FC) in expression was > 1.5 with a *p*-value < 0.5. From the data set of miRNAs and mRNAs for which expression was significantly altered between serum exosomes isolated from young and old rats, the potential regulation of a mRNA by a particular miRNA was predicted with TargetScan (http://www.targetscan.org/, MIT, MA, USA). The paired miRNAs and mRNAs were used for further analysis.

Hierarchical cluster analysis (HCA) is an algorithmic approach to identify groups with varying degrees of (dis)similarity in a data set represented by a (dis)similarity matrix. This analysis was carried out with the Pheatmap package (https://CRAN.R-project.org/package=pheatmap, Estonia). The volcano plot is a type of scatter-plot that can quickly identify changes in individual data in large data-sets composed of replicate data; ggplot2 package (http://ggplot2.org, New Zealand) was used for this purpose.

4.7. Gene Ontology (GO) and Pathway Enrichment Analysis

The biological function of each protein was annotated with Blastp (Blast2GO version 5, BioBam Bioinformatics, Spain) using the entire gene-expression database, and subsequent mapping was carried out with the GO database (www.geneontology.org/, Gene ontology resource, USA). To further understand the biological significance of differentially expressed exosome proteins, pathway analysis was carried out with the KEGG Orthology-Based Annotation System (http://kobas.cbi.pku.edu.cn, China). The association of proteins with different pathways was computed using the KEGG database (www.genome.jp/kegg, Japan). EuKaryotic Orthologous Groups (KOG) Analysis was based on the phylogenetic classification of proteins encoded in the complete genomes (www.ncbi.nlm.nih.gov/COG/, NCBI, MD, USA) project. IPA (QIAGEN, Germany) was used for additional functional annotation, including top canonical pathway, top disease and function and molecular and cellular functions, prediction of upstream, regulator effectors, and miRNA–mRNA relationship and interaction network analysis.

4.8. Statistical Analysis

For the GO analysis, statistically significant alterations in functions of differentially expressed exosome proteins were assessed with Fisher's exact test in Blast2GO with an adjusted *p*-value (false discovery rate, FDR) of <0.05 and fold change > 1.5. The statistical significance of changes in pathways identified with IPA was assessed with the right-tailed Fisher's Exact test. A *p*-value of < 0.05 implies that the relationship of a set of targeted molecules and a process/pathway/transcription was randomly matched. A Z score of ≥ 2 or ≤ -2 indicated significant activation or significant inhibition, respectively. For all analysis, the difference was considered significant for *p* < 0.05.

Supplementary Materials: Supplementary materials can be found at http://www.mdpi.com/1422-0067/21/6/1908/s1.

Author Contributions: Formal analysis, H.Z.; Investigation, K.J.; Methodology, H.Z. Project administration, K.J.; Writing—original draft, H.Z.; Writing—review and editing, K.J. All authors have read and agreed to the published version of the manuscript.

Funding: American Heart Association: 18PRE34020126; National Institutes of Health: 5R21-NS094859-02.

Conflicts of Interest: The authors declare no conflict of interest.

References

1. Tinetti, M.E.; Speechley, M.; Ginter, S.F. Risk factors for falls among elderly persons living in the community. *N. Engl. J. Med.* **1988**, *319*, 1701–1707. [CrossRef]
2. Hebert, L.E.; Scherr, P.A.; Bienias, J.L.; Bennett, D.A.; Evans, D.A. Alzheimer disease in the us population: Prevalence estimates using the 2000 census. *Arch. Neurol.* **2003**, *60*, 1119–1122. [CrossRef] [PubMed]

3. Jin, K. Modern biological theories of aging. *Aging Dis.* **2010**, *1*, 72–74. [PubMed]
4. Jin, K. A microcirculatory theory of aging. *Aging Dis.* **2019**, *10*, 676–683. [CrossRef] [PubMed]
5. Conboy, I.M.; Conboy, M.J.; Wagers, A.J.; Girma, E.R.; Weissman, I.L.; Rando, T.A. Rejuvenation of aged progenitor cells by exposure to a young systemic environment. *Nature* **2005**, *433*, 760–764. [CrossRef] [PubMed]
6. Brack, A.S.; Conboy, M.J.; Roy, S.; Lee, M.; Kuo, C.J.; Keller, C.; Rando, T.A. Increased wnt signaling during aging alters muscle stem cell fate and increases fibrosis. *Science* **2007**, *317*, 807–810. [CrossRef] [PubMed]
7. Loffredo, F.; Steinhauser, M.L.; Jay, S.M.; Gannon, J.; Pancoast, J.R.; Yalamanchi, P.; Sinha, M.; Dall'Osso, C.; Khong, D.; Shadrach, J.L.; et al. Growth differentiation factor 11 is a circulating factor that reverses age-related cardiac hypertrophy. *Cell* **2013**, *153*, 828–839. [CrossRef]
8. Zhang, H.; Cherian, R.; Jin, K. Systemic milieu and age-related deterioration. *Geroscience* **2019**, *41*, 275–284. [CrossRef]
9. Sebastiani, P.; Thyagarajan, B.; Sun, F.; Schupf, N.; Newman, A.B.; Montano, M.; Perls, T. Biomarker signatures of aging. *Aging Cell* **2017**, *16*, 329–338. [CrossRef]
10. Xia, X.; Chen, W.; McDermott, J.; Han, J.J. Molecular and phenotypic biomarkers of aging. *F1000Research* **2017**, *6*, 860. [CrossRef]
11. Urbanelli, L.; Magini, A.; Buratta, S.; Brozzi, A.; Sagini, K.; Polchi, A.; Tancini, B.; Emiliani, C. Signaling pathways in exosomes biogenesis, secretion and fate. *Genes* **2013**, *4*, 152–170. [CrossRef] [PubMed]
12. Zomer, A.; Vendrig, T.; Hopmans, E.S.; van Eijndhoven, M.; Middeldorp, J.M.; Pegtel, D.M. Exosomes: Fit to deliver small rna. *Commun. Integr. Biol.* **2010**, *3*, 447–450. [CrossRef] [PubMed]
13. Zhang, Y.; Kim, M.S.; Jia, B.; Yan, J.; Zuniga-Hertz, J.P.; Han, C.; Cai, D. Hypothalamic stem cells control ageing speed partly through exosomal mirnas. *Nature* **2017**, *548*, 52–57. [CrossRef]
14. Lee, B.R.; Kim, J.H.; Choi, E.S.; Cho, J.H.; Kim, E. Effect of young exosomes injected in aged mice. *Int. J. Nanomed.* **2018**, *13*, 5335–5345. [CrossRef] [PubMed]
15. Pulliam, L.; Sun, B.; Mustapic, M.; Chawla, S.; Kapogiannis, D. Plasma neuronal exosomes serve as biomarkers of cognitive impairment in hiv infection and alzheimer's disease. *J. Neurovirol.* **2019**, *25*, 702–709. [CrossRef] [PubMed]
16. Sun, B.; Dalvi, P.; Abadjian, L.; Tang, N.; Pulliam, L. Blood neuron-derived exosomes as biomarkers of cognitive impairment in hiv. *AIDS* **2017**, *31*, F9–F17. [CrossRef]
17. Rani, A.; O'Shea, A.; Ianov, L.; Cohen, R.A.; Woods, A.J.; Foster, T.C. Mirna in circulating microvesicles as biomarkers for age-related cognitive decline. *Front. Aging Neurosci.* **2017**, *9*, 323. [CrossRef]
18. Alvarez-Erviti, L.; Seow, Y.; Yin, H.; Betts, C.; Lakhal, S.; Wood, M.J. Delivery of sirna to the mouse brain by systemic injection of targeted exosomes. *Nat. Biotechnol.* **2011**, *29*, 341–345. [CrossRef]
19. Matsumoto, J.; Stewart, T.; Banks, W.A.; Zhang, J. The transport mechanism of extracellular vesicles at the blood-brain barrier. *Curr. Pharm. Des.* **2017**, *23*, 6206–6214. [CrossRef]
20. Yang, T.; Martin, P.; Fogarty, B.; Brown, A.; Schurman, K.; Phipps, R.; Yin, V.P.; Lockman, P.; Bai, S. Exosome delivered anticancer drugs across the blood-brain barrier for brain cancer therapy in danio rerio. *Pharm. Res.* **2015**, *32*, 2003–2014. [CrossRef]
21. Qu, M.; Lin, Q.; Huang, L.; Fu, Y.; Wang, L.; He, S.; Fu, Y.; Yang, S.; Zhang, Z.; Zhang, L.; et al. Dopamine-loaded blood exosomes targeted to brain for better treatment of parkinson's disease. *J. Control Release* **2018**, *287*, 156–166. [CrossRef] [PubMed]
22. Eitan, E.; Green, J.; Bodogai, M.; Mode, N.A.; Bæk, R.; Jørgensen, M.M.; Freeman, D.W.; Witwer, K.W.; Zonderman, A.B.; Biragyn, A.; et al. Age-related changes in plasma extracellular vesicle characteristics and internalization by leukocytes. *Sci. Rep.* **2017**, *7*, 1342. [CrossRef] [PubMed]
23. Lewis, B.P.; Shih, I.H.; Jones-Rhoades, M.W.; Bartel, D.P.; Burge, C.B. Prediction of mammalian microrna targets. *Cell* **2003**, *115*, 787–798. [CrossRef]
24. Bartel, D.P. Micrornas: Target recognition and regulatory functions. *Cell* **2009**, *136*, 215–233. [CrossRef] [PubMed]
25. Tan, C.L.; Plotkin, J.; Venø, M.T.; Von Schimmelmann, M.; Feinberg, P.; Mann, S.; Handler, A.; Kjems, J.; Surmeier, D.J.; O'Carroll, D.; et al. Microrna-128 governs neuronal excitability and motor behavior in mice. *Science* **2013**, *342*, 1254–1258. [CrossRef]

26. Pedersen, M.E.; Snieckute, G.; Kagias, K.; Nehammer, C.; Multhaupt, H.A.; Couchman, J.R.; Pocock, R. An epidermal microrna regulates neuronal migration through control of the cellular glycosylation state. *Science* **2013**, *341*, 1404–1408. [CrossRef]

27. Gomez, G.G.; Volinia, S.; Croce, C.M.; Zanca, C.; Li, M.; Emnett, R.; Gutmann, D.H.; Brennan, C.; Furnari, F.B.; Cavenee, W.K. Suppression of microrna-9 by mutant egfr signaling upregulates foxp1 to enhance glioblastoma tumorigenicity. *Cancer Res.* **2014**, *74*, 1429–1439. [CrossRef]

28. Mushtaq, G.; Greig, N.H.; Anwar, F.; A Zamzami, M.; Choudhry, H.; Shaik, M.M.; Tamargo, I.A.; Kamal, M.A. Mirnas as circulating biomarkers for alzheimer's disease and parkinson's disease. *Med. Chem.* **2016**, *12*, 217–225. [CrossRef]

29. Schwarzenbach, H.; Nishida, N.; Calin, G.A.; Pantel, K. Clinical relevance of circulating cell-free micrornas in cancer. *Nat. Rev. Clin. Oncol.* **2014**, *11*, 145–156. [CrossRef]

30. Sheinerman, K.S.; Tsivinsky, V.G.; Abdullah, L.; Crawford, F.; Umansky, S.R. Plasma microrna biomarkers for detection of mild cognitive impairment: Biomarker validation study. *Aging* **2013**, *5*, 925–938. [CrossRef]

31. Cossetti, C.; Iraci, N.; Mercer, T.R.; Leonardi, T.; Alpi, E.; Drago, D.; Alfaro-Cervello, C.; Saini, H.K.; Davis, M.; Schaeffer, J.; et al. Extracellular vesicles from neural stem cells transfer ifn-gamma via ifngr1 to activate stat1 signaling in target cells. *Mol. Cell* **2014**, *56*, 193–204. [CrossRef]

32. Dragovic, R.A.; Gardiner, C.; Brooks, A.S.; Tannetta, D.S.; Ferguson, D.J.; Hole, P.; Carr, B.; Redman, C.W.; Harris, A.L.; Dobson, P.J.; et al. Sizing and phenotyping of cellular vesicles using nanoparticle tracking analysis. *Nanomedicine* **2011**, *7*, 780–788. [CrossRef]

33. Feng, L.-B.; Pang, X.-M.; Zhang, L.; Li, J.; Huang, L.-G.; Su, S.-Y.; Zhou, X.; Li, S.-H.; Xiang, H.-Y.; Chen, C.-Y.; et al. Microrna involvement in mechanism of endogenous protection induced by fastigial nucleus stimulation based on deep sequencing and bioinformatics. *BMC Med. Genom.* **2015**, *8*, 79. [CrossRef] [PubMed]

34. Kanehisa, M.; Furumichi, M.; Tanabe, M.; Sato, Y.; Morishima, K. Kegg: New perspectives on genomes, pathways, diseases and drugs. *Nucleic Acids Res.* **2017**, *45*, D353–D361. [CrossRef]

35. Hatse, S.; Brouwers, B.; Dalmasso, B.S.; Laenen, A.; Kenis, C.; Schöffski, P.; Wildiers, H. Circulating micrornas as easy-to-measure aging biomarkers in older breast cancer patients: Correlation with chronological age but not with fitness/frailty status. *PLoS ONE* **2014**, *9*, e110644. [CrossRef] [PubMed]

36. Xiao, H.; Huang, R.; Diao, M.; Li, L.; Cui, X. Integrative analysis of microrna and mrna expression profiles in fetal rat model with anorectal malformation. *PeerJ* **2018**, *6*, e5774. [CrossRef]

37. Kim, J.-H.; Lee, B.-R.; Choi, E.-S.; Lee, K.-M.; Choi, S.-K.; Cho, J.H.; Jeon, W.B.; Kim, E. Reverse expression of aging-associated molecules through transfection of mirnas to aged mice. *Mol. Ther. Nucleic Acids* **2017**, *6*, 106–115. [CrossRef] [PubMed]

38. Lee, B.P.; Buric, I.; George-Pandeth, A.; Flurkey, K.; Harrison, D.E.; Yuan, R.; Peters, L.; Kuchel, G.A.; Melzer, D.; Harries, L.W. Micrornas mir-203-3p, mir-664-3p and mir-708-5p are associated with median strain lifespan in mice. *Sci. Rep.* **2017**, *7*, 44620. [CrossRef]

39. Zhao, Y.; Li, C.; Wang, M.; Su, L.; Qu, Y.; Li, J.; Yu, B.; Yan, M.; Yu, Y.; Liu, B.; et al. Decrease of mir-202-3p expression, a novel tumor suppressor, in gastric cancer. *PLoS ONE* **2013**, *8*, e69756. [CrossRef]

40. Dluzen, D.F.; Noren Hooten, N.; De, S.; Wood, W.H., 3rd; Zhang, Y.; Becker, K.G.; Zonderman, A.B.; Tanaka, T.; Ferrucci, L.; Evans, M.K. Extracellular rna profiles with human age. *Aging Cell* **2018**, *17*, e12785. [CrossRef]

41. Tsai, H.P.; Huang, S.F.; Li, C.F.; Chien, H.T.; Chen, S.C. Differential microrna expression in breast cancer with different onset age. *PLoS ONE* **2018**, *13*, e0191195. [CrossRef] [PubMed]

42. Noren Hooten, N.; Fitzpatrick, M.; Wood, W.H., 3rd; De, S.; Ejiogu, N.; Zhang, Y.; Mattison, J.A.; Becker, K.G.; Zonderman, A.B.; Evans, M.K. Age-related changes in microrna levels in serum. *Aging* **2013**, *5*, 725–740. [CrossRef]

43. Noren Hooten, N.; Abdelmohsen, K.; Gorospe, M.; Ejiogu, N.; Zonderman, A.B.; Evans, M.K. Microrna expression patterns reveal differential expression of target genes with age. *PLoS ONE* **2010**, *5*, e10724. [CrossRef] [PubMed]

44. Inukai, S.; de Lencastre, A.; Turner, M.; Slack, F. Novel micrornas differentially expressed during aging in the mouse brain. *PLoS ONE* **2012**, *7*, e40028. [CrossRef] [PubMed]

45. Hara, N.; Kikuchi, M.; Miyashita, A.; Hatsuta, H.; Saito, Y.; Kasuga, K.; Murayama, S.; Ikeuchi, T.; Kuwano, R. Serum microrna mir-501-3p as a potential biomarker related to the progression of alzheimer's disease. *Acta Neuropathol. Commun.* **2017**, *5*, 10. [CrossRef] [PubMed]

46. Kovanda, A.; Leonardis, L.; Zidar, J.; Koritnik, B.; Dolenc-Grošelj, L.; Kovacic, S.R.; Curk, T.; Rogelj, B. Differential expression of micrornas and other small rnas in muscle tissue of patients with als and healthy age-matched controls. *Sci. Rep.* **2018**, *8*, 5609. [CrossRef] [PubMed]

47. Chung, H.Y.; Kim, D.H.; Lee, E.K.; Chung, K.W.; Chung, S.; Lee, B.; Seo, A.Y.; Chung, J.H.; Jung, Y.S.; Im, E.; et al. Redefining chronic inflammation in aging and age-related diseases: Proposal of the senoinflammation concept. *Aging Dis.* **2019**, *10*, 367–382. [CrossRef]

48. Lafferty-Whyte, K.; Cairney, C.J.; Jamieson, N.B.; Oien, K.A.; Keith, W.N. Pathway analysis of senescence-associated mirna targets reveals common processes to different senescence induction mechanisms. *Biochim. Biophys. Acta* **2009**, *1792*, 341–352. [CrossRef]

49. Carlson, M.E.; Silva, H.S.; Conboy, I.M. Aging of signal transduction pathways, and pathology. *Exp. Cell Res.* **2008**, *314*, 1951–1961. [CrossRef]

50. Newgard, C.B.; Pessin, J.E. Recent progress in metabolic signaling pathways regulating aging and life span. *J. Gerontol. Ser. A Biol. Sci. Med. Sci.* **2014**, *69*, S21–S27. [CrossRef]

51. Pan, H.; Finkel, T. Key proteins and pathways that regulate lifespan. *J. Biol. Chem.* **2017**, *292*, 6452–6460. [CrossRef] [PubMed]

52. Dharap, A.; Vemuganti, R. Ischemic pre-conditioning alters cerebral micrornas that are upstream to neuroprotective signaling pathways. *J. Neurochem.* **2010**, *113*, 1685–1691. [CrossRef] [PubMed]

53. Locke, J.M.; da Silva Xavier, G.; Dawe, H.R.; Rutter, G.A.; Harries, L.W. Increased expression of mir-187 in human islets from individuals with type 2 diabetes is associated with reduced glucose-stimulated insulin secretion. *Diabetologia* **2014**, *57*, 122–128. [CrossRef] [PubMed]

54. Jones, A.; Danielson, K.M.; Benton, M.C.; Ziegler, O.; Shah, R.; Stubbs, R.S.; Das, S.; Macartney-Coxson, D. Mirna signatures of insulin resistance in obesity. *Obesity* **2017**, *25*, 1734–1744. [CrossRef]

55. Tatar, M.; Bartke, A.; Antebi, A. The endocrine regulation of aging by insulin-like signals. *Science* **2003**, *299*, 1346–1351. [CrossRef]

56. Markowska, A.L.; Mooney, M.; Sonntag, W.E. Insulin-like growth factor-1 ameliorates age-related behavioral deficits. *Neuroscience* **1998**, *87*, 559–569. [CrossRef]

57. Vijg, J.; Campisi, J. Puzzles, promises and a cure for ageing. *Nature* **2008**, *454*, 1065–1071. [CrossRef]

58. Holzenberger, M.; Dupont, J.; Ducos, B.; Leneuve, P.; Géloën, A.; Even, P.; Cervera, P.; Le Bouc, Y. Igf-1 receptor regulates lifespan and resistance to oxidative stress in mice. *Nature* **2003**, *421*, 182–187. [CrossRef]

59. Kennedy, B.K.; Lamming, D.W. The mechanistic target of rapamycin: The grand conductor of metabolism and aging. *Cell Metab.* **2016**, *23*, 990–1003. [CrossRef]

60. Henderson, S.T.; Bonafe, M.; Johnson, T.E. Daf-16 protects the nematode caenorhabditis elegans during food deprivation. *J. Gerontol. Ser. A Biol. Sci. Med. Sci.* **2006**, *61*, 444–460. [CrossRef]

61. Hansen, M.; Taubert, S.; Crawford, D.; Libina, N.; Lee, S.J.; Kenyon, C. Lifespan extension by conditions that inhibit translation in caenorhabditis elegans. *Aging Cell* **2007**, *6*, 95–110. [CrossRef] [PubMed]

62. Syntichaki, P.; Troulinaki, K.; Tavernarakis, N. Eif4e function in somatic cells modulates ageing in caenorhabditis elegans. *Nature* **2007**, *445*, 922–926. [CrossRef] [PubMed]

63. Kapahi, P.; Zid, B.M.; Harper, T.; Koslover, D.; Sapin, V.; Benzer, S. Regulation of lifespan in drosophila by modulation of genes in the tor signaling pathway. *Curr. Biol.* **2004**, *14*, 885–890. [CrossRef] [PubMed]

64. Harrison, D.E.; Strong, R.; Sharp, Z.D.; Nelson, J.F.; Astle, C.M.; Flurkey, K.; Nadon, N.L.; Wilkinson, J.E.; Frenkel, K.; Carter, C.S.; et al. Rapamycin fed late in life extends lifespan in genetically heterogeneous mice. *Nature.* **2009**, *460*, 392–395. [CrossRef]

65. Miller, R.A.; Harrison, D.E.; Astle, C.M.; Fernandez, E.; Flurkey, K.; Han, M.; Javors, M.A.; Li, X.; Nadon, N.L.; Nelson, J.F.; et al. Rapamycin-mediated lifespan increase in mice is dose and sex dependent and metabolically distinct from dietary restriction. *Aging Cell* **2014**, *13*, 468–477. [CrossRef]

66. Greer, E.L.; Brunet, A. Signaling networks in aging. *J. Cell Sci.* **2008**, *121*, 407–412. [CrossRef]

67. Tait, I.S.; Li, Y.; Lu, J. Pten, longevity and age-related diseases. *Biomedicines* **2013**, *1*, 17–48. [CrossRef]

68. Iwanami, A.; Cloughesy, T.F.; Mischel, P.S. Striking the balance between pten and pdk1: It all depends on the cell context. *Genes Dev.* **2009**, *23*, 1699–1704. [CrossRef]

International Journal of
Molecular Sciences

Article

Exceptionally Long-Lived Individuals (ELLI) Demonstrate Slower Aging Rate Calculated by DNA Methylation Clocks as Possible Modulators for Healthy Longevity

Danielle Gutman [1], **Elina Rivkin** [2], **Almog Fadida** [2], **Lital Sharvit** [1], **Vered Hermush** [3,4], **Elad Rubin** [5], **Dani Kirshner** [4,5], **Irina Sabin** [4,5], **Tzvi Dwolatzky** [4,5] and **Gil Atzmon** [1,6,*]

[1] Department of Human Biology, Faculty of Natural Sciences, University of Haifa, Haifa 3498838, Israel;
    dgutman546@gmail.com (D.G.); lsharvit@univ.haifa.ac.il (L.S.)
[2] Faculty of Public Health, University of Haifa, Haifa 3498838, Israel; duelina2705@gmail.com (E.R.);
    almogfadida@gmail.com (A.F.)
[3] Department of Geriatrics and Skilled Nursing, Laniado Medical Center, Netanya 4244916, Israel;
    vhermush@laniado.org.il
[4] Ruth and Bruce Rappaport Faculty of Medicine, Technion–Israel Institute of Technology, Haifa 3200003,
    Israel; kirshner.dani@gmail.com (D.K.); i_sabin@rambam.health.gov.il (I.S.);
    t_dwolatzky@rambam.health.gov.il (T.D.)
[5] Department of Geriatrics, Rambam Health Care Campus, Haifa 3109601, Israel; eladrob@gmail.com
[6] Departments of Genetics and Medicine, Division of endocrinology, Institute for Aging Research and the
    Diabetes Research Center, Albert Einstein College of Medicine, Bronx, New York, NY 10461, USA
* Correspondence: gatzmon@univ.haifa.ac.il; Tel.: +972-4664-7927

Received: 25 November 2019; Accepted: 15 January 2020; Published: 17 January 2020

Abstract: Exceptionally long-lived individuals (ELLI) who are the focus of many healthy longevity studies around the globe are now being studied in Israel. The Israeli Multi-Ethnic Centenarian Study (IMECS) cohort is utilized here for assessment of various DNA methylation clocks. Thorough phenotypic characterization and whole blood samples were obtained from ELLI, offspring of ELLI, and controls aged 53–87 with no familial exceptional longevity. DNA methylation was assessed using Illumina MethylationEPIC Beadchip and applied to DNAm age online tool for age and telomere length predictions. Relative telomere length was assessed using qPCR T/S (Telomere/Single copy gene) ratios. ELLI demonstrated juvenile performance in DNAm age clocks and overall methylation measurement, with preserved cognition and relative telomere length. Our findings suggest a favorable DNA methylation profile in ELLI enabling a slower rate of aging in those individuals in comparison to controls. It is possible that DNA methylation is a key modulator of the rate of aging and thus the ELLI DNAm profile promotes healthy longevity.

Keywords: healthy aging; DNA methylation; epigenetic clocks; telomere length; centenarians

1. Introduction

Healthy aging is usually characterized by preserved cognitive and motor functions. A unique group of aging individuals termed centenarians serves as a healthy aging model, outliving the age of 100, with mostly intact cognition and physical health [1–3]. Such exceptionally long-lived individuals (ELLI) are the focus of many studies around the world [4–12], and this group is now being studied in Israel as well. Our newly established cohort of ELLI is part of the Israeli Multi-Ethnic Centenarian Study (IMECS), which aims to elucidate the mechanisms of their healthy aging process.

Two of the most-studied hallmarks of aging [13] are DNA methylation and telomere attrition. Telomere shortening has long been documented to have inverse correlation with age [14–17], with mean telomere length (TL) considered a marker for cellular senescence and aging [18–20]. Alongside this inverse correlation, mean TL has also been strongly correlated with several age-associated diseases [21–25], adding significance to the negative outcomes of telomere shortening. That said, longer TL has been associated with exceptional longevity [2,26] through several potential mechanisms [27]. Telomere length is commonly measured by southern blot or by quantitative PCR. The latter method has gained popularity for its ease of use and robustness [2,28–34].

The other hallmark of aging, DNA methylation, increases with age, mostly through a phenomenon termed epigenetic drift [35]. The DNA methylation of centenarians, however, seems to be slightly lower, hinting at a mechanism promoting healthy aging. A study performed on semi-supercentenarians (ages 105–109 years) and their offspring demonstrated that the semi-supercentenarians and their offspring displayed younger "epigenetic age" (calculated on DNA methylation values) with age-matched controls (to the offspring) displaying same "epigenetic age" as actual age [36]. There are several such "epigenetic age" estimators which are mostly developed and utilized using standardized DNA methylation data [35,37–41]. There are various methods for measuring DNA methylation, with the most recently developed Illumina MethylationEPIC beadchip array serving as a thorough, genome-wide, standardized method. Recently, Lu et al. developed two clocks, one for telomere length and one for age, based on DNA methylation levels measured using the Illumina arrays [42,43]. To this date, the DNAmTL or DNAmGrimAge have not been used on DNA methylation data of ELLI. DNAmTL uses 140 CpG sites to estimate telomere length in Kb, while DNAmGrimAge uses 12 sub-DNAm-measures, alongside age and gender, to estimate physiological age with an addition of an estimate of time-to-death termed DNAmAccelGrim. Prior to the development of DNAmGrimAge, the same team developed DNAmPhenoAge, a DNA methylation-based aging biomarker that utilizes 513 CpGs to predict the phenotypic age of an individual [44].

The current study utilizes the IMECS cohort (consisting of ELLI, offspring of ELLI, and controls aged 53–87 with no familial exceptional longevity) to compare between the different DNAm clocks and actual phenotypic measures (such as relative telomere length measurements, cognitive performance, and actual age) from the IMECS cohort. We hypothesize that the DNAm age biomarkers and molecular phenotype of ELLI do not differ from those measures in the much younger offspring and control populations. These efforts aim to add knowledge on the phenotype of exceptional longevity and perhaps point at potential therapeutic avenues that might aid in cognitive and physical health preservation or even improvement (as suggested by Fahy et al. [45]).

2. Results

2.1. DNA Methylation

DNA methylation raw data of all 70 IMECS participants (described in Tables A1 and A2) were normalized using Noob normalization, and beta values of all CpG sites passing QC filtering were used to calculate mean beta for each sample, as a measure for global DNA methylation. As can be seen in Figure 1A, the mean beta value for the centenarian group is slightly lower than that of the control group, however this difference was not significant. Lack of significance in this value surprisingly shows great similarity in whole-genome methylation percentage between the groups, hinting at a juvenile methylation profile for ELLI, seeing as global DNA methylation is known to increase with age [35,46]. This similarity is continued, as expected, in the offspring group, demonstrating slightly lower average beta value compared to control as well. In Figure 1B the decrease in methylation with age is easily visible, and contradicting the increase reported by Hannum et al. [35].

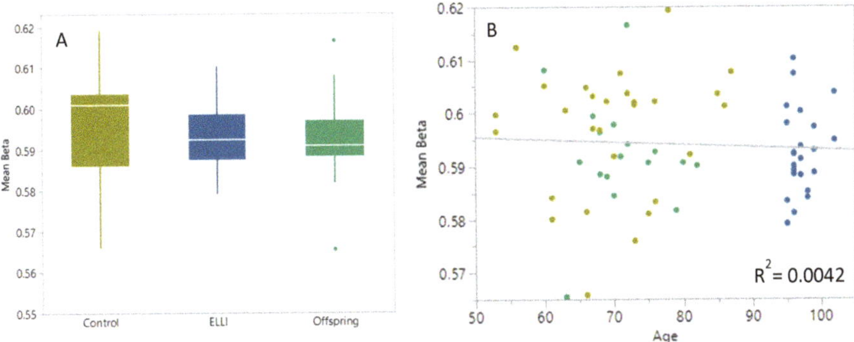

Figure 1. Mean methylation by group. Methylation levels measured via Illumina MethylationEPIC beadchip, converted to beta values with preprocessNoob to remove background read signal through Minfi R package. $N_{control}$ = 28, N_{ELLI} = 24, and $N_{offspring}$ = 17 (one outlier removed). Difference between groups found non-significant by one-way Kruskal–Wallis analysis. (**A**) Mean beta values per group, $mean_{control}$ = 0.596 ± 0.012, $mean_{ELLI}$ = 0.593 ± 0.008, and $mean_{offspring}$ = 0.592 ± 0.011. (**B**) Linear regression of mean beta values over age, non-significant.

2.2. DNAm Age Clocks

We used two recently developed epigenetic clocks, DNAmPhenoAge [44] and DNAmGrimAge [42], both developed by the Horvath group at UCLA. In short, DNAmPhenoAge is an age clock based on beta values of 513 CpG sites established via correlation with clinical markers, and DNAmGrimAge, the more recent and accurate clock, relies on 12 different sub-DNAm-estimators alongside actual age and gender. Both clocks aim to describe health and lifespan predictions through clinical and phenotypic measurements. Both clocks predict younger age of our groups (Figure 2), with DNAmGrimAge outperforming DNAmPhenoAge, and the ELLI estimations are the most juvenile (differences between actual age and clocks is largest). The differences between chronological age and DNAmGrimAge in the control and offspring groups were very slight (Tables A3 and A4), whereas the DNAmPhenoAge consistently underestimated the ages of control and offspring participants. This performance is consistent with the DNAmGrimAge performance in the validation data used by the developers, yet is the first to be reported in ELLI, whose ages were calculated to be younger by DNAmGrimAge.

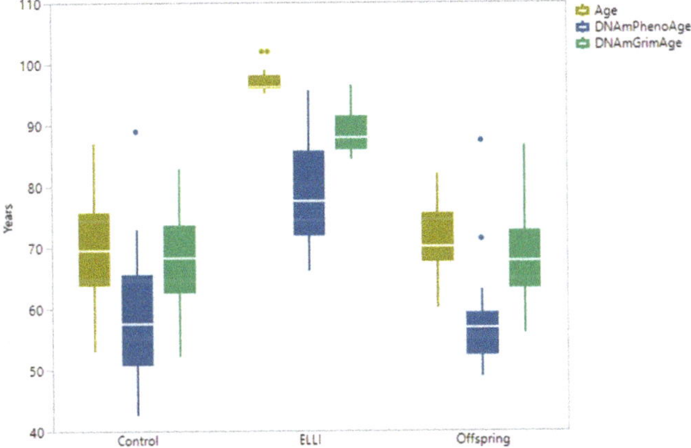

Figure 2. Comparison of actual age and two age clocks. DNAmGrimAge and DNAmPhenoAge calculated by applying methylation beta values to DNAm online tool [37]. $N_{control}$ = 28, N_{ELLI} = 24, and $N_{offspring}$ = 18.

Further, there is a high correlation between chronological age and both DNAm clocks (Figure 3), with DNAmGrimAge outperforming DNAmPhenoAge in actual age prediction. Though DNAmGrimAge is more closely related to chronological age (especially due to the use of actual age as a parameter of DNAmGrimAge), DNAmPhenoAge was originally designed to capture a phenotypic age (rather than chronological age). As depicted by our results, the phenotypic age prediction was lower than the chronological age especially for ELLI, indicating a juvenile phenotype of this group. With this high correlation in mind, we proceeded to examine correlation between age and DNAm clocks with cognitive state of IMECS participants. For this extent, we used the Mini-Mental State Exam (MMSE) questionnaire score as a measure for the cognitive impairment of participants. This assessment revealed no significant correlation between MMSE score and age in neither group (Figure 4).

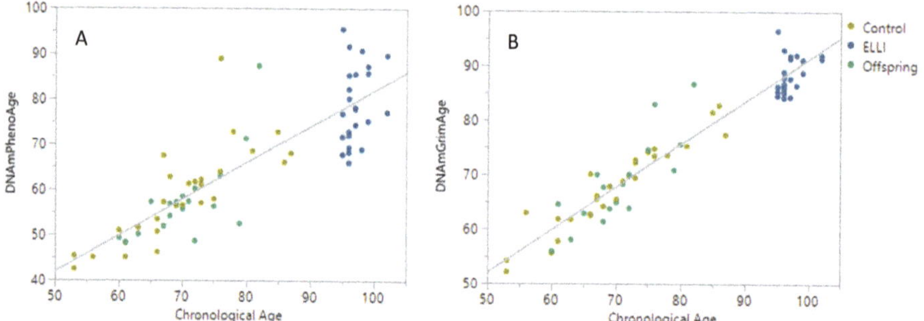

Figure 3. Chronological age vs. DNAm epigenetic age clocks. $N_{control}$ = 28, N_{ELLI} = 24, and $N_{offspring}$ = 18. (**A**) DNAmPhenoAge as a function of chronological age, R^2 = 0.716, p < 0.001. (**B**) DNAmGrimAge as a function of chronological age, R^2 = 0.919, p < 0.001. Linear regressions performed and plotted using JMP 14 (SAS Institute Inc., Cary, NC, USA).

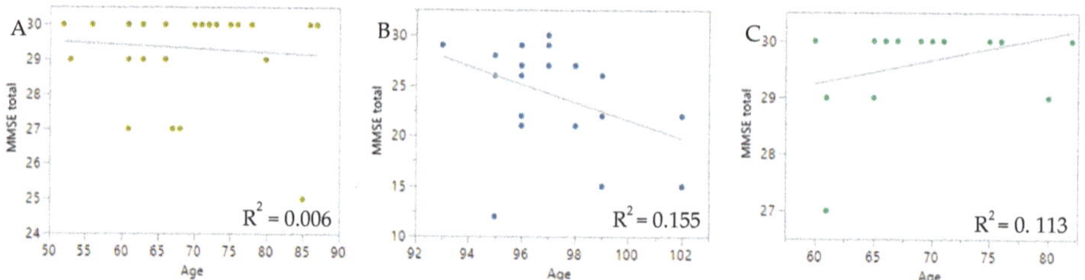

Figure 4. Cognition vs. age. Mini-Mental State Exam scores used for measuring cognitive impairment. (**A**) MMSE score of controls as a function of chronological age, N = 28, p = 0.7021. (**B**) MMSE score of ELLI as a function of chronological age, N = 21, p = 0.0776. (**C**) MMSE score of offspring as a function of chronological age, N = 16, p = 0.2021.

2.3. Telomere length

Finally, we turned to telomere length measurement using qPCR and the DNAm estimator of telomere length, DNAmTL. Our qPCR results did not demonstrate different T/S ratios between the three groups (Figure 5 and Table A5). However, the DNAmTL estimator found the telomeres of ELLI to be approximately 500 bp shorter compared to the control and offspring groups (Figure 6 and Table A6). When comparing T/S ratio and DNAmTL (Figure A1), there is no correlation between the two TL measures. Interestingly, when T/S ratio is tested between ELLI and controls with adjustment by DNAmGrimAge, it approaches significant correlation (p = 0.0508), hinting at a masking effect of the physiological age (representing juvenile methylation levels of centenarians) on the T/S ratio obtained with qPCR.

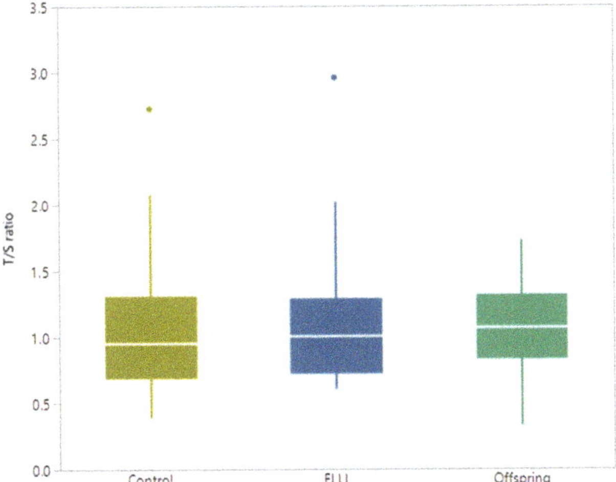

Figure 5. Average T/S ratio as measured by qPCR. T/S ratio obtained by dividing concentration of telomeric reaction by concentration of SCG (Single Copy Gene) reaction, as calculated using standard curve reactions. N_{ELLI} = 12, $N_{control}$ = 17, and $N_{offspring}$ = 12. All pair comparison (Dunn Joint Ranking) non-significant.

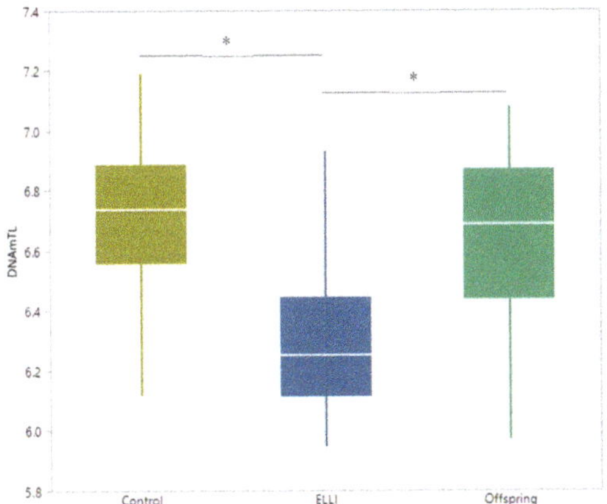

Figure 6. DNAmTL calculated using the online tool. DNAmTL was calculated by applying methylation beta values of 140 CpGs to online tool [37]. $N_{control}$ = 28, N_{ELLI} = 24, and $N_{offspring}$ = 18. * Significant differences, p <0.05.

3. Discussion

Many studies are aimed at biomarker discovery and improvement for aging [38,41,42,44,47–52]. The need for such characterization is of upmost importance in light of efforts to achieve longer health and lifespans across the world. Such biomarker detection would enable tracking and even reversal [45] of aging processes and allow for drug targeting and development to benefit the already graying population. Molecular and genomic biomarkers for aging are still sparse and inaccurate with the exception of the very recent development of DNAmGrimAge [42]. This DNA methylation biomarker outperforms all previously reported methylation age estimators and serves as a very accurate estimate of chronological age. Although this is expected due to the use of chronological as a surrogate for the age prediction, DNAmGrimAge, as DNAmPhenoAge, also serve as an evaluation of health status, indicative of the rate of epigenetic aging. Use of such biomarkers as indication of rate of age acceleration could promote better understanding of the processes underlying progression of aging and replace use of chronological age in clinical assessments relating to those conditions.

That said, the centenarian DNAm still remains elusive, even to the most accurate DNAmGrimAge. We show here that although accurate in offspring of ELLI and unrelated controls, DNAmGrimAge, along with DNAmPhenoAge, underestimates the chronological age of our IMECS ELLI participants, predicting a younger epigenetic age. We believe that this represents a slower rate of aging processes occurring in ELLI, and enabling them to reach such exceptional chronological age. This is in agreement with the methylation profile of semi-supercentenarians and their offspring, described by Horvath et al. [36], and replicates their results in our independent cohort.

The juvenile DNAm profile demonstrated in our cohort together with mostly intact cognition add support to the idea that ELLI age at a slower rate. Even though there was a small decline in the MMSE scores of the ELLI, this decline was not statistically significant, indicating intact cognition in the majority of the ELLI participants.

Further, DNAmTL estimated telomere length compared to T/S ratio of qPCR measurement showed no correlation with each other, until adjusted by DNAmGrimAge, at which the correlation approached significance. This masking effect of the physiological age (measured by DNAmGrimAge) adds support to the slower rate of aging. Telomere length has long been argued for and against use as an age indicator, but it is well-established to be decreased with age. Our qPCR measurements are consistent with previous observations of longer telomeres in ELLI [2]. While T/S ratio of the ELLI was expected to shorten in respect to offspring and controls because of their relatively advanced age, it remained unchanged, indicating a similar telomere length despite almost 30 years average age difference between group participants, demonstrating once again, a decreased aging rate. Taken together with the juvenile methylation rates in ELLI, we suggest that ELLI age slower than the general population through a beneficial methylation profile that may affect telomere length and other aspects of the hallmarks of aging.

To further draw conclusions, there is a need for bigger sample size and thorough molecular validation. We acknowledge that these are limitations in our current study and are already planning to pursue various directions for validation of our results. In addition, since the work presented here is part of an ongoing study, new IMECS participants are recruited and new recruitment centers should be established to increase ease and rate of recruitment. We believe that with adequate sample size and further validation in primary cells from participants we will be able to obtain more information on the juvenile epigenetic profile of ELLI and their offspring.

4. Materials and Methods

4.1. Ethics Statement

All IMECS participants gave their informed consent for inclusion prior to participation and blood collection. The study was conducted in accordance with the Declaration of Helsinki, and the protocol was approved by the Ethics Committee of Clinical Trials Department, Ministry of Health, Israel (project 109-2014) and by the Institutional Review Board at the Rambam Health Care Center in Haifa, Israel (project RMB 0312-14). Any person over 95 years of age was included in the study with the exclusion of persons cognitively unable to sign informed consent and cognitively impaired persons with no legal guardian. For the offspring group, people with one or more parent outliving the age of 95, either alive or deceased at the time of recruitment, and cognitively able to sign informed consent were included. For the control group any person in the range of 50–90 years of age both of whose parents did not reach 95 years of age and cognitively able to sign informed consent were included.

4.2. Sample Collection and Preparation

All IMECS participants (for demographic information see Tables A1 and A2) underwent physical and cognitive assessment including family history, general health questions, functional assessment including Instrumental Activities of Daily Living (IADL) and Basic Activities of Daily Living (BADL), Mini-Mental State Examination (MMSE), and the 12-item Short Form Health Survey (SF-12). Following this assessment, 20 mL whole blood was drawn from each participant. White blood cells were separated using lymphocyte separation medium (LSM, MP Biomedicals, LLC, CA, USA) and DNA extracted using High Pure PCR Template Preparation Kit (Roche Diagnostics GmbH, Mannheim, Germany) according to manufacturer instructions and quantified using dsDNA High Sensitivity Kit for Qubit (Life technologies Corporation, Eugene, OR, USA) and Qubit (Life Technologies Corporation, Carlsbad, CA, USA).

4.3. DNA Methylation Analysis

For detection of DNA methylation, 500 ng DNA were subjected to bisulfite sequencing and hybridized to Illumina MethylationEPIC beadchip at the Technion Biomedical Core Facilities, Rappaport Faculty of Medicine, Haifa, Israel. Raw data underwent an analysis pipeline using Minfi [53] and BumpHunter [54] (Bioconductor R packages) for quality control and statistical analyses. PreprocessNoob was used for background correction through dye-bias normalization. DNAmTL and all age estimators were obtained using the online tool developed by Lu et al. (https://dnamage.genetics.ucla.edu/home) [37].

4.4. Quantitative PCR for Relative Telomere Length Assessment

Average relative telomere length was measured as previously described [2,28,31,32] with modifications. IFNB1 was used as a single copy gene as published by Vasilishina et al (primer sequences detailed in Table A7). qPCR reactions mixes were prepared according to Table A8 with a standard curve performed for each run. qPCR run in LightCycler 480 II (Roche Diagnostics International Ltd, Rotkreuz, Switzerland) under following conditions: pre-incubation at 95 °C for 10 min, 35 cycles of 95 °C for 15 s, 60 °C for 60 s, and 72 °C for 10 s, followed by melting curve at 95 °C for 5 s, and 65 °C for 60 s. Triplicates for each sample were performed and concentration of sample calculated according to same-run standard curve and averaged for each sample. Relative telomere length (T/S ratio) of each sample was calculated as the ratio between average concentration of telomere reactions to the average concentration of the single copy gene reactions.

4.5. Statistical Analyses

All analyses and plots generated and analyzed using JMP 14 (SAS Institute Inc., Cary, NC, USA). For all analyses, p-values < 0.05 were considered significant.

Author Contributions: Conceptualization, D.G., T.D. and G.A.; methodology, D.G. and G.A.; sample collection, D.G., E.R. (Elina Rivkin), A.F., V.H., E.R. (Elad Rubin), D.K., I.S., T.D.; software, D.G. and G.A.; validation, D.G., E.R. (Elina Rivkin), A.F., L.S., V.H., E.R. (Elad Rubin), D.K., I.S., T.D., G.A.; formal analysis, D.G. and G.A.; investigation, D.G., E.R. (Elina Rivkin), A.F., L.S., V.H., E.R. (Elad Rubin), D.K., I.S., T.D., G.A.; resources, D.G., E.R. (Elina Rivkin), A.F., L.S., V.H., E.R. (Elad Rubin), D.K., I.S., T.D., G.A.; data curation, D.G., E.R. (Elina Rivkin), A.F., L.S., V.H., E.R. (Elad Rubin), D.K., I.S., T.D., G.A.; writing—original draft preparation, D.G.; writing—review and editing, T.D. and G.A.; visualization, T.D. and G.A.; supervision, G.A.; project administration, G.A.; funding acquisition, T.D. and G.A. All authors have read and agreed to the published version of the manuscript.

Funding: This research was funded by the Israel Science Foundation (ISF), grant number 193/16.

Acknowledgments: We would like to thank Ake T. Lu and Steve Horvath for their dedicated efforts with the DNAmAge online tool troubleshooting. We would also like to thank all the IMECS participants for their time and donation to our study.

Conflicts of Interest: The authors declare no conflict of interest.

Abbreviations

IMECS	Israeli Multi-Ethnic Centenarian Study
ELLI	Exceptionally Long-Lived Individuals
DNAm	DNA methylation
qPCR	Quantitative Polymerase Chain Reaction
MMSE	Mini-Mental State Exam
CpG	Cytosine Guanine dinucleotide
TL	Telomere Length

Appendix A

Table A1. Age distribution of Israeli Multi-Ethnic Centenarian Study (IMECS) participants.

Group	Number of Participants	Mean Age	STDEV	Min Age	Max Age
ELLI	24	97.1	2.04	95	102
Offspring	18	70.4	6.25	60	82
Control	28	69.7	9.1	53	87

Table A2. IMECS demography.

Group	Gender F	Gender M	% Smokers	Education level Elementary	Education level High School	Education level College Degree	Education level NA
ELLI	12	12	57%	4	10	7	3
Offspring	12	6	37%	1	3	12	2
Control	20	8	39%	0	7	21	0

Table A3. One-way analysis of each age measurement through groups.

Age Measure	Group Pairs	Kruskal-Wallis *P*-Value	Wilcoxon Post-Hoc
DNAm PhenoAge	ELLI—Offspring	<0.0001	<0.0001
DNAm PhenoAge	ELLI—Control	<0.0001	<0.0001
DNAm PhenoAge	Control—Offspring	<0.0001	0.5969
DNAm GrimAge	ELLI—Offspring	<0.0001	<0.0001
DNAm GrimAge	ELLI—Control	<0.0001	<0.0001
DNAm GrimAge	Control—Offspring	<0.0001	0.9731

Table A4. Repeated measures test within and among groups.

Measurement Pairs	Within Pairs *P*-Value	Among Pairs *P*-Value
DNAmPhenoAge—actual age	<0.0001	<0.0001
DNAmGrimAge—actual age	<0.0001	<0.0001
DNAmGrimAge—DNAmPhenoAge	0.7812	<0.0001

Table A5. Distribution of average T/S ratio.

Group	Median Average T/S	Minimum Average T/S	Maximum Average T/S
ELLI	1.01	0.61	5.91
Offspring	1.11	0.33	4.23
Control	0.95	0.39	2.73

Table A6. One-way analysis of DNAmTL measurements.

Group Pairs	ANOVA *P*-Value	Tukey Post-Hoc
ELLI—Offspring		0.0003
ELLI—Control	<0.0001	<0.0001
Control—Offspring		0.3388

Table A7. Telomere length qPCR primer information.

Primer Name	Primer Sequence	Working Concentration
Telo-F	CGGTTTGTTTGGGTTTGGGTTTGGGTTTGGGTTTGGGTT	10 μM
Telo-R	GGCTTGCCTTACCCTTACCCTTACCCTTACCCTTACCCT	10 μM
IFNB1-F	GGTTACCTCCGAAACTGAAGA	7.5 μM
IFNB1-R	CCTTTCATATGCAGTACATTAGCC	7.5 μM

Table A8. Telomere length qPCR reaction volumes per 1 reaction.

Reagent	Telo Reaction	IFNB1 Reaction
Ultra-Pure Water (Bio-Lab ltd, Jerusalem, Israel)	6 μL	4 μL
SYBR green (Thermo Fisher Scientific Baltics, Vilnius, Lithuania)	10 μL	10 μL
F primer (Sigma-Aldrich Israel ltd, Rehovot, Israel)	0.2 μL	0.6 μL
R primer (Sigma-Aldrich Israel ltd, Rehovot, Israel)	1.8 μL	1.4 μL
Genomic DNA (5 ng/μL)	2 μL	4 μL

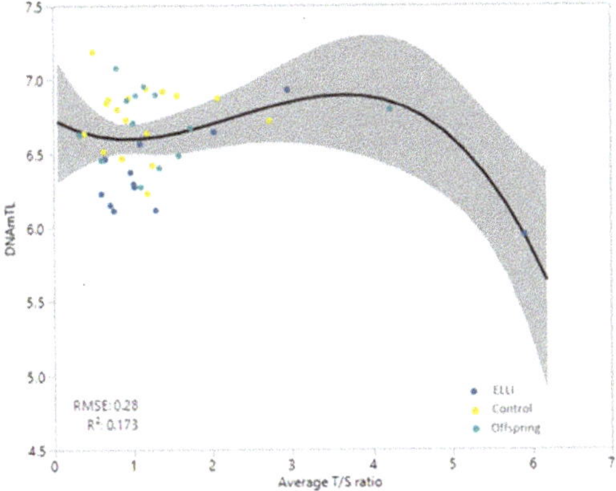

Figure A1. T/S ratio vs. DNAmTL. DNAmTL as a function of T/S ratio (telomeric concentration divided by SCG concentration).

References

1. Yashin, A.; De Benedictis, G.; Vaupel, J.; Tan, Q.; Andreev, K.; Iachine, I.; Bonafe, M.; Valensin, S.; De Luca, M.; Carotenuto, L.; et al. Genes and longevity: Lessons from studies of centenarians. *J. Gerontol. Ser. A Boil. Sci. Med. Sci.* **2000**, *55*, B319–B328. [CrossRef] [PubMed]

2. Atzmon, G.; Cho, M.; Cawthon, R.M.; Budagov, T.; Katz, M.; Yang, X.; Siegel, G.; Bergman, A.; Huffman, D.M.; Schechter, C.B.; et al. Evolution in health and medicine Sackler colloquium: Genetic variation in human telomerase is associated with telomere length in Ashkenazi centenarians. *Proc. Natl. Acad. Sci. USA* **2010**, *107*, 1710–1717. [CrossRef] [PubMed]

3. Freudenberg-Hua, Y.; Freudenberg, J.; Vacic, V.; Abhyankar, A.; Emde, A.-K.; Ben-Avraham, D.; Barzilai, N.; Oschwald, D.; Christen, E.; Koppel, J.; et al. Disease variants in genomes of 44 centenarians. *Mol. Genet. Genom. Med.* **2014**, *2*, 438–450. [CrossRef] [PubMed]

4. Rubino, G.; Bulati, M.; Aiello, A.; Aprile, S.; Gambino, C.M.; Gervasi, F.; Caruso, C.; Accardi, G. Sicilian centenarian offspring are more resistant to immune ageing. *Aging Clin. Exp. Res.* **2018**, *31*, 125–133. [CrossRef] [PubMed]

5. Stevenson, M.; Bae, H.; Schupf, N.; Andersen, S.; Zhang, Q.; Perls, T.; Sebastiani, P. Burden of disease variants in participants of the long life family Study. *Aging* **2015**, *7*, 123–132. [CrossRef] [PubMed]

6. Giuliani, C.; Sazzini, M.; Pirazzini, C.; Bacalini, M.G.; Marasco, E.; Ruscone, G.A.G.; Fang, F.; Sarno, S.; Gentilini, D.; Di Blasio, A.M.; et al. Impact of demography and population dynamics on the genetic architecture of human longevity. *Aging* **2018**, *10*, 1947–1963. [CrossRef]

7. Teixeira, L.; Araújo, L.; Jopp, D.; Ribeiro, O. Centenarians in Europe. *Maturitas* **2017**, *104*, 90–95. [CrossRef]

8. Puca, A.A.; Spinelli, C.; Accardi, G.; Villa, F.; Caruso, C. Centenarians as a model to discover genetic and epigenetic signatures of healthy ageing. *Mech. Ageing Dev.* **2018**, *174*, 95–102. [CrossRef]

9. Milman, S.; Barzilai, N. Dissecting the Mechanisms Underlying Unusually Successful Human Health Span and Life Span. *Cold Spring Harb. Perspect. Med.* **2015**, *6*, a025098. [CrossRef]

10. Atzmon, G.; Rincon, M.; Schechter, C.B.; Shuldiner, A.R.; Lipton, R.B.; Bergman, A.; Barzilai, N. Lipoprotein genotype and conserved pathway for exceptional longevity in humans. *PLoS Boil.* **2006**, *4*, e113. [CrossRef]

11. Barzilai, N.; Atzmon, G.; Schechter, C.; Schaefer, E.J.; Cupples, A.L.; Lipton, R.; Cheng, S.; Shuldiner, A.R. Unique Lipoprotein Phenotype and Genotype Associated With Exceptional Longevity. *JAMA* **2003**, *290*, 2030–2040. [CrossRef] [PubMed]

12. Ben-Avraham, D.; Govindaraju, D.R.; Budagov, T.; Fradin, D.; Durda, P.; Liu, B.; Ott, S.; Gutman, D.; Sharvit, L.; Kaplan, R.; et al. The GH receptor exon 3 deletion is a marker of male-specific exceptional longevity associated with increased GH sensitivity and taller stature. *Sci. Adv.* **2017**, *3*, e1602025. [CrossRef] [PubMed]

13. López-Otín, C.; Blasco, M.A.; Partridge, L.; Serrano, M.; Kroemer, G. The hallmarks of aging. *Cell* **2013**, *153*, 1194–1217. [CrossRef]

14. Austriaco, N.R., Jr.; Guarente, L.P. Changes of telomere length cause reciprocal changes in the lifespan of mother cells in Saccharomyces cerevisiae. *Proc. Natl. Acad. Sci. USA* **1997**, *94*, 9768–9772. [CrossRef]

15. Teixeira, M.T. Saccharomyces cerevisiae as a Model to Study Replicative Senescence Triggered by Telomere Shortening. *Front. Oncol.* **2013**, *3*, 101. [CrossRef]

16. Espejel, S.; Klatt, P.; Murcia, J.M.-D.; Martiín-Caballero, J.; Flores, J.M.; Taccioli, G.; De Murcia, G.; Blasco, M.A. Impact of telomerase ablation on organismal viability, aging, and tumorigenesis in mice lacking the DNA repair proteins PARP-1, Ku86, or DNA-PKcs. *J. Cell Boil.* **2004**, *167*, 627–638. [CrossRef]

17. Honig, L.S.; Schupf, N.; Lee, J.H.; Tang, M.X.; Mayeux, R. Shorter telomeres are associated with mortality in those with APOE ε4 and dementia. *Ann. Neurol.* **2006**, *60*, 181–187. [CrossRef]

18. Vera, E.; De Jesus, B.B.; Foronda, M.; Flores, J.M.; Blasco, M.A. The Rate of Increase of Short Telomeres Predicts Longevity in Mammals. *Cell Rep.* **2012**, *2*, 732–737. [CrossRef]

19. Simons, M.J. Questioning causal involvement of telomeres in aging. *Ageing Res. Rev.* **2015**, *24*, 191–196. [CrossRef]

20. Harley, C.B.; Futcher, A.B.; Greider, C.W. Telomeres shorten during ageing of human fibroblasts. *Nature* **1990**, *345*, 458–460. [CrossRef]

21. Aviv, A.; Shay, J.W. Reflections on telomere dynamics and ageing-related diseases in humans. *Philos. Trans. R. Soc. B Boil. Sci.* **2018**, *373*, 20160436. [CrossRef]

22. Sanders, J.L.; Newman, A.B. Telomere Length in Epidemiology: A Biomarker of Aging, Age-Related Disease, Both, or Neither? *Epidemiologic Rev.* **2013**, *35*, 112–131. [CrossRef] [PubMed]

23. Xu, Z.; Duc, K.D.; Holcman, D.; Teixeira, M.T. The length of the shortest telomere as the major determinant of the onset of replicative senescence. *Genetics* **2013**, *194*, 847–857. [CrossRef] [PubMed]

24. Savage, S.A. Beginning at the ends: Telomeres and human disease. *F1000Research* **2018**, *7*. [CrossRef]

25. Levy, D.; Neuhausen, S.L.; Hunt, S.C.; Kimura, M.; Hwang, S.-J.; Chen, W.; Bis, J.C.; Fitzpatrick, A.L.; Smith, E.; Johnson, A.D.; et al. Genome-wide association identifies OBFC1 as a locus involved in human leukocyte telomere biology. *Proc. Natl. Acad. Sci. USA* **2010**, *107*, 9293–9298. [CrossRef] [PubMed]

26. Terry, D.F.; Nolan, V.G.; Andersen, S.L.; Perls, T.T.; Cawthon, R. Association of longer telomeres with better health in centenarians. *Journals Gerontol. Ser. A: Boil. Sci. Med Sci.* **2008**, *63*, 809–812. [CrossRef] [PubMed]

27. Gutman, D.; Sharvit, L.; Atzmon, G. Possible Mechanisms for Telomere Length Maintenance in Extremely Old People. *Hered. Genet.* **2014**, *3*, 1–2.

28. Cawthon, R.M. Telomere measurement by quantitative PCR. *Nucleic Acids Res.* **2002**, *30*, e47. [CrossRef]

29. Pfaffl, M.W. A new mathematical model for relative quantification in real-time RT-PCR. *Nucleic Acids Res.* **2001**, *29*, 45. [CrossRef]

30. Cawthon, R.M.; Smith, K.R.; O'Brien, E.; Sivatchenko, A.; Kerber, R.A. Association between telomere length in blood and mortality in people aged 60 years or older. *Lancet* **2003**, *361*, 393–395. [CrossRef]

31. Shekhidem, H.A.; Sharvit, L.; Leman, E.; Manov, I.; Roichman, A.; Holtze, S.; Huffman, D.M.; Cohen, H.Y.; Hildebrandt, T.B.; Shams, I.; et al. Telomeres and Longevity: A Cause or an Effect? *Int. J. Mol. Sci.* **2019**, *20*, 3233. [CrossRef] [PubMed]

32. Vasilishina, A.; Kropotov, A.; Spivak, I.; Bernadotte, A. Relative human telomere length quantification by real-time PCR. *Methods Mol. Biol.* **2019**, *1896*, 39–44. [PubMed]

33. Axelrad, M.D.; Budagov, T.; Atzmon, G. Telomere Length and Telomerase Activity; A Yin and Yang of Cell Senescence. *J. Vis. Exp.* **2013**, *75*, e50246. [CrossRef] [PubMed]

34. Montpetit, A.J.; Alhareeri, A.A.; Montpetit, M.; Starkweather, A.R.; Elmore, L.W.; Filler, K.; Mohanraj, L.; Burton, C.W.; Menzies, V.S.; Lyon, D.E.; et al. Telomere length: A review of methods for measurement. *Nurs. Res.* **2014**, *63*, 289–299. [CrossRef]

35. Hannum, G.; Guinney, J.; Zhao, L.; Zhang, L.; Hughes, G.; Sadda, S.; Klotzle, B.; Bibikova, M.; Fan, J.-B.; Gao, Y. Genome-wide methylation profiles reveal quantitative views of human aging rates. *Mol. Cell* **2013**, *49*, 359–367. [CrossRef]

36. Horvath, S.; Pirazzini, C.; Bacalini, M.G.; Gentilini, D.; Di Blasio, A.M.; Delledonne, M.; Mari, D.; Arosio, B.; Monti, D.; Passarino, G.; et al. Decreased epigenetic age of PBMCs from Italian semi-supercentenarians and their offspring. *Aging* **2015**, *7*, 1159–1170. [CrossRef]

37. Horvath, S. DNA methylation age of human tissues and cell types. *Genome Boil.* **2013**, *14*, R115. [CrossRef]

38. Chen, B.H.; Marioni, R.E.; Colicino, E.; Peters, M.J.; Ward-Caviness, C.K.; Tsai, P.-C.; Roetker, N.S.; Just, A.C.; Demerath, E.W.; Guan, W.; et al. DNA methylation-based measures of biological age: Meta-analysis predicting time to death. *Aging* **2016**, *8*, 1844–1859. [CrossRef]

39. Horvath, S.; Oshima, J.; Martin, G.M.; Lu, A.T.; Quach, A.; Cohen, H.; Felton, S.; Matsuyama, M.; Lowe, D.; Kabacik, S.; et al. Epigenetic clock for skin and blood cells applied to Hutchinson Gilford Progeria Syndrome and ex vivo studies. *Aging* **2018**, *10*, 1758. [CrossRef]

40. Armstrong, N.J.; Mather, K.A.; Thalamuthu, A.; Wright, M.J.; Trollor, J.N.; Ames, D.; Brodaty, H.; Schofield, P.R.; Sachdev, P.S.; Kwok, J.B. Aging, exceptional longevity and comparisons of the Hannum and Horvath epigenetic clocks. *Epigenomics* **2017**, *9*, 689–700. [CrossRef]

41. Ryan, J.; Wrigglesworth, J.; Loong, J.; Fransquet, P.D.; Woods, R.L.; Anderson, R. A systematic review and meta-analysis of environmental, lifestyle and health factors associated with DNA methylation age. *J. Gerontol. Ser. A Boil. Sci. Med. Sci.* **2019**. [CrossRef] [PubMed]

42. Lu, A.T.; Quach, A.; Wilson, J.G.; Reiner, A.P.; Aviv, A.; Raj, K.; Hou, L.; Baccarelli, A.A.; Li, Y.; Stewart, J.D.; et al. DNA methylation GrimAge strongly predicts lifespan and healthspan. *Aging* **2019**, *11*, 303–327. [CrossRef] [PubMed]

43. Lu, A.T.; Seeboth, A.; Tsai, P.-C.; Sun, D.; Quach, A.; Reiner, A.P.; Kooperberg, C.; Ferrucci, L.; Hou, L.; Baccarelli, A.A.; et al. DNA methylation-based estimator of telomere length. *Aging* **2019**, *11*, 5895–5923. [CrossRef]

44. Levine, M.E.; Lu, A.T.; Quach, A.; Chen, B.H.; Assimes, T.L.; Bandinelli, S.; Hou, L.; Baccarelli, A.A.; Stewart, J.D.; Li, Y.; et al. An epigenetic biomarker of aging for lifespan and healthspan. *Aging* **2018**, *10*, 573–591. [CrossRef]

45. Fahy, G.M.; Brooke, R.T.; Watson, J.P.; Good, Z.; Vasanawala, S.S.; Maecker, H.; Leipold, M.D.; Lin, D.T.S.; Kobor, M.S.; Horvath, S. Reversal of epigenetic aging and immunosenescent trends in humans. *Aging Cell* **2019**, *18*, e13028. [CrossRef]

46. Xiao, F.-H.; Wang, H.-T.; Kong, Q.-P. Dynamic DNA Methylation During Aging: A "Prophet" of Age-Related Outcomes. *Front. Genet.* **2019**, *10*, 107. [CrossRef]

47. Bernadotte, A.; Mikhelson, V.M.; Spivak, I.M. Markers of cellular senescence. Telomere shortening as a marker of cellular senescence. *Aging* **2016**, *8*, 3–11. [CrossRef]

48. Lin, Q.; Weidner, C.I.; Costa, I.G.; Marioni, R.E.; Ferreira, M.R.P.; Deary, I.J.; Wagner, W. DNA methylation levels at individual age-associated CpG sites can be indicative for life expectancy. *Aging* **2016**, *8*, 394–401. [CrossRef]

49. Shiels, P.G.; Stenvinkel, P.; Kooman, J.P.; McGuinness, D. Circulating markers of ageing and allostatic load: A slow train coming. *Pract. Lab. Med.* **2017**, *7*, 49–54. [CrossRef]

50. Olivieri, F.; Capri, M.; Bonafè, M.; Morsiani, C.; Jung, H.J.; Spazzafumo, L.; Viña, J.; Suh, Y. Circulating miRNAs and miRNA shuttles as biomarkers: Perspective trajectories of healthy and unhealthy aging. *Mech. Ageing Dev.* **2017**, *165*, 162–170. [CrossRef]

51. Deelen, J.; Akker, E.B.V.D.; Trompet, S.; Van Heemst, D.; Mooijaart, S.P.; Slagboom, P.E.; Beekman, M. Employing biomarkers of healthy ageing for leveraging genetic studies into human longevity. *Exp. Gerontol.* **2016**, *82*, 166–174. [CrossRef] [PubMed]

52. Barral, S.; Singh, J.; Fagan, E.; Cosentino, S.; Andersen-Toomey, S.L.; Wojczynski, M.K.; Feitosa, M.; Kammerer, C.M.; Schupf, N.; Long Life Family Study. Age-Related Biomarkers in LLFS Families With Exceptional Cognitive Abilities. *J. Gerontol. Ser. A Boil. Sci. Med Sci.* **2017**, *72*, 1683–1688.

53. Aryee, M.J.; Jaffe, A.E.; Corrada-Bravo, H.; Ladd-Acosta, C.; Feinberg, A.P.; Hansen, K.D.; Irizarry, R.A. Minfi: A flexible and comprehensive Bioconductor package for the analysis of Infinium DNA methylation microarrays. *Bioinformatics* **2014**, *30*, 1363–1369. [CrossRef] [PubMed]

54. Jaffe, A.E.; Murakami, P.; Lee, H.; Leek, J.T.; Fallin, M.D.; Feinberg, A.P.; Irizarry, R.A. Bump hunting to identify differentially methylated regions in epigenetic epidemiology studies. *Int. J. Epidemiol.* **2012**, *41*, 200–209. [CrossRef]

MDPI

St. Alban-Anlage 66

4052 Basel

Switzerland

Tel. +41 61 683 77 34

Fax +41 61 302 89 18

www.mdpi.com

International Journal of Molecular Sciences Editorial Office

E-mail: ijms@mdpi.com

www.mdpi.com/journal/ijms